Growth Control During Cell Aging

Editors

Eugenia Wang, Ph.D.

Director
The Bloomfield Centre for Research in Aging
Lady Davis Institute for Medical Research
The Sir Mortimer B. Davis - Jewish General Hospital and McGill University
Montreal, Quebec, Canada

Huber R. Warner, Ph.D.

Chief
Molecular and Cell Biology Branch
National Institute on Aging
Bethesda, Maryland

CRC Press, Inc.
Boca Raton, Florida

Library of Congress Cataloging-in-Publication Data

Growth control during cell aging/editors, Eugenia Wang, Huber
Warner.
    p.   cm.
   Includes bibliographies and index.
   ISBN 0-8493-4580-4
   1. Cells—Growth—Regulation. 2. Cell proliferation. 3. Cells—
Aging. I. Wang, E. (Eugenia), 1945-   . II. Warner, Huber R.
   [DNLM: 1. Aging. 2. Cell Division. 3. Cell Survival. 4. Growth
Substances—physiology.   WT 102 G884]
QH604.G76 1989
574.87'61—dc19
DNLM/DLC
for Library of Congress   88-36740
                                                 CIP

Direct all inquiries to CRC Press, Inc., 2000 Corporate Blvd., N.W., Boca Raton, Florida, 33431.

© 1989 by CRC Press, Inc.

International Standard Book Number 0-8493-4580-4

Library of Congress Card Number 88-36740
Printed in the United States

PREFACE

Nontransformed cells grown in culture have a definite life span, as shown by the early work of Leonard Hayflick and his colleagues. After a rather specific number of doublings these cells enter a senescent state, cease dividing, and eventually die. This state resembles the quiescent state, which is defined as the G_0 stage of the normal cell cycle, but differs in several important ways. One obvious, important difference is the inability to stimulate these cells to divide by adding serum or mitogens.

A growing cadre of investigators is turning its attention to the question, ''Why can senescent cells no longer be stimulated to divide?'' To obtain a complete answer to this question, it will first be necessary to understand what happens when cultured cells leave G_0, pass through G_1, and enter S phase. It will be necessary to understand how and why cells become committed to DNA synthesis and what controls this process. Finally, it will be necessary to understand how quiescent and senescent cells differ.

This volume is an attempt to address these questions from the standpoint of what is currently known about control of cell proliferation (including growth factors, oncogenes and their products, inhibitors, transforming growth factors, etc.), and it derives from a workshop held in Bethesda, Maryland on December 11 and 12, 1985. Progress is being made in this area, and this information will not only be useful in understanding aging *in vitro,* but will also increase our understanding of cell transformation, wound healing, tissue regeneration, and abnormal cell growth.

Huber R. Warner
October 1988

THE EDITORS

Eugenia Wang, Ph.D., is Director of the Bloomfield Centre for Research in Aging at the Lady Davis Institute for Medical Research of the Sir Mortimer B. Davis - Jewish General Hospital and McGill University Teaching Hospital, in Montreal, Quebec, Canada. She is also an Associate Professor in the Departments of Anatomy and Medicine of the School of Medicine at McGill University.

Dr. Wang received her B.S. degree from National Taiwan University, Taipei, Taiwan in 1966 and her M.A. degree in 1969 from Northern Michigan University, Marquette. In 1974, she received her Ph.D. in cell biology from Case Western Reserve University, Cleveland, Ohio. Following that she was a postdoctoral fellow in virology at the Rockefeller University, New York, where she was later appointed Research Associate and then Assistant Professor during the period of 1976 to 1987. In 1987, she moved to the McGill University community to assume her present position in establishing The Bloomfield Centre for Research in Aging.

Dr. Wang is a member of the American Association for the Advancement of Science, the American Association of Cell Biology, the Gerontological Society of America, the New York Academy of Sciences, and the Honorary Harvey Society. Furthermore, she has been Senior Editor for the 455 Annals of the New York Academy of Sciences, titled ''Intermediate Filaments''. She is a member of the scientific reviewing section for the National Institute on Aging, and she is serving as the advisory member for cell biology for the Institute of Biomedical Sciences, Academia Sinica, Taipei, Taiwan, Republic of China. She has received the Distinguished Scientist Award from the Medical Research Council (MRC) of Canada and has been the recipient of many research grants from the National Institutes of Health and the MRC.

Dr. Wang is the author of more than 50 papers and has co-authored two books. Her current major research interests relate to the investigation of the molecular mechanisms governing the cessation of proliferation during the *in vitro* aging process of human fibroblasts.

Huber Warner, Ph.D., is Chief, Molecular and Cell Biology Branch, National Institute on Aging (NIA), National Institutes of Health, Bethesda, Maryland.

Dr. Warner graduated in 1958 from the Massachusetts Institute of Technology (MIT), Cambridge, with a B.S. degree in chemical engineering, and obtained his Ph.D. degree in biochemistry in 1962 from the University of Michigan, Ann Arbor. From 1962 to 1964, he was postdoctoral fellow at MIT, and then he was on the faculty of the Department of Biochemistry, University of Minnesota, St. Paul, from 1964 to 1984. In 1984, he moved to the NIA.

Dr. Warner is a member of the American Society for Biochemistry and Molecular Biology and the American Association for the Advancement of Science. He has published over 60 research papers in the areas of virus replication and DNA repair, and he served for 1 year as the Program Director for the Biochemistry Program at the National Science Foundation. He currently serves as Program Administrator for the Molecular Biology and Genetics Programs at the National Institute on Aging.

CONTRIBUTORS

Renato Baserga, M.D.
Department of Pathology and
  Fels Research Institute
Temple University Medical School
Philadelphia, Pennsylvania

Sarah A. Bruce, Ph.D.
Division of Biophysics
School of Hygiene and Public Health
The Johns Hopkins University
Baltimore, Maryland

David L. Busbee, Ph.D.
Department of Anatomy and Cell Biology
College of Veterinary Medicine
Texas A & M University
College Station, Texas

Stanley Cohen, M.D.
Department of Pathology
Hahnemann University School
  of Medicine
Philadelphia, Pennsylvania

V. J. Cristofalo, Ph.D.
The Wistar Institute
Philadelphia, Pennsylvania

Geoffrey M. Curtin, Ph.D.
Department of Anatomy and Cell Biology
College of Veterinary Medicine
Texas A & M University
College Station, Texas

David T. Denhardt, Ph.D.
Cancer Research Laboratory
University of Western Ontario
London, Ontario, Canada

Kerin L. Fresa, Ph.D.
Department of Pathology
Hahnemann University School
  of Medicine
Philadelphia, Pennsylvania

G. Gerhard, M.D.
The Wistar Institute
Philadelphia, Pennsylvania

Luis Glaser, Ph.D.
Departments of Biology and Biochemistry
University of Miami
Coral Gables, Florida

Samuel Goldstein, M.D.
Departments of Medicine and
  Biochemistry & Molecular Biology
University of Arkansas for Medical
  Sciences and McClellan V. A.
  Memorial Hospital
Little Rock, Arkansas

Bruce H. Howard, M.D.
Laboratory of Molecular Biology
Division of Cancer Biology and
  Diagnosis
National Cancer Institute
Bethesda, Maryland

Brian Lathrop, Ph.D.
Department of Biology
Boston University School of Medicine
Boston, Massachusetts

Adam Lerner, M.D.
Department of Pathology
Boston University School of Medicine
Boston, Massachusetts

John Macauley, B.A.
Department of Biochemistry
Boston University School of Medicine
Boston, Massachusetts

George M. Martin, M.D.
Departments of Pathology and Genetics
University of Washington
Seattle, Washington

Michael B. Mathews, Ph.D.
Cold Spring Harbor Laboratory
Cold Spring Harbor, New York

Richard A. Miller, M.D., Ph.D.
Departments of Pathology and
  Biochemistry
Boston University School of Medicine
Boston, Massachusetts

Elena Moerman, B. A.
Department of Medicine
University of Arkansas for Medical
  Sciences
Little Rock, Arkansas

Audrey L. Muggleton-Harris, F.I. Biol.
MRC Experimental Embryology and
  Teratology Unit
St. George's Hospital Medical School
London, United Kingdom

Craig L. J. Parfett, Ph.D.
Cancer Research Laboratory
University of Western Ontario
London, Ontario, Canada

Olivia M. Pereira-Smith, Ph.D.
Roy M. and Phyllis Gough Huffington
  Center on Aging
Division of Molecular Virology
  and Department of Medicine
Baylor College of Medicine
Houston, Texas

P. D. Phillips, Ph.D.
The Wistar Institute
Philadelphia, Pennsylvania

Ben Philosophe, B.S.
Department of Pathology
Boston University School of Medicine
Boston, Massachusetts

Susan R. Rittling, Ph.D.
Agriculture Research Division
American Cyanamid Company
Princeton, New Jersey

Robert J. Shmookler Reis, Ph.D.
Departments of Medicine and
  Biochemistry & Molecular Biology
University of Arkansas for Medical
  Sciences and McClellan V.A.
  Memorial Hospital
Little Rock, Arkansas

James R. Smith, Ph.D.
Roy M. and Phyllis Gough Huffington
  Center on Aging
Division of Molecular Virology
  and Departments of Medicine
  and Cell Biology
Baylor College of Medicine
Houston, Texas

T. Sorger, Ph.D.
The Wistar Institute
Philadelphia, Pennsylvania

Andrea L. Spiering, Ph.D.
Roy M. and Phyllis Gough Huffington
  Center on Aging
Division of Molecular Virology
Baylor College of Medicine
Houston, Texas

Gretchen H. Stein, Ph.D.
Department of Molecular, Cellular, and
  Developmental Biology
University of Colorado
Boulder, Colorado

Victor L. Sylvia, Ph.D.
Department of Anatomy and Cell Biology
College of Veterinary Medicine
Texas A & M University
College Station, Texas

Quan Sun, Ph.D.
Department of Biochemistry
Michigan State University
East Lansing, Michigan

Patricia G. Voss, M.S.
Department of Biochemistry
Michigan State University
East Lansing, Michigan

Eugenia Wang, Ph.D.
The Bloomfield Centre for Research
 in Aging
Lady Davis Institute for Medical
 Research
The Sir Mortimer B. Davis - Jewish
 General Hospital and McGill University
Montreal, Quebec, Canada

John L. Wang, Ph.D.
Department of Biochemistry
Michigan State University
East Lansing, Michigan

Huber R. Warner, Ph.D.
National Institute on Aging
National Institutes of Health
Bethesda, Maryland

Burton Wice, B.A.
Department of Biochemistry and
 Molecular Biophysics
Washington University School
 of Medicine
St. Louis, Missouri

TABLE OF CONTENTS

I. Introduction

II. Positive Control of Cell Proliferation

III. Negative Control of Cell Proliferation

Section I. Introduction

INTRODUCTION

George M. Martin

This volume reports on the proceedings of the Workshop on Control of Cell Proliferation in Senescent Cells, which was held at the National Institute on Aging in Bethesda, MD on December 12, 1985. Virtually all the original discussants have contributed to this publication. (My close colleague Tom Norwood is an important exception; a summary of the studies of T. H. Norwood, W. R. Pendergrass, and their colleagues on the relationships of DNA polymerase alpha concentration, cell volume, and rate of initiation of DNA synthesis will appear elsewhere.) We are all grateful to Huber Warner, Chief of the Molecular and Cell Biology Branch of the National Institute on Aging, and to Eugenia Wang for editing this volume and for bringing together this congenial group of investigators, some old and some new to the field. I should say, parenthetically, that this field has been characterized by a high degree of collegiality and a ready exchange of information amongst its practitioners. The group in fact held a follow-up conference at the Fourth International Congress of Cell Biology in Montreal in August 1988. Readers should be alert for the publication of the proceedings of that conference. A glance at the diagram in Dr. Warner's useful summary of this volume immediately tells us that there is a great deal more mystery to be unraveled!

The questions which are being addressed are indeed fascinating. In vertebrate organisms, somatic cell lineages are apparently mortal, with the exception of *populations* of neoplastic or transformed cells. (Individual *clones* of neoplastic or transformed cells may, in fact, exhibit finite proliferative potentials due to a variety of mechanisms, such as terminal differentiation and other epigenetic events, chromosomal segregation, mitotic recombination, and various types of point mutation.) Germ-line lineages are, of course, immortal.

There have been numerous challenges to the inevitability of the "Hayflick limit", some of them cited in the recent historical review by Witkowski.[1] (For a more comprehensive timely review, with 575 reference citations, see Stanulis-Praeger.[2]) The most recent challenge cites evidence of an apparent indefinite proliferation of murine embryonic diploid cells established and maintained for multiple passages in the absence of serum.[3] It remains to be seen, however, to what extent such behavior relates to selection, from a large population of cells, of unusual transformed variants (diploid, but not necessarily euploid). Another apparent exception is perhaps less surprising, as it involves cultures derived from multipotent embryonal stem cells.[4]

This book, however, is not entirely preoccupied with the model of *in vitro* clonal senescence. Particularly notable has been the progress in characterizing proliferative behaviors of subsets of peripheral blood lymphocytes derived from donors of varying ages (see especially Chapter 5). Although some unifying principles may well emerge, the reader should take care to avoid general conclusions based upon a mix of evidence from these different model systems. It is unfortunate that there has not been a collective focus of research upon *in vivo/in vitro* comparative studies with identical cell types.

A nice feature of the present volume is that it provides a vehicle for stimulating speculative discussions beyond what is traditionally permissive in journal publications. A good example is the contribution by Bruce Howard (Chapter 12). I look forward to learning more about the possible role of interspersed repetitive elements in cell senescence. I also look forward to progress in elucidating the role of alterations in site-specific DNA methylations, as discussed by Robert Shmookler Reis, Elena Moerman, and Samuel Goldstein (Chapter 13). There is certainly a growing interest among gerontologists in heritable epigenetic alterations in gene expression.

A number of authors outline, and sometimes diagram, interesting models. There appears to be a trend toward embracing the idea of multiple restriction points in the G_0/G_1 stage(s) of the cell cycle, as well as a growing emphasis on the importance of negative regulation. The chapters by Olivia Pereira-Smith and her colleagues (Chapter 8), Gretchen Stein (Chapter 9), Eugenia Wang (Chapter 11), and others bring one up to date on the important new advances concerning such regulation.

While there have been theoretical constructs that attribute species-specific life spans to genetically controlled differential rates of clonal attenuation in key cell populations,[5] it is likely that a variety of additional mechanisms contribute to such constitutional limits of longevity, including differential stabilities of postreplicative cell populations. Among mammals, however, the phenotype of senescent organisms is characterized by a striking loss of proliferative homeostasis, including multifocal hyperplasias.[6] Thus, the concern of a number of the contributors for the problem of the interface between clonal senescence and oncogenesis is quite important. The group was fortunate to have had the participation of Renato Baserga and Vincent Cristofalo, who have devoted long and productive careers investigating mitotic cell cycle control in the context of cancer and aging. A dissection of both positive and negative regulation of the mitotic cell cycle is clearly the right approach, although I suspect that we are seeing only the tip of the iceberg in terms of complexity. In yeast, there are at least 70 different cell cycle division genes, and many more are likely to be discovered that influence cell cycle function.[7] A number of different loci have been implicated in the control of the ''start'' signals. In mammals, even at the single cell level, many growth factors can have both stimulatory and inhibitory effects or can stimulate cell cycle traverse in one cell type but inhibit it in another.[8] Moreover, it will be essential to consider cell-cell interactions, as emphasized by Audrey Muggleton-Harris. Perhaps we need to be cell sociologists as well as molecular biologists!

REFERENCES

1. **Witkowski, J. A.** Cell aging *in vitro*: a historical perspective, *Exp. Gerontol.,* 22, 231, 1987.
2. **Stanulis-Praeger, B. M.,** Cellular senescence revisited: a review, *Mech. Ageing Dev.,* 38, 1, 1987.
3. **Loo, D. T., Fuquay, J. I., Rawson, C. L., and Barnes, D. W.,** Extended culture of mouse embryo cells without senescence: inhibition by serum, *Science,* 236, 200, 1987.
4. **Suda, Y., Suzuki, M., Ikawa, Y., and Aizawa, S.,** Mouse embryonic stem cells exhibit indefinite proliferative potential, *J. Cell. Physiol.,* 133, 197, 1987.
5. **Juckett, D. A.,** Cellular aging (the Hayflick limit) and species longevity: a unification model based on clonal succession, *Mech. Ageing Dev.,* 38, 49, 1987.
6. **Martin, G. M.,** Proliferative homeostasis and its age-related aberrations, *Mech. Ageing Dev.,* 9, 385, 1979.
7. **Pringle, J. R. and Hartwell, L. H.,** The *Saccharomyces cerevisiae* cell cycle, in *Molecular Biology of the Yeast Saccharomyces: Life Cycle and Inheritance,* Strathern, J. N., Jones, E. W., and Broach, J. R., Eds., Cold Spring Harbor Laboratory, Cold Spring Harbor, NY, 1981, 97.
8. **Sporn, M. B. and Roberts, A. B.,** Peptide growth factors are multifunctional, *Nature,* 332, 217, 1988.

Section II. Positive Control of Cell Proliferation

Chapter 1

THE REGULATION OF CELL SENESCENCE

V. J. Cristofalo, P. D. Phillips, T. Sorger, and G. Gerhard

Our studies of the mechanisms of cell aging have focused on two areas. One has been to characterize cell cycle timing and arrest. These studies have shown that during the course of serial subcultivation, the percentage of cells participating in DNA synthesis decreases, the average cell cycle time increases, primarily due to an increase in G_1, and senescent cells become blocked with a 2C-DNA content, presumably in G_1.[1-3] Furthermore, when senescent cells are stimulated with serum or a mixture of growth factors, the activity of the enzyme thymidine kinase increases[4,5] and the thymidine triphosphate pool expands. Both of these events are associated with the late G_1 period and occur similarly in mitogen-stimulated young and old cells. In addition, the nuclear fluorescence pattern of stimulated senescent cells stained with quinacrine dihydrochloride resembles that of proliferating young cells in late G_1.[6] Finally, following growth factor stimulation, senescent cells continue to express mRNAs for several cell cycle-dependent genes, some of which are characteristic of late G_1.[7] All of these observations are consistent with the idea that senescent cells are blocked in late G_1 or at the G_1/S border.

The only way known to overcome this late G_1 block in senescent cells is by infection with SV40 virus.[8] Since both viral and host DNA replication depend on cell-encoded enzymes, and since a temperature-sensitive mutant of SV40 was used at the restrictive temperature, these data suggest that the cell DNA-synthesizing machinery is not damaged but rather is turned off in senescent cells.

One apparent "switching" mechanism for the initiation of DNA synthesis is growth factor stimulation. In the presence of growth factors, young cells will initiate the events which lead to DNA replication and then replicate the DNA, while senescent cells, although carrying out some G_1 events, do not synthesize DNA. A question of major interest is why and how the triggering action of specific growth factors for DNA synthesis fails in senescence.

Several years ago, we began a systematic study of the growth factor requirements of normal human fibroblast cell line WI-38. As a result of these studies, we devised a serum-free, growth factor-supplemented medium which would support multiple rounds of cell proliferation to an extent equivalent to that supported by 10% serum-supplemented medium. Our serum-free medium was composed of basal medium MCDB-104 plus platelet-derived growth factor (PDGF), epidermal growth factor (EGF), insulin (INS), transferrin (TRS), and dexamethasone (DEX).[9,10] Supplementing the basal medium with freshly prepared $FeSO_4$ eliminates the need for TRS. INS actually works as a mitogen in this system due to its ability to bind to the insulin-like growth factor-I (IGF-I) receptor. IGF-I is a potent mitogen at much lower concentrations (100 ng/ml) than INS (5 μg/ml).[11] Furthermore, IR-3, a monoclonal antibody which blocks the IGF-I receptor, also blocks the mitogenic effect of insulin.

In this chapter, we will examine some of our recent work focusing on growth factor responsiveness and various aspects of growth factor-cell interactions throughout the proliferative life span of WI-38 cells.

Figure 1 is a cartoon of a cell which shows a diagrammatic summary of some of the growth factors we are studying, their metabolism, and the putative second messengers and regulators associated with them. We have been systematically working from the outside of the cell inward to determine how these systems operate and to identify age-associated changes. We have examined cell responsiveness, receptor binding, and receptor autophos-

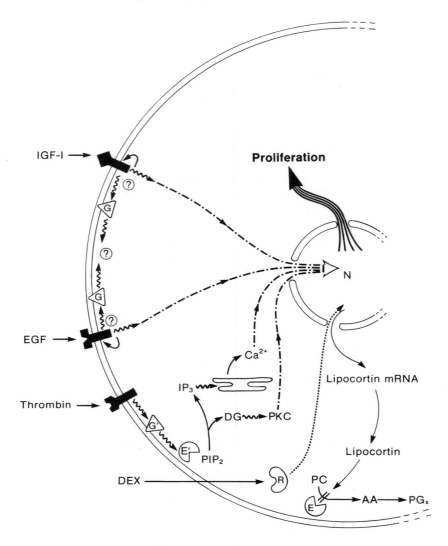

FIGURE 1.   A characterization of growth factor-cell interactions. G = G proteins; AA = arachidonic acid; PC = phosphatidylcholine; PG = prostaglandin; IP₃ = inositol triphosphate; PIP₂ = phosphatidyl inositol bisphosphate; PKC = protein kinase C; DG = diacylgylcerol; E′ = phospholipase C; E = phospholipase A.

phorylation for several growth factors. We have also studied the products of arachidonic acid (AA) metabolism as potential modulators of growth.

Table 1 shows cell responsiveness to various combinations of growth factors. Low-density, mitogen-deprived young cells were refed with the combinations of growth factors shown, along with ^{3}H-TdR. After 24 h, cells were fixed and prepared for autoradiography, and the percent labeled nuclei was scored by standard techniques.[1] Combining EGF, fibroblast growth factor (FGF), PDGF, thrombin (THR), IGF-I, and DEX gives a proliferative response equivalent to serum (compare experiments 1 and 13, Table 1). Equally effective, however, are IGF-I and DEX in combination with EGF (experiment 2), FGF (experiment 3), PDGF (experiment 4), or THR (experiment 5). However, by using EGF as a representative of these last four factors and testing various combinations of EGF, IGF-I, and DEX, we see that when any one is left out, the stimulation is barely more than basal medium alone (compare

TABLE 1
The Mitogenic Functional Equivalency of EGF, FGF, PDGF, and Thrombin

	EGF (25 ng/ml)	FGF (100 ng/ml)	PDGF (6 ng/ml)	THR (500 ng/ml)	IGF-I (100 ng/ml)	DEX (55 ng/ml)	L.N. (%)
1	+	+	+	+	+	+	58
2	+	−	−	−	+	+	51
3	−	+	−	−	+	+	60
4	−	−	+	−	+	+	53
5	−	−	−	+	+	+	60
6	+	−	−	−	+	−	19
7	+	−	−	−	−	+	20
8	−	−	−	−	+	+	20
9	+	−	−	−	−	−	18
10	−	−	−	−	+	−	18
11	−	−	−	−	−	+	16
12	−	−	−	−	−	−	9
13			10% Fetal bovine serum				59

Note: Low-density cultures of young cells (>50% life span completed) were made quiescent by refeeding with MCDB-104. After 48 h, they were refed as shown in the presence of 1 μCi/ml ^{3}H-TdR. After 24 h, they were fixed and prepared for autoradiography, and triplicate coverslips were scored for percent labeled nuclei (L.N.[%]).

TABLE 2
Mitogen Classification for WI-38 Cells

Class I	Class II[a]	Class III[b]
EGF[c]	IGF-I	HC[d]
FGF[c]	IGF-II	DEX
PDGF[c]	MSA[e]	
THR	INS	

[a] Function mitogenically through the IGF-I receptor system.
[b] Function mitogenically through the glucocorticoid receptor system.
[c] Separate Class I receptor systems.
[d] Hydrocortisone.
[e] Multiplication-stimulating activity.

experiments 6 to 8 with 12). This points to a synergistic effect. Note that in addition to the previously mentioned growth factors, FGF, THR, and IGF-I are also potent mitogens.

From these and related experiments we have constructed a classification scheme. The factors which we studied that stimulate cell proliferation in WI-38 cells can be placed into three classes, as shown in Table 2. The Class I mitogens include EGF, FGF, PDGF, and THR. These growth factors act via their own separate cell surface receptor systems.[12-15] The Class II mitogens are IGF-I (also known as somatomedin c), IGF-II (or the rat homolog multiplication-stimulating activity), and INS. These structurally related factors, however, all act by their varying abilities to bind to the IGF-I receptor.[16] Binding to their own receptors on these cells, however, does not mediate cell proliferation. The Class III mitogens are made up of hydrocortisone and the synthetic analog DEX. Both of these steroids operate through the glucocorticoid receptor system in WI-38 cells.[17]

TABLE 3
Young Cell Growth Supported
by Insulin or IGF-I

Growth factors used			Cells/cm² on day 7
EGF	INS	DEX	7.4×10^4
EGF	IGF-I	DEX	7.6×10^4
EGF		DEX	2.0×10^4

Note: Cultures of middle-aged cells were seeded at 1×10^4 cells/cm² into each of the growth factor combinations in MCDB-104. After 7 d, the cell number on triplicate wells for each set was determined by means of a Coulter Counter®.

The Class I mitogens are functionally equivalent in that when any one of them is combined with a Class II mitogen and a Class III mitogen, DNA synthesis is stimulated to an extent equivalent to that following the addition of fetal bovine serum. This means that for the maximum proliferative response, we must activate the glucocorticoid receptor system, the IGF-I receptor system, and any one of several other receptor systems.

The optimal concentrations for these growth factors are EGF at 25 ng/ml, FGF at 100 ng/ml, PDGF at 6.6 ng/ml, partially purified THR at 500 to 1000 ng/ml, IGF-I at 100 ng/ml, IGF-II at 400 ng/ml, INS at 5 μg/ml, and DEX at 55 ng/ml. In our experiments we routinely use EGF as the Class I mitogen, IGF-I as the Class II mitogen, and DEX as the Class III mitogen. In fact, these three factors support multiple rounds of cell proliferation. Table 3 shows that IGF-I, EGF, and DEX in combination are as effective as the trio of INS, EGF, and DEX in supporting growth and are significantly better than EGF and DEX alone.

We have examined all three growth factor receptor systems in some detail. There is no age-associated change in the binding of EGF[18] or IGF-I[11] to their receptors. There is also no change in the binding affinities of either growth factor-receptor complex. There is some decrease in the amount of DEX binding;[17] whether this is sufficient to account for the decreased responsiveness, however, is still unknown. There are no changes in any of the ligand-receptor affinities with age. The EGF and IGF-I receptors are known to be tyrosine-specific autocatalytic protein kinases.[19,20] For EGF receptors in membrane preparations, this enzyme activity appears to be unchanged with age.[21,22] However, when the EGF receptor is purified by detergent solubilization and immunoprecipitation, the autocatalytic kinase activity in senescent cell preparations is greatly reduced and nearly absent.[23] At present, we are pursuing the cause of this age-associated increased enzyme lability, but at this time neither its basis nor its functional significance is understood. We have only recently begun studies of the IGF-I receptor's tyrosine kinase activity and have not drawn any conclusions at this point. We have also begun a series of studies designed to characterize the PDGF receptor system in cultures of young and senescent cells. We have found that old cells can bind at least twice as much ^{125}I-PDGF as young cells. This is similar to what we have observed for EGF- and IGF-I-specific binding and presumably is accounted for by the increase in size of older cells. The K_d of the PDGF-receptor complex is approximately $2 \times 10^{-9}\ M$ and does not change with age. This is also similar to what we have observed for the EGF and IGF-I receptor systems. The PDGF binding data are summarized in Table 4.

We have also asked whether the PDGF receptor becomes phosphorylated on its tyrosine residues in response to PDGF in membranes prepared from young and old cells. Under these

TABLE 4

**^{125}I-PDGF-Specific Binding to Young and Old
Cells**

	Young cells	Old cells
Cpm specifically bound per 10^5 cells	450	900
50% displacement of ^{125}I-PDGF	2.8 nM	2.4 nM

Note: Confluent cultures of young (<50% life span completed)
and old (>90% life span completed) cells were incubated
at 4°C with 1 ng/ml of ^{125}I-PDGF and increasing concentrations of unlabeled PDGF. Nonspecific binding was determined in the presence of a 100-fold excess of unlabeled
PDGF.

TABLE 5

**Summary of Age-Associated Changes in the EGF, IGF-I,
PDGF, and DEX Receptor Systems**

	Receptor systems			
	EGF	IGF-I	PDGF	DEX
Binding sites per cell	I	I	I	D
Apparent K_d	NC	NC	NC	NC
Receptor phosphorylation in membranes	NC	ND	NC	NA
Receptor phosphorylation in solution	D	ND	ND	NA

Note: D = decreases with age; I = increases with age; NA = not applicable;
NC = no change with age; ND = not done.

conditions; the PDGF-stimulated phosphorylation of the receptor appears to be equivalent
in membranes from young and old cells. This is similar to what we have observed for the
EGF receptor.[22] We are now attempting to determine if there is an age-associated loss in
the autophosphorylating activity of PDGF receptors, similar to what we have observed for
solubilized EGF receptors. The properties of the EGF, IGF-I, PDGF, and DEX receptor
systems which we have studied are summarized in Table 5. In general, aging changes do
not seem to occur at the receptor level, although there is clear evidence from the EGF
receptor that molecular or regulatory changes have occurred in senescent cells.

As we described earlier, there are a variety of G_0/G_1 events which occur in response to
mitogen stimulation in both young and old cells. These include late G_1 events like increased
thymidine kinase activity, thymidine triphosphate pool expansion, and histone H3 gene
expression. Hence, at least some of these pathways appear complete or nearly so. Given
such similarities, it is potentially important to know if any sequential or temporal action
differences exist among the three growth factors. If growth factors act at different times, it
may be possible to dissect age-associated changes in responsiveness to specific growth
factors. This should make it possible to isolate the pathways involved in the loss of proliferative capacity. With this in mind, we have examined the timed addition of each of the
three growth factors and have monitored the entry of cells into DNA synthesis. Table 6
shows the results of such an experiment. Quiescent, mitogen-deprived young cells were
stimulated with EGF, IGF-I, and DEX at various times in the presence of ^{3}H-TdR. At the
times indicated, the cells, which were growing on coverslips, were fixed and prepared for

TABLE 6
The Effect of the Timed Addition of EGF, IGF-I, and DEX on the Entry of Cells into S Phase

Time of growth factor addition						Entry into S phase (h)
0	(h)		6	9	12	
EGF	IGF-I	DEX				9—12
	IGF-I	DEX	EGF			15—18
	IGF-I	DEX		EGF		21—24
	IGF-I	DEX			EGF	21—24
EGF		DEX	IGF-I			12—15
EGF		DEX		IGF-I		18—21
EGF		DEX			IGF-I	21—24
EGF	IGF-I		DEX			9—12
EGF	IGF-I			DEX		9—12
EGF	IGF-I				DEX	12—15

Note: Low-density cultures of young cells were made quiescent by refeeding with MCDB-104. After 48 h they were refed as shown and were also given 1 μCi/ml ^{3}H-TdR. At 3-h intervals, duplicate coverslips for each experimental condition were fixed and prepared for autoradiography, and percent labeled nuclei were scored. At the times shown, EGF, IGF-I, or DEX was added.

autoradiography and then scored for percent labeled nuclei. When EGF, IGF-I, and DEX were added at time 0, the cells began to enter DNA synthesis after about 12 h. When IGF-I and DEX were added at time 0 and EGF then added at 6, 9, or 12 h, there was a delay in the entry into DNA synthesis which was approximately equal to the time for which EGF was withheld. There was a similar pattern when EGF and DEX were added at time 0 and IGF-I at 6, 9, or 12 h. Although there is a suggestion that some cells enter S phase on time, there is a clear delay for the majority of entering cells which is approximately equal to the time for which IGF-I was withheld. The pattern is different, however, for DEX. DEX can be added to EGF and IGF-I at 6, 9, or 12 h without delaying the time of entry or affecting the magnitude of the response. DEX appears to act as a "trigger" or "gate" to allow cells which have otherwise progressed through G_1 to initiate DNA synthesis. Unlike either EGF or IGF-I, DEX does not have to be present continuously, only near the G_1/S boundary. This is, of course, a very exciting development in our view of a late G_1 arrest of senescent cells because it provides a probe for that late G_1 period.

Calcium is recognized as an important element in the regulation of cell proliferation. In light of this, we investigated the effects of various concentrations of extracellular calcium on the growth of WI-38 cells in serum-free medium with and without the addition of exogenous growth factors. At 5 mM CaCl$_2$, WI-38 cells seeded at low density without serum or hormone supplementation showed up to a 12-fold increase in cell number at saturation density over that obtained at day 1. Saturation densities were comparable when either 5 mM CaCl$_2$ or EGF (plus 1 mM CaCl$_2$) was used in the presence of INS and DEX. Combining suboptimal doses of EGF and CaCl$_2$ resulted in an additive effect on saturation density. Thus, normal human cells are capable of substantial growth in serum-free, growth factor-free medium. In contrast, confluent cultures refed with the same medium formulation are not responsive to elevated CaCl$_2$. In fact, elevated CaCl$_2$ inhibits the proliferative response of confluent cultures to EGF, but it enhances their response to the combination of INS and DEX.[24]

Lithium ion (Li$^+$) at concentrations of 2 to 20 mM has been reported to stimulate the

TABLE 7
Li⁺ Acts in a Synergistic Manner to Stimulate DNA Synthesis in the Presence of EGF and Insulin

Treatment	CPM (±SEM)	Stimulation (%)
No GFs[a]	1,850 (268)	—
No GFs + Li⁺	1,995 (355)	8
EGF + INS	9,111 (656)	—
EGF + INS + Li⁺	13,498 (762)	48
EGF + INS + DEX	13,298 (1,134)	46

Note: WI-38 cells which had been serum starved for 2 d were refed with fresh basal medium (MCDB-104). After 2 h, either no growth factors or EGF (25 ng/ml) and INS (5 μg/ml) were added to sets of four cultures, with or without LiCl (2 mM) or DEX (55 ng/ml). ³H-TdR (1 μCi/ml) was also added at this time, and the incorporation of this label into TCA-precipitable material was determined at the end of 24 h.

[a] GFs = growth factors.

proliferation of a variety of cell types, including lymphocytes,[25] mammary epithelial cells,[26-28] and 3T3 fibroblasts.[29] Several years ago, it was proposed that the growth-promoting effects of Li⁺ may be related to its ability to interfere with the phosphoinositide cycle,[30] in particular, to amplify the rapid increases in the levels of inositol triphosphate (IP3)[31-33] and diacylglycerol (DG)[34] caused by calcium-mobilizing agonists. These second messengers are thought to mediate the intracellular actions of certain mitogens, such as THR, but not others, such as EGF and INS.[35-37] However, the ability of Li⁺ to amplify the agonist-dependent generation of IP3 and DG has yet to be demonstrated in the same cells in which Li⁺ promotes growth.

We have investigated the action of Li⁺ on WI-38 cells in growth factor-supplemented medium. As described above, growth comparable to that seen in the presence of serum can be obtained in this medium using only EGF, IGF-I (or INS), and DEX. We found that Li⁺ was functionally equivalent to DEX in three important respects. Like DEX, Li⁺ stimulated DNA synthesis in a synergistic manner with EGF and INS (Table 7). When we examined the temporal requirement during the prereplicative phase in order for Li⁺ to exert this effect, we found that Li⁺, like DEX, needed to be present only during late G_1. This observation is difficult to reconcile with a postulated action of Li⁺ on phosphoinositide turnover early during the response to growth factors.

Both Li⁺ and DEX also caused a shift in the kinetics of DNA synthesis, preferentially stimulating replication at times *later* than the time of peak DNA synthesis in control cultures stimulated only by EGF and INS. This finding is consistent with a model in which both Li⁺ and DEX can recruit into the cell cycle a subpopulation of cells with a long prereplicative phase relative to the control population.

Finally, we asked whether the ability of Li⁺ to mimic the action of DEX in these respects was dependent on the glucocorticoid receptor. Using the specific glucocorticoid receptor antagonist RU486,[38] we found that Li⁺ stimulated DNA synthesis even at inhibitor concentrations which completely blocked the action of DEX. The interesting conclusion from these results is that Li⁺ may exert its effects on growth by activating the glucocorticoid response pathway at a site distal to hormone binding and translocation of the receptor.

The metabolism of AA (Figure 2) has received increasing attention since the demonstration

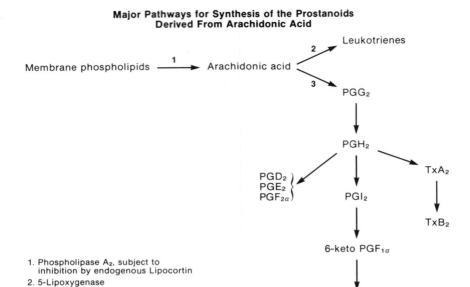

FIGURE 2.   Major pathways for synthesis of the prostanoids derived from arachidonic acid.

that two mitogens, THR and PDGF, can elicit the rapid mobilization of AA from cellular phospholipid and its conversion to prostaglandins (PG).[39,40] However, a number of observations make it appear unlikely that there is a general role for prostanoids in the control of proliferation. First, not all of the cell types in which AA metabolism is stimulated by PDGF respond mitogenically to this factor.[39] Moreover, other growth factors, such as EGF and INS, do not appear to activate this pathway.[40a] Finally, the effects of PGs on cell growth can vary, depending on the particular PG, the target cell, and the presence of other growth factors. For example, PG_{E_2} (at concentrations of 10^{-7} M or less) (1) stimulates the proliferation of mammary epithelial cells,[41] (2) has no effect alone, but blocks the stimulation of DNA synthesis by $PG_{F_{2\alpha}}$ in endometrial cells,[42] and (3) inhibits the proliferation of granulocyte/macrophage stem cells and lung fibroblasts.[43,44]

Several studies have shown that there is a selective increase in the production of PG_{E_2} as lung fibroblasts and other types of cells are subcultivated *in vitro*. These findings have led us to consider that late-passage cells may be refractory to mitogenic stimulation, in part because of the negative autocrine control of growth by PG_{E_2}. After reviewing the evidence for increased PG_{E_2} synthesis as a function of *in vitro* age, we shall describe some of the experiments that we are currently conducting to test this hypothesis.

Murota et al.,[45] using homogenates prepared from early-, mid-, and late-passage human embryonic lung fibroblasts, demonstrated a progressive decrease in the conversion of radiolabeled AA into 6-keto $PG_{F_{1\alpha}}$ (a stable metabolite of PG_{I_2} or prostacylin), with a corresponding increase in the appearance of PG_{E_2}, $PG_{F_{2\alpha}}$, and thromboxane B2 (TX_{B_2}).[45] These authors also reported that clones derived from single cells were capable of elaborating both PG_{I_2} and TX_{B_2}, suggesting that the shift in the distribution of PGs in aging cells reflected a switching off of PG_{I_2} synthesis and diversion of the precursor PG_{h_2} into the competing biosynthetic pathways rather than selection against a subpopulation of PG_{I_2}-producing cells[45] (see Figure 2).

TABLE 8
Inhibition of Proliferation by PG_{E_2}

PG_{E_2} (ng/ml)	Cells × $10^4/cm^2$ (±SD)	Inhibition (%)
0	6.14 (0.12)	—
1	5.26 (0.13)	14
10	3.55 (0.08)	42
100	2.71 (0.04)	56

Note: WI-38 cells seeded at $1 \times 10^4/cm^2$ and serum starved for 48 h were refed with fresh basal medium supplemented with EGF, INS, and DEX. Duplicate cultures received various concentrations of PG_{E_2} at this time, and the number of cells in each culture was determined 1 week later. The differences between final cell densities at successive concentrations of PG_{E_2} were all significant ($p < 0.02$).

A similar change in the pattern of prostanoid metabolism was found as a result of *in vivo* aging in homogenates of rat aortic smooth muscle cells which had grown out in culture and had been subcultivated fewer than five times.[46] In both of these studies, total PG production by young and old cells was similar.[45,46]

Based on these and other published data, it appears that in the course of repeated subcultivation, lung fibroblasts and vascular smooth muscle cells generally lose the capacity to elaborate PG_{I_2}, while the activity of one or both of the competing pathways of prostanoid biosynthesis may be enhanced.

In our own studies of AA metabolism in WI-38 cells, we began by looking at the sensitivity of growth to PG_{E_2} and the basal rate of PG_{E_2} output in serum-free medium MCDB-104.[47] When this medium was supplemented with EGF, INS, and DEX, it supported the rapid proliferation of WI-38 cells. PG_{E_2} added at the same time as the growth factors inhibited proliferation in a dose-dependent manner, an effect which was small (-14%) but highly significant even at 1.0 ng/ml (Table 8).

The same concentration of PG_{E_2} (1.0 ng/ml) was more effective at inhibiting DNA synthesis in a short-term assay. This action could be reproduced by adding 0.33 ng/ml PG_{E_2} three times, at 12-h intervals (Table 9). These results are consistent with the notion that the effective concentration of added PG_{E_2} during a prolonged incubation is reduced by cellular uptake and/or metabolism.

We then determined the relationship between *in vitro* age and the output of PG_{E_2} into serum-free medium. As shown in Table 10, there was an expected decline in the cell density achieved by cultures at higher population doubling levels. On the other hand, the production of PG_{E_2} increased as a function of population doubling level, so that the concentration of PG_{E_2} in the medium of cells at population doubling level 60 was almost three times that found in the medium of cells at population doubling level 32, corresponding to a ninefold increase in PG_{E_2} per cell (Table 10).

Since the release of AA is thought to be rate limiting for the synthesis of PGs, we compared the rates of release of ^3H from early- and late-passage cultures which had been labeled with ^3H-AA. As can be seen in Table 11, in the first 8 h after the cultures were rinsed and refed

TABLE 9

Inhibition of DNA Synthesis by

PG_{E_2}

PG_{E_2} (ng/ml)	CPM ($\pm$ SD)	Inhibition (%)
0	9962 (300)	—
0.33	7082 (55)[a]	19
1.00	4812 (473)[b]	52
0.33 × 3	4745 (180)[b]	52

Note: WI-38 cells which had been serum starved for 48 h were stimulated to initiate DNA synthesis by the addition of EGF, INS, and DEX in fresh defined medium. At this time (0 h), duplicate cultures received 0, 0.33, or 1.00 ng PG_{E_2} in 1 μl ethanol. One group received 0.33 ng PG_{E_2} again at 12 and 24 h, while the other groups received 1 μl ethanol at these times. The incorporation of ³H-TdR into TCA-precipitable material was determined 40 h after addition of growth factors.

[a] Significantly different from the control value ($p < 0.01$).
[b] Significantly different from the value at 0.33 ng/ml ($p < 0.05$).

TABLE 10

PG_{E_2} Output as a Function of *In Vitro* Age

PDL[a]	PG_{E_2} (pg/ml)	Cells × 10^4/cm²	PG_{E_2} (pg/10^4 cells)
32	102 ± 18	12.0 ± 0.5	8.5
45	106 ± 36[b]	7.0 ± 0.4[b]	23.7
60	202 ± 26[c]	3.1 ± 0.2[c]	91.0

Note: WI-38 cells at different PDLs were allowed to grow for 1 week in serum-containing medium and were then serum starved for 48 h. The cultures were then rinsed and refed with fresh defined medium (MCDB-104). The number of cells and the PG_{E_2} concentration in the medium were determined for each culture 24 h later. The data represent the mean ± SD of four cultures per group.

[a] PDL = population doubling level.
[b] Significantly different from values at PDL 32 ($p < 0.02$).
[c] Significantly different from values at PDL 45 ($p < 0.005$).

with fresh medium, senescent cells released ³H (predominantly ³H-AA and metabolites) at three times the rate of release from young cells.

When cultures were treated with indomethacin to inhibit the enzyme cyclo-oxygenase and prevent the conversion of AA into PGs (Figure 2), the release of AA from young cells was stimulated threefold, to a rate similar to that seen with untreated senescent cells. The latter showed a much weaker response to indomethacin (Table 11). These data are consistent with

TABLE 11
Effects of *In Vitro* Age and Indomethacin on the Release of ^{3}H-AA

PDL	+ Indomethacin[a]	− Indomethacin[a]	Stimulation (%)
29	248 ± 79	899 ± 13[b]	262
64	806 ± 158[c]	1318 ± 43	63

Note: Early- and late-passage WI-38 cells, which had been serum starved for 48 h and labeled with ^{3}H-AA, were rinsed and refed with fresh defined medium. Indomethacin (2 μg/ml) was added to half of the cultures, and the release of ^{3}H into the medium was determined after 8 h. The data represent the mean values (± SD) of duplicate cultures.

[a] cpm/8 h/10^4 cells.
[b] Significantly different from the control value at PDL 29 ($p < 0.005$).
[c] Significantly different from the control value at PDL 29 ($p < 0.05$).

TABLE 12
Lipocortin Content of Membranes from Early- and Late-Passage Cells

	^{125}I-Protein A specifically bound (cpm/μg membrane protein)		Change (%)
	PDL 32	PDL 60	
Lipocortin I	700 ± 192	654 ± 86	−7
Lipocortin II	416 ± 64	262 ± 25	−35

Note: Crude membrane preparations were resolved by polyacrylamide gel electrophoresis, transferred to nitrocellulose, and blotted with antisera directed against lipocortins I and II.[51] Following incubation with ^{125}I-protein A and autoradiography, the bands corresponding to lipocortin I (36 kDa) and lipocortin II (35 kDa) were excised and counted. The data represent the mean values (± SD) of duplicate cultures at each PDL. The antisera were generously provided by Dr. Blake Pipinsky, Biogen Corp., Cambridge, MA.

the notions that a product of cyclo-oxygenase acts by negative feedback to inhibit the release of AA and that late-passage cells fail either to produce or to respond to this factor.

Control of the release of AA is also exerted by endogenous inhibitors of the enzyme phospholipase-A_2, collectively referred to as lipocortin.[48-50] It is conceivable that the accelerated release of AA from senescent cells is related to a deficiency of lipocortin. However, Western blot analysis of crude membrane preparations from mid- and late-passage cells revealed no change in the level of lipocortin I and only a 35% decrease in the level of lipocortin II, the two being predominant phospholipase-A_2 inhibitors (Table 12).

Taken together, these findings suggest an altered regulation of AA metabolism in late-passage cells. Under the serum-free conditions of our experiments, senescent cells elaborate substantially more PG_{E_2} than do young cells, and the concentrations achieved (approximately 0.3 ng/ml) are in the range of the effective concentrations for the inhibition of growth of early-passage cells when this compound is added exogenously (see Table 9).

Our studies have led us to conclude that senescent cells respond to growth factors in part in much the same way as young cells. The receptor systems are largely unchanged with age, although some subtle modifications do occur. Furthermore, many of the early growth factor-initiated events occur in a similar way in both young and old cells. This has led us to theorize that senescent cells are not arrested in the same way as mitogen-deprived young cells, but rather become blocked at a new arrest point in late G_1 just prior to entry into DNA synthesis.

REFERENCES

1. **Cristofalo, V. J. and Sharf, B. B.,** Cellular senescence of DNA synthesis. Thymidine incorporation as a measure of population age in human diploid cells, *Exp. Cell Res.,* 76, 419, 1973.
2. **Yanishevsky, R., Mendelsohn, M., Mayall, B., and Cristofalo, V. J.,** Proliferative capacity and DNA content in aging human diploid cell cultures: a cytophotometric and autoradiographic analysis, *J. Cell. Physiol.,* 84, 165, 1974.
3. **Grove, G. L. and Cristofalo, V. J.,** Characterization of cell cycle of cultured human diploid cells: effects of aging and hydrocortisone, *J. Cell. Physiol.,* 90, 415, 1977.
4. **Cristofalo, V. J.,** Cellular senescence: factors modulating cell proliferation *in vitro, INSERM,* 27, 65, 1973.
5. **Olashaw, N. E., Kress, E. D., and Cristofalo, V. J.,** Thymidine triphosphate synthesis in senescent WI-38 cells, *Exp. Cell Res.,* 149, 547, 1983.
6. **Gorman, S. D. and Cristofalo, V. J.,** Analysis of the G_1 arrest position of senescent WI-38 cells by quinacrine dihydrochloride nuclear fluorescence: evidence for a later G_1 arrest, *Exp. Cell Res.,* 167, 87, 1986.
7. **Rittling, S. R., Brooks, K. M., Cristofalo, V. J., and Baserga, R.,** Expression of cell cycle-dependent genes in young and senescent WI-38 fibroblasts, *Proc. Natl. Acad. Sci. U.S.A.,* 83, 3316, 1986.
8. **Gorman, S. D. and Cristofalo, V. J.,** Reinitiation of cellular DNA synthesis in BrdU-selected nondividing senescent WI-38 cells by simian virus 40 infection, *J. Cell Physiol.,* 125, 122, 1985.
9. **Phillips, P. D. and Cristofalo, V. J.,** A procedure for the serum-free growth of normal human fibroblasts, *J. Tissue Cult. Meth.,* 6, 123, 1980.
10. **Phillips, P. D. and Cristofalo, V. J.,** Growth regulation of WI-38 cells in a serum-free medium, *Exp. Cell Res.,* 134, 297, 1981.
11. **Phillips, P. D., Pignolo, R. J., and Cristofalo, V. J.,** Insulin-like growth factor-I: specific binding to high and low affinity sites and mitogenic action throughout the lifespan of WI-38 cells, *J. Cell. Physiol.,* 133, 135, 1987.
12. **Carney, D. H., Steirnberg, J., and Fentor, J. W.,** Initiation of proliferative events by human-thrombin requires both receptor binding and enzymatic activity, *J. Cell. Biochem.,* 26, 181, 1984.
13. **Carpenter, G. and Cohen, S.,** ^{125}I-Labeled human epidermal growth factor, *J. Cell Biol.,* 71, 159, 1976.
14. **Heldin, C.-H., Westermark, B., and Wasteson, A.,** Specific receptors for platelet-derived growth factor on cells derived from connective tissue and glia, *Proc. Natl. Acad. Sci. U.S.A.,* 78, 3664, 1981.
15. **Schreiber, A. B., Kenney, J., Kowalski, W. J., Friesel, R., Mehlman, T., and Maciag, T.,** Interaction of endothelial cell growth factor with heparin: characterization by receptor and antibody recognition, *Proc. Natl. Acad. Sci. U.S.A.,* 82, 6138, 1985.
16. **Van Wyk, J. J., Graves, D. C., Casella, S. J., and Jacobs, S.,** Evidence from monoclonal antibody studies that insulin stimulates deoxyribonucleic acid synthesis through the type I somatomedin receptor, *J. Clin. Endocrinol. Metab.,* 61, 639, 1985.
17. **Rosner, B. A. and Cristofalo, V. J.,** Changes in specific dexamethasone binding during aging in WI-38 cells, *Endocrinology,* 108, 1965, 1981.
18. **Phillips, P. D., Kuhnle, E., and Cristofalo, V. J.,** [^{125}I]EGF binding ability is stable throughout the replicative lifespan of WI-38 cells, *J. Cell. Physiol.,* 114, 311, 1983.
19. **Carpenter, G., King, L., and Cohen, S.,** Epidermal growth factor stimulated the phosphorylation in membrane preparations *in vitro, Nature,* 276, 409, 1978.
20. **Jacobs, S., Kull, F. C., Earp, H. S., Suoboda, M. E., Van Wyk, J. J., and Cuatrecasas, P.,** Somatomedin-C stimulates the phosphorylation of the α-subunit of its own receptor, *J. Biol. Chem.,* 258, 9581, 1983.
21. **Chua, C. C., Geiman, D. E., and Ladda, R. L.,** Receptor for epidermal growth factor retains normal structure and function in aging cells, *Mech. Ageing Dev.,* 34, 35, 1986.

22. **Brooks, K. M., Phillips, P. D., Carlin, C. R., Knowles, B. B., and Cristofalo, V. J.,** EGF-dependent phosphorylation of the EGF receptors in plasma membranes isolated from young and senescent WI-38 cells, *J. Cell. Physiol.,* in press.

23. **Carlin, C. R., Phillips, P. D., Knowles, B. B., and Cristofalo, V. J.,** Diminished *in vitro* tyrosine kinase activity of the EGF receptor of senescent human fibroblasts, *Nature,* 306, 617, 1981.

24. **Praeger, F. C. and Cristofalo, V. J.,** The growth of WI-38 cells in a serum-free medium with elevated calcium concentration, *In Vitro,* 22, 355, 1986.

25. **Hart, D. A.,** Evidence that lithium ions can modulate lectin stimulation of lymphoid cells by multiple mechanisms, *Cell. Immunol.,* 58, 372, 1981.

26. **Hori, C. and Oka, T.,** Induction by lithium ions of multiplication of mammary epithelium in culture, *Proc. Natl. Acad. Sci. U.S.A.,* 76, 2823, 1979.

27. **Ptashne, K., Stockdale, F. E., and Conkon, S.,** Initiation of DNA synthesis in mammary epithelium and mammary tumors by lithium ions, *J. Cell. Physiol.,* 103, 44, 1980.

28. **Tomooka, Y., Imagawa, W., Nandi, S., and Bern, H. A.,** Growth effect of lithium on mouse mammary epithelial cells in serum-free collagen gel culture, *J. Cell. Physiol.,* 117, 290, 1983.

29. **Rybak, S. M. and Stockdale, F. E.,** Growth effects of lithium chloride in BALB/c3T3 fibroblasts and Madin-Darby canine kidney epithelial cells, *Exp. Cell Res.,* 136, 263, 1981.

30. **Berridge, M. J. and Irvine, R. F.,** Inositol triphosphate, a novel second messenger in cellular signal transduction, *Nature,* 312, 315, 1984.

31. **Drummond, A. H., Bushfield, M., and Macphee, C. H.,** Thyrotropin-releasing hormone-stimulated [^{3}H]-inositol metabolism in GH3 pituitary tumor cells, *Mol. Pharmacol.,* 25, 201, 1984.

32. **Rubin, R. P., Godfrey, P. P., Chapman, D. A., and Putney, J. W.,** Secretagogue-induced formation of inositol phosphates in rat exocrine pancreas, *Biochem. J.,* 219, 655, 1984.

33. **Thomas, A. P., Alexander, J., and Williamson, J. R.,** Relationship between inositol polyphosphate production and the increase of cytosolic free Ca^{++} induced by vasopressin in isolated hepatocytes, *J. Biol. Chem.,* 259, 5574, 1984.

34. **Drummond, A. H. and Raeburn, C. A.,** The interaction of lithium with thyrotropin-releasing hormone-stimulated lipid metabolism in GH3 pituitary tumor cells, *Biochem. J.,* 224, 129, 1984.

35. **Besterman, J. M., Watson, S. P., and Cuatrecasas, P.,** Lack of association of epidermal growth factor, insulin and serum induced mitogenesis with stimulation of phosphoinositide degradation in BALB/c3T3 fibroblasts, *J. Biol. Chem.,* 261, 723, 1986.

36. **Chambard, J. C., Paris, S., L'Allemain, G., and Pouyssegur, J.,** Two growth factor signalling pathways in fibroblasts distinguished by pertussis toxin, *Nature,* 326, 800, 1987.

37. **L'Allemain, G. and Pouyssegur, J.,** EGF and insulin action in fibroblasts, *FEBS Lett.,* 197, 344, 1986.

38. **Moguilewsky, M. and Philibert, D.,** RU 38486: potent antiglucocorticoid activity related in the strong binding to the cytosolic glucocorticoid receptor followed by an impaired activation, *J. Steroid Biochem.,* 20, 271, 1984.

39. **Coughlin, S. R., Moskowitz, M. A., Zetter, R. R., Antoniades, H. N., and Levine, C.,** Platelet-dependent stimulation of prostacyclin synthesis by platelet-derived growth factor, *Nature,* 288, 600, 1980.

40. **Hong, S.-C. L. and Levine, L.,** Stimulation of prostaglandin synthesis by bradykinin and thrombin and their mechanisms of action on MC5-5 fibroblasts, *J. Biol. Chem.,* 251, 5814, 1976.

40a. **Sorger, T. and Cristofalo, V. J.,** unpublished data.

41. **Rudland, P. S., Davies, A. C. T., and Tsao, S.-W.,** Rat mammary preadipocytes in culture produce a trophic agent for mammary epithelia:prostaglandin E$_2$, *J. Cell. Physiol.,* 120, 364, 1984.

42. **Orlicky, D. J., Lieberman, R., and Gerschenson, L. E.,** Prostaglandin F$_{2\alpha}$ and E$_1$ regulation of proliferation in primary cultures of rabbit endometrial cells, *J. Cell. Physiol.,* 127, 55, 1986.

43. **Freundlich, B., Bomalaski, J. S., Neilson, E., and Jiminez, S. A.,** Regulation of fibroblast proliferation and collagen synthesis by cytokines, *Immunol. Today,* 7, 303, 1986.

44. **Pelus, L. M.,** Prostaglandin E: biphasic control of hematopoiesis, in *Biological Protection with Prostaglandins,* Vol. 1, Cohen, M. M., Ed., CRC Press, Boca Raton, FL, 1985, 45.

45. **Murota, S.-I., Mitsui, Y., and Kawamura, M.,** Effect of in vitro aging on 6-ketoprostaglandin F$_{1\alpha}$-producing activity in cultured human diploid lung fibroblasts, *Biochim. Biophys. Acta,* 574, 351, 1979.

46. **Chang, W.-C., Murota, S.-I., Nakao, J., and Orimo, H.,** Age-related decrease in prostacyclin biosynthetic activity in rat aortic smooth muscle cells, *Biochim. Biophys. Acta,* 620, 159, 1980.

47. **McKeehan, W. C., McKeehan, K. A., Hammond, S. C., and Ham, R. G.,** Improved medium for clonal growth of human diploid fibroblasts at low concentrations of serum protein, *In Vitro,* 13, 399, 1977.

48. **Brugge, J. S.,** The p35/p36 substrates of protein-tyrosine kinase as inhibitors of phospholipase A2, *Cell,* 46, 149, 1986.

49. **Flower, R. J., Wood, J. N., and Parente, L.,** Macrocortin and the mechanism of action of the glucocorticoids, *Adv. Inflammation Res.,* 7, 61, 1984.

50. **Hirata, F.,** Roles of lipomodulin: a phospholipase inhibitory protein in immunoregulation, *Adv. Inflammation Res.,* 7, 71, 1984.

51. **Huang, K. S., Wallner, B. P., Mattaliano, R. J., Tizard, R., Burne, C., Frey, A., Hession, C., McGray, P., Sinclair, L. K., Chow, E. P., Browning, J. C., Ramachandran, K. C., Tang, J., Smart, J. E., and Pepinsky, R. B.,** Two human 35 kd inhibitors of phospholipase A_2 are related to substrates of pp60$^{v\text{-}src}$ and of the epidermal growth factor receptor/kinase, *Cell,* 46, 191, 1986.

Chapter 2

EXPRESSION OF GROWTH-RELATED GENES IN SENESCENT HUMAN DIPLOID FIBROBLASTS

Susan R. Rittling and Renato Baserga

TABLE OF CONTENTS

I. INTRODUCTION

It is now generally agreed that reproductive failure is the fundamental characteristic of cellular senescence in cultures of human diploid fibroblasts. Therefore, an understanding of normal growth and the factors that control it is the prerequisite for understanding cellular senescence *in vitro*. It is then not surprising that one of the major questions that biologists in the field of aging have been asking for the past few years is, "What regulates the extent of cell proliferation?".

We can look at this problem in several ways. From a cell biologist's point of view, populations of cells, whether in tissue culture or in multicellular organisms, consist of a mixture of three different subpopulations:

1. Cells that continuously proliferate, i.e, cycling cells that go from one mitosis to the next one
2. G_0 cells, i.e., nondividing, noncycling cells still capable of reentering the cell cycle upon the application of an appropriate stimulus
3. Terminally differentiated cells that are destined to die without dividing again

The number of cells produced at any given time depends on the length of the cell cycle as well as on the proportion of cells that are in the cell cycle. Therefore, growth of a population of cells depends on the respective proportions of these three subpopulations, the rate of cell death, and the rapidity with which cycling cells proliferate. In the adult animal, in which there is no net growth, the number of cells that die in a given time equals the number of cells that are produced in the same time (for a review see Baserga[1]). Whenever the number of cells produced exceeds the number of cells that die, there is growth. Conversely, whenever the number of cells that die exceeds the number of cells that are produced, the tissue or cell population undergoes atrophy and the cell population eventually disappears. In cultures of senescent human diploid fibroblasts, cells continue to die and they are not replaced by newly produced cells, resulting eventually in the extinction of the cultured population. We can now rephrase the previous question ("What regulates cell proliferation?") and ask, *"What regulates the length of the cell cycle, the proportion of cells that are cycling, and the ability of the G_0 cells to reenter the cell cycle?"*. There are no answers to these questions yet, but today it is possible to reduce the study of cell proliferation to a reasonably simple proposition, namely that the extent of cell proliferation depends on (1) the environmental signals (i.e., the stimulatory and inhibitory growth factors in the environment that regulate cell proliferation) and (2) the genes and gene products that interact with and respond to these growth factors. By studying cells in culture, where we can control the environment by using defined media, we reduce the problem further, to the genes and gene products required for cell proliferation. Thus, as we will see below, we can use what we know about the genes which respond to growth factors to begin to address the three questions posed above. First we will discuss the genes themselves, then their role in these processes. We shall use for these genes the general term "growth-related genes", to include genes that may play a role in the regulation of cell proliferation in animal cells as well as genes that are simply growth regulated.

II. GROWTH-RELATED GENES

Growth-related genes can be divided into three classes based on their method of identification. There are three different approaches to the identification of genes controlling cell reproduction in animal cells, namely

1. The identification of proto-oncogenes, i.e., the cellular equivalents of retroviral transforming genes
2. The isolation and molecular cloning of those genes that can complement defects in temperature-sensitive mutants of the cell cycle
3. The identification of sequences that are growth regulated, i.e., genes preferentially expressed in a specific phase of the cell cycle or inducible by growth factors

This last systematic approach usually requires differential screening of cDNA libraries.

These three approaches identify genes which, in the first two cases, have a demonstrated role and, in the third case, have a potential role in cell proliferation.

A. CELLULAR ONCOGENES

It seems reasonable to assume that cellular oncogenes must have something to do with the regulation of animal cell proliferation, because when they are changed by point mutations, deletions, or insertions into a retroviral vector, or when they are overexpressed, they can transform cells in culture. We can say, therefore, that an alteration in a cellular oncogene can result in an alteration in the regulation of cell growth. A discussion of oncogenes and of the mechanisms by which they are activated requires a separate review, and the reader is referred, for this purpose, to other sources.[2,3] At the moment, we will limit ourselves to stating that the identification of cellular oncogenes is a very important and crucial approach to our understanding of cell proliferation. It is not, however, directly related to the objective of this review, and so the discussion of oncogenes will be limited to the expression of cellular oncogenes in senescent WI-38 cells.

B. TEMPERATURE-SENSITIVE (*ts*) MUTANTS OF THE CELL CYCLE

The usefulness of conditionally lethal mutants for studying structure and function in biological materials has been amply demonstrated. In the past 20 years, a large number of cell cycle-specific mutants have been isolated and partially characterized, both in yeasts and in animal cells. Animal cell *ts* mutants of G_0, G_1, S phase, G_2, and even of mitosis have been isolated. (For reviews see Basilico,[4] Marcus et al.,[5] and Baserga.[1]) The *ts* mutants of the cell cycle offer a very interesting set of cell lines which one can use to identify and molecularly clone genes that are necessary for cell cycle progression. It is obvious that these genes also have something to do with cell proliferation, since when their products are defective the cell stops at a precise point in the cell cycle. Molecular cloning of the wild-type alleles of the genes complementing these *ts* mutants is therefore of extreme importance, since it will identify a class of genes that is essential for cell cycle progression. Unfortunately, very little is known about the wild-type alleles of cell cycle *ts* genes in animal cells, and nothing is known about their expression during the cell cycle in young and senescent cells. Therefore, they will not be considered further in this review.

C. GROWTH-REGULATED GENES

Growth-regulated genes refer to those genes whose expression is regulated by growth factors or by the cell cycle. A number of genes have been shown to be cell cycle dependent, i.e., to be expressed in a cell cycle-dependent manner. A partial list of these genes is given in Table 1. Several growth-regulated genes have been identified as cDNA clones by differential screening of cDNA libraries. The strategy is straightforward and has been followed by a number of laboratories. There are several pitfalls and criticisms to this approach, but the importance of growth-related genes in cell proliferation is strengthened by the finding that several bona fide oncogenes are expressed in a cell cycle-dependent manner.[8-11] Table 1 provides a list of the oncogenes whose expression has been shown to be regulated by growth factors. The fact that these oncogenes are inducible by growth factors does not

TABLE 1
Cell Cycle-Dependent Genes

Growth-regulated genes	Growth-regulated oncogenes
Core histones	c-*fos*
DNA topoisomerase I	
Thymidine kinase	c-*myc*
DNA polymerases	
Thymidylate synthase	c-*ras*
Ribonucleotide reductase	
Dihydrofolate reductase	c-*fgr*
Cyclin	
Interleukin-2 receptor	p53
Transferrin receptor	
Ornithine decarboxylase	
Major excreted protein	
Mitogen-regulated protein	
Calmodulin	
Interferons	
Vimentin	
β-Actin	
Four glycolytic enzymes	
Collagen and collagenase	
ADP/ATP carrier	
Calcyclin	
Transin	

Note: Appropriate references can be found in books and reviews by Baserga,[1] Kaczmarek,[6] and Denhardt et al.[7]

necessarily mean that they regulate the cell cycle. However, these oncogenes, when properly modified as mentioned above, are known to transform cells, and therefore they must have some influence on growth regulation. One can turn this statement around and say that if oncogenes can be induced by growth factors, then one can look at other growth factor-inducible cellular genes and perhaps identify possible candidates for potential new oncogenes or, at any rate, genes that play a role in the control of cell proliferation. This actually turned out to be true for the p53 gene, which was first demonstrated by Reich and Levine[12] to be growth regulated and subsequently was identified as a bona fide oncogene that can replace c-*myc* in transformation experiments.[13,14] Interestingly, no p53 sequence has ever been found in a retroviral transforming gene, not even a remotely homologous sequence. It is therefore clear that some potential oncogenes have never been captured by retroviruses and that growth-regulated genes form another set of cellular genes that are of great interest in our understanding of animal cell growth.

As mentioned above, several laboratories have used this approach and have identified, by differential screening of cDNA libraries, many clones whose cognate RNAs are either growth factor inducible or cell cycle regulated. Often these genes are still inducible in the presence of concentrations of cycloheximide that completely suppress protein synthesis,[15] indicating that they are primary responders. The genes that have been identified by this approach represent an interesting mixture. For instance, Linzer and Nathans[16] have isolated a cDNA that has a 70% homology with prolactin, which is a growth factor, and this cDNA turns out to be identical to a mitogen-regulated protein described previously.[17] A number of growth-regulated genes in several laboratories turned out to have repetitive sequences,[18] but the meaning of these genes, which contain highly repetitive sequences and are inducible by growth factors, is at present unclear. Matrisian et al.[19] identified a group of cDNA clones whose cognate RNAs were inducible by growth factors. Of these cDNAs, one was β-actin and four were glycolytic enzymes (lactate dehydrogenase, glyceraldehyde triphosphate dehydrogenase, enolase, and triosephosphate isomerase). It has been known for a long time that triosephosphate isomerase is strongly growth regulated. We should also add that the RAS-2 gene of the budding yeast is also required for glucose metabolism.[20]

Of the three cDNA clones isolated by Hirschhorn et al.[21] as growth regulated, one is an ATP/ADP translocase[22] which is the major energy carrier of the cell. Another is a cytoskeletal protein, vimentin,[23] while the third one, which was called 2A9, turned out to code for a calcium-binding protein, specifically a gene that has a strong homology to the S100 protein and other calcium-binding proteins and has since been renamed calcycline.[24] Other cDNA clones isolated by differential screening of cDNA libraries have also turned out to code for proteins like cysteine protease, an inhibitor of metalloproteins, or cathepsin B.[7]

Despite this strange assortment of growth-regulated genes, there is no question that the identification of such genes is an important approach to our understanding of cell proliferation. There are three main reasons for studying growth-related genes. The first reason is to identify genes which control cell proliferation and which define the intracellular pathway from growth factor to DNA synthesis. Second, the mechanisms of regulation of any growth-related genes can give insight into this pathway. Thirdly, the expression of growth-related genes represents a series of intracellular events that can be used as markers of cell progression from G_0 to S. Perhaps we will not find among these sets of genes the regulatory genes in which we are fundamentally interested, but undoubtedly our knowledge of how these genes are growth regulated and how they participate in the mitogenic process will help us throw considerable light on the mechanisms by which the extent of cell proliferation is regulated in animal cells. In addition, we can use the expression of these genes as an assay for intracellular events occurring in response to growth factors. We shall therefore examine, together with the celluar oncogenes, several of these growth-regulated genes and explore how they are expressed in young and senescent WI-38 cells in order to elucidate the mechanism of the block to cell proliferation which exists in the senescent population.

III. GENE EXPRESSION AND CELLULAR SENESCENCE

The WI-38 cell system allows us to address two of the questions posed earlier: what regulates the length of the cell cycle, and what regulates the ability of G_0 cells to reenter the cell cycle. By leaving WI-38 cells quiescent for different times, we can experimentally manipulate the length of the cell cycle. By studying WI-38 cultures of different ages, we are in effect altering the ability of G_0 cells to reenter the cell cycle. Because these experiments are done in a controlled way, so that the environment of the cells is constant, we know that the differences between those populations do not reside at the level of the environmental signals; rather, differences in the genes and gene products which respond to these signals must be responsible for the differences between these populations of cells. What follows is a review of what is known about the expression of growth-related genes in these different experimental conditions.

WI-38 human diploid fibroblasts were originally isolated by Hayflick and Moorhead.[25] They have a limited life span, and they usually stop proliferating after 50 to 60 population doublings.[26] These cells have been extensively studied both in the laboratory of Vincent Cristofalo[27] and in our own. Cultures of WI-38 cells are capable of a prolonged period of high proliferative activity, followed by a gradual decrease in growth rate and an increase in the fraction of cells arrested in the nonreplicative phase.[27] The cells were derived from the lung of a single human embryo, and the *in vitro* aging of these cells is characterized by both an exponential decline in the percentage of the population capable of synthesizing DNA and an increase in cell cycle time.[27,28] However, DNA synthesis can still be induced in senescent human diploid fibroblasts if they are infected with the DNA oncogenic virus, SV40.[29,30]

WI-38 cells can be made quiescent by simply plating them in growth medium containing 10% fetal calf serum and then leaving them alone. Usually by the fifth day after plating the cells are confluent and cell proliferation has all but ceased. The addition of serum at this point induces quiescent WI-38 cells to reenter the cell cycle.[31,32] If WI-38 cells are left quiescent for a prolonged period of time, they remain quiescent; in fact, the fraction of cells that are labeled by a 24-h pulse of ^{3}H-thymidine decreases even further, to levels of about 0.01%. When WI-38 cells are kept quiescent for a prolonged period of time, there is a lengthening of the prereplicative phase; i.e., there is a decrease in the fraction of cells reentering S phase as well as a delay in the time required for the cells to reenter S phase after addition of serum.[33] For instance, when confluent monolayers of WI-38 cells are stimulated by a change of medium containing 10% serum at 5 d after plating, the prereplicative phase is relatively short (8 h). If the same monolayers are stimulated 9 d after plating, the prereplicative phase is considerably longer (14 h). In fact, the cells apparently go into a deeper G_0 and the prereplicative phase becomes very long (up to 22 h) when the cells are stimulated 17 or 18 d after plating.[33,34] This is rather interesting, since it has been known that age affects liver regeneration after partial hepatectomy.[35] Thus, rapidly growing young rats restore tissue mass and cell population number more rapidly than adults, while the older animals lack new cell production. DNA synthesis peaks at 22 h after partial hepatectomy in weanling rats, at 25 h in young adults, and at 32 h in 1-year-old rats.[35] In addition, there is in weanlings a second peak at 35 h which is absent in other animals. Similar findings were reported by Adelman et al.[36] in the isoproterenol-stimulated salivary gland of rats. In a 2-month-old rat, a single injection of isoproterenol caused stimulation of DNA synthesis, which began at about 18 h after injection, reaching a peak at about 28 h. A similar treatment of 12-month-old rats produced a wave of DNA synthesis which began to increase very slowly 24 h after the injection of isoproterenol and reached a peak between 36 and 38 h after stimulation.

It seems, therefore, from these data that the length of the prereplicative phase in quiescent cells stimulated to proliferate by an appropriate stimulus is lengthened in old animals or in cells left quiescent for increasing periods of time. The increased length of the prereplicative phase is not due to the type of cells, but it seems to be due to the length of time the cell has been quiescent, i.e., age in adult animals or a longer time in a nonstimulated condition in tissue culture.

Since aging in the adult animal and aging or prolonged quiescence in tissue culture causes an increase in the length of the prereplicative phase and a decrease in the fraction of cells capable of responding to a proliferative stimulus, a legitimate question that we may now ask is whether the events that are known to occur in quiescent cells stimulated to proliferate are altered by aging or prolonged quiescence. We review here the data concerning one type of event, the expression of growth-related genes in sensecent cells and in cells after prolonged quiescence.

A. WI-38 CELLS AFTER PROLONGED QUIESCENCE (DIFFERENT LENGTH OF CELL CYCLE)

As mentioned above, when WI-38 cells are plated in medium containing 10% fetal calf serum, they become quiescent upon reaching confluence, at least in terms of DNA synthesis and in mitosis. Recently, however, Ferrari et al.[37] have shown that the expression of certain growth-related genes (notably, c-*myc*, ornithine decarboxylase, and the p53 gene) remains elevated even after the cells have become quiescent. Nonetheless, by the 15th day after plating, the expression of growth-regulated genes is no longer detectable. Interestingly, as long as the mRNAs for c-*myc*, p53, and ornithine decarboxylase are still elevated, WI-38

TABLE 2
Expression of Growth-Related Genes in Senescent
WI-38 Human Diploid Fibroblasts[41]

Gene	Time of maximum expression in young cells (h)	Time of maximum expression in senescent cells (h)
Vimentin	4	17
c-*myc*	4	4
JE-3	16	16
ATP/ADP translocase	17	17—24
Ornithine decarboxylase	16	4
Calcyclin	4	17—24
p53	16	4
c-Ha-*ras*	16	24
β-Actin	16	4
Thymidine kinase	17	24
Histone H3	16	24

cells respond, not only to serum but also to platelet-poor plasma. When these mRNAs are no longer detectable, WI-38 cells no longer respond to platelet-poor plasma, although they can still be induced to enter S phase by serum.

Addition of serum to WI-38 cells in prolonged quiescence still elicits a quick increase in the RNA levels of c-*fos* and c-*myc,* despite the fact that the entry into S phase is delayed.[38] Thus, prolonged quiescence (up to 34 d after plating) does not seem to greatly affect the inducibility of growth-regulated genes by growth factors in WI-38 cells.[37,38] Late-G_1 or S-phase genes, however, are delayed.[38a]

B. SENESCENT WI-38 CELLS

As mentioned above, there are several genes whose expression is known to be growth regulated (Table 1). The term "expression" is being used here in one of its accepted usages, i.e., as in steady-state levels of cytoplasmic mRNA. Growth-regulated genes include at least six bona fide oncogenes, several well-characterized cellular proteins (such as ornithine decarboxylase, thymidine kinase, or histones), and several genes isolated as cDNA clones in our laboratories and others. Most of the genes are induced in different cell types stimulated by different mitogens, and their mRNA levels can be used as markers of a cell's progression through the cell cycle. (See reviews by Kaczmarek[6] and Denhardt et al.[7])

Several lines of evidence have suggested that senescent cells are blocked in late G_1 rather than in G_0 phase, as are quiescent cells. First, the level of thymidine kinase activity in senescent, slowly proliferating cultures is similar to that of young, rapidly dividing populations. Second, the expansion of the thymidine trisphosphate pool prior to DNA synthesis is similar in young and senescent cells.[39] Finally, the nuclear fluorescence pattern of senescent cells, after staining with quinacrine dihydrochloride, is typical of cells that are blocked in late G_1 or at the G_1-S boundary.[40]

We therefore investigated the expression of several growth-regulated genes in senescent WI-38 cells and compared them to the results obtained in young cells.[41] The growth-regulated genes that were investigated in this study are listed in Table 2. The number of cells entering S phase after quiescence and stimulation was constantly monitored throughout these experiments by the addition of ^{3}H-thymidine. In stimulated cultures, the cells begin to enter S phase after about 12 to 15 h and the maximum number of DNA-synthesizing cells plateaus at 24 to 30 h. The maximum percentage of cells are labeled at this plateau stage if the labeling is continuous. In a representative experiment, the maximum percentage of labeled cells for the young cells was 62%, while only 12% of the senescent cells eventually entered DNA synthesis. The time at which half the maximum percentage of cells was labeled was slightly later in senescent cells (21 h) than in young cells (17 h), confirming the longer average prereplicative phase of still-cycling senescent cells.

1. Expression of Early-G$_1$ Genes

The early-G$_1$ genes that we studied were c-*myc*, vimentin, ADP/ATP translocase, and the cDNA clone, JE-3. All of these genes were expressed when quiescent young cells were stimulated to proliferate with serum. Similarly, they were all expressed when senescent cells were stimulated with serum. There were quantitative but no qualitative differences between young and senescent cells. In general, it seemed that the expression of these genes in senescent cells increased somewhat later than in the young cells and remained higher for a longer period of time.

2. Expression of Mid-G$_1$ Genes

We examined the expression of several mid-G$_1$ genes, including calcyclin, c-Ha-*ras*, and the gene encoding ornithine decarboxylase (ODC). Included in this group were also the genes encoding β-actin and the p53 antigen. Again, all of the mid-G$_1$ genes were expressed at least as well, if not slightly better, in senescent cells as in young cells. This applied to four genes, but there was one exception; c-Ha-*ras* did not follow the same pattern of cell cycle-dependent expression in senescent cells as in young cells. However, the c-HA-*ras* gene was still expressed in senescent cells at levels similar to those in young cells.

3. Expression of G$_1$-S Genes

Lastly, we examined the expression of two genes known to be preferentially expressed at the G$_1$-S boundary, i.e., the genes encoding thymidine kinase[42] and histone H3.[43,44] Both of these genes were well expressed in young cells as well as in senescent cells when stimulated by serum.

We can conclude that senescent WI-38 cells fully express not only the early and mid-G$_1$ genes examined, but also the G$_1$-S genes. In particular, WI-38 cells fully express the genes encoding both thymidine kinase and histone H3, which are usually tightly coupled to DNA synthesis. A detailed comparison of the expression in young and senescent cells is complicated by the slightly different length of the G$_1$ period in the two cell types. However, the levels of expression are similar in young and senescent cells, and that indicates that it must be due to the noncycling cells in the senescent population and not just the 12% of senescent cells that do enter S phase.

Similar studies were carried out by Paulsson et al.[45] They investigated the response of human fibroblasts to platelet-derived growth factor when they become senescent. They found that senescent human diploid fibroblasts responded to platelet-derived growth factors with increased levels of c-*fos* and c-*myc* mRNA, similar to growth-arrested fibroblasts. However, the expression of the nuclear antigen K67, which in young cells is induced in the S phase and continues to be expressed throughout the cell cycle, is not induced in senescent cells in response to platelet-derived growth factor. The exact nature of this K67 antigen is unknown, but it is absent in G$_0$-arrested cells, appears in S phase, and peaks in G$_2$. Slightly different results were obtained by Kihara et al.[46] with TIG-1 cells, a different strain of human diploid fibroblast. These authors found that early events were not inhibited in senescent cells, but late events (including thymidine uptake) were completely suppressed. Finally, Tsuji et al.[47] found that heat-shock proteins were still inducible in senescent human diploid fibroblasts.

IV. CONCLUSIONS

It therefore seems reasonably well established that senescent human diploid fibroblasts respond to growth factors with the expression of growth-regulated genes in a manner that is not different from that of young cells. Early-G_1 genes, mid-G_1 genes, and even those genes that are generally expressed at the G_1-S boundary are expressed at similar levels in young and senescent cells. This is different from what happens in G_1-specific temperature-sensitive mutants of the cell cycle. In these G_1 *ts* mutants, stimulation at the restrictive temperature induces some of the early- and mid-G_1 genes, but genes like thymidine kinase and histone H3 that are expressed at the G_1-S boundary are not inducible.[44,48] It seems, therefore, that a G_1 block really results in the complete suppression of S phase-related genes. By making a comparison of G_1 *ts* mutants with senescent WI-38 cells, one is tempted to conclude that the block in senescent WI-38 cells is at the G_1-S boundary, in agreement with our data. It seems that the block is actually in the S phase itself, and one is tempted on the basis of the reports in the literature (see elsewhere in this monograph) to speculate that this block may be due to the presence of an inhibitor of DNA synthesis itself.

REFERENCES

1. **Baserga, R.,** *The Biology of Cell Reproduction,* Harvard University Press, Cambridge, MA, 1985.
2. **Bishop, J. M.,** Cellular oncogenes and retroviruses, *Annu. Rev. Biochem.,* 54, 301, 1983.
3. **Bishop, J. M.,** The molecular genetics of cancer, *Science,* 235, 305, 1987.
4. **Basilico, C.,** Temperature-sensitive mutations in animal cells, *Adv. Cancer Res.,* 24, 223, 1977.
5. **Marcus, M., Fainrod, A., and Diamond, G.,** The genetic analysis of mammalian cell cycle mutants, *Annu. Rev. Genet.,* 19, 389, 1985.
6. **Kaczmarek, L.,** Proto-oncogene expression during the cell cycle, *Lab. Invest.,* 54, 365, 1986.
7. **Denhardt, D. T., Edwards, D. R., and Parfett, C. L. J.,** Gene expression during the mammalian cell cycle, *Biochim. Biophys. Acta,* 865, 83, 1986.
8. **Kelly, B. K., Cochran, B. H., Stiles, C. D., and Leder, P.,** Cell specific regulation of the c-*myc* gene by lymphocyte mitogens and platelet-derived growth factor, *Cell,* 35, 603, 1983.
9. **Greenberg, M. E. and Ziff, E. B.,** Stimulation of 3T3 cells induces transcription of the c-*fos* proto-oncogene, *Nature,* 311, 433, 1984.
10. **Campisi, J., Gray, H. E., Paradee, A. B., Dean, M., and Sonenshein, G. E.,** Cell cycle control of c-*myc* and not c-*ras* expression is lost following chemical transformation, *Cell,* 36, 241, 1984.
11. **Torelli, G., Selleri, L., Donelli, A., Ferrari, S., Emilia, G., Venturelli, D., Morelli, L., and Torelli, U.,** Activation of c-*myb* expression by phytohemagglutinin stimulation in normal and human T-lymphocytes, *Mol. Cell. Biol.,* 5, 2874, 1985.
12. **Reich, N. C. and Levine, A. J.,** Growth regulation of a cellular tumor antigen p53 in nontransformed cells, *Nature,* 308, 199, 1984.
13. **Eliyahu, D., Raz, A., Gruss, P., Givol, D., and Oren, M.,** Participation of p53 cellular tumour antigen in transformation of normal embryonic cells, *Nature,* 312, 646, 1984.
14. **Parada, L. F., and Land, H., Weinberg, R. A., Wolf, D., and Rotter, V.,** Cooperation between encoding p53 tumour antigen and *ras* in cellular transformation, *Nature,* 312, 649, 1984.
15. **Rittling, S. R., Gibson, C. W., Ferrari, S., and Baserga, R.,** The effect of cycloheximide on the expression of cell cycle-dependent genes, *Biochem. Biophys. Res. Commun.,* 132, 327, 1985.
16. **Linzer, D. I. H. and Nathans, D.,** Nucleotide sequence of a growth-regulated mRNA encoding a member of the prolactin growth hormone family, *Proc. Natl. Acad. Sci. U.S.A.,* 81, 4255, 1984.
17. **Parfett, C. L. J., Hamilton, R. T., Howell, B. W., Edwards, D. R., Nilsen-Hamilton, M., and Denhardt, D. T.,** Characterization of a cDNA clone coding murine mitogen-regulated protein: regulation of mRNA levels in mortal and immortal cell lines, *Mol. Cell. Biol.,* 5, 3289, 1985.
18. **Edwards, D. R., Parfett, C. L. J., and Denhardt, D. T.,** Transcriptional regulation of two serum induced RNAs in mouse fibroblasts. Equivalents of one species of B2 repetitive elements, *Mol. Cell. Biol.,* 5, 3280, 1985.

19. **Matrisian, L. M., Rautman, G., Magun, B. E., and Breathnach, R.**, Epidermal growth factor or serum stimulation of rat fibroblasts induced an elevation in mRNA levels for lactate dihydrogenase and other glycolytic enzymes, *Nucleic Acids Res.*, 13, 711, 1985.

20. **Tatchell, K., Robinson, L. C., and Breitenbach, M.**, RAS2 of *Saccharomyces cerevisiae* is required for gluconeogenic growth and proper response to nutrient limitation, *Proc. Natl. Acad. Sci. U.S.A.*, 82, 3785, 1985.

21. **Hirschhorn, R. R., Aller, P., Yuan, Z. A., Gibson, C. W., and Baserga, R.**, Cell cycle specific cDNAs from mammalian cells temperature-sensitive for growth, *Proc. Natl. Acad. Sci., U.S.A.*, 81, 6004, 1984.

22. **Battini, R., Ferrari, S., Kaczmarek, L., Calabretta, B., Chen, S. T., and Baserga, R.**, Molecular cloning of a cDNA for a human ADP/ATP carrier which is growth-regulated, *J. Biol. Chem.*, 262, 4355, 1987.

23. **Ferrari, S., Battini, R., Kaczmarek, L., Rittling, S., Calabretta, B., deRiel, J. K., Philiponis, V., Wei, J.-F., and Baserga, R.**, Coding sequence and growth regulation of the human vimentin gene, *Mol. Cell. Biol.*, 6, 3614, 1986.

24. **Calabretta, B., Battini, R., Kaczmarek, L., deRiel, J. K., and Baserga, R.**, Molecular cloning of the cDNA for a growth factor inducible gene with strong homology to S-100, a calcium binding protein, *J. Biol. Chem.*, 261, 1261, 1986.

25. **Hayflick, L. and Moorhead, P. S.**, The serial cultivation of human diploid cell strains, *Exp. Cell Res.*, 25, 585, 1961.

26. **Hayflick, L.**, The limited *in vitro* lifetime of human diploid cell strains, *Exp. Cell Res.*, 37, 614, 1965.

27. **Cristofalo, V. J. and Sharf, B. B.**, Cellular senescence and DNA synthesis, *Exp. Cell Res.*, 76, 419, 1973.

28. **Grove, G. L. and Cristofalo, G. A.**, Characterization of the cell cycle of cultured human diploid cells. Effects of aging and hydrocortisone, *J. Cell. Physiol.*, 90, 415, 1977.

29. **Ide, T., Tsuji, Y., Ishibashi, S., and Mitsui, Y.**, Reinitiation of host DNA synthesis in senescent human diploid cells by infection with simian virus 40, *Exp. Cell Res.*, 143, 343, 1983.

30. **Gorman, S. D. and Cristofalo, V. J.**, Reinitiation of cellular DNA synthesis in BrdU-selected nondividing senescent WI-38 cells by Simian virus 40 infection, *J. Cell. Physiol.*, 125, 122, 1985.

31. **Wiebel, F. and Baserga, R.**, Early alterations in amino acid pools and protein synthesis of diploid fibroblasts stimulated to synthesize DNA by addition of serum, *J. Cell. Physiol.*, 74, 191, 1969.

32. **Bombik, B. M. and Baserga, R.**, Increased RNA synthesis in nuclear monolayers of WI38 cells stimulated to proliferate, *Proc. Natl. Acad. Sci. U.S.A.*, 71, 2038, 1974.

33. **Augenlicht, L. F. and Baserga, R.**, Changes in the G_0 state of WI-38 fibroblasts at different times after confluence, *Exp. Cell Res.*, 89, 225, 1974.

34. **Rossini, M., Lin, J. C., and Baserga, R.**, Effects of prolonged quiescence on nuclei and chromatin of WI-38 fibroblasts, *J. Cell. Physiol.*, 88, 1, 1976.

35. **Bucher, M. L. R.**, Regeneration of mammalian liver, *Int. Rev. Cytol.*, 15, 241, 1963.

36. **Adelman, R. C., Stein, G., Roth, G. S., and Englander, D.**, Age dependent regulation of mammalian DNA synthesis and cell proliferation *in vivo*, *Mech. Ageing Dev.*, 1, 49, 1973.

37. **Ferrari, S., Calabretta, B., Battini, R., Cosenza, S. C., Owen, T. A., Soprano, K. J., and Baserga, R.**, Expression of c-*myc* and induction of DNA synthesis by platelet-poor plasma in human diploid fibroblasts, *Exp. Cell Res.*, 174, 25, 1988.

38. **Owen, T. A., Cosenza, S. C., Soprano, D. R., and Soprano, K. J.**, Effects of prolonged quiescence on c-*fos* and c-*myc* expression in human diploid fibroblasts, *J. Biol. Chem.*, 262, 15111, 1987.

38a. **Soprano, K.**, personal communication.

39. **Olashaw, N. E., Kress, E. D., and Cristofalo, V. J.**, Thymidine triphosphate synthesis in senescent WI38 cells, *Exp. Cell Res.*, 149, 547, 1983.

40. **Gorman, S. D. and Cristofalo, V. J.**, Analysis of the G_1 arrest position of senescent WI-38 cells by quinacrine dihydrochloride nuclear fluorescence, *Exp. Cell Res.*, 167, 87, 1986.

41. **Rittling, S. R., Brooks, K. M., Cristofalo, V. J., and Baserga, R.**, Expression of cell cycle dependent genes in young and senescent WI38 fibroblasts, *Proc. Natl. Acad. Sci. U.S.A.*, 83, 3316, 1986.

42. **Liu, H.-T., Gibson, C. W., Hirschhorn, R. R., Rittling, S., Baserga, R., and Mercer, W. E.**, Expression of thymidine kinase and dihydrofolate reductase genes in mammalian *ts* mutants of the cell, *J. Biol. Chem.*, 260, 3269, 1985.

43. **Plumb, M., Stein, J., and Stein, G.**, Co-ordinate regulation of multiple histone mRNAs during the cell cycle in HeLa cells, *Nucleic Acids Res.*, 11, 2391, 1983.

44. **Hirschhorn, R. R., Marashi, F., Baserga, R., Stein, J., and Stein, G.**, Expression of histone genes in a G_1 specific temperature-sensitive mutant of the cell cycle, *Biochemistry*, 23, 3731, 1984.

45. **Paulsson, Y., Bywater, M., Pfeifer-Ohlsson, S., Paulsson, R., Nilsson, S., Heldin, C. H., Westermark, B., and Betsholtz, C.**, Growth factors induce early prereplicative changes in senescent human fibroblasts, *EMBO J.*, 5, 2157, 1986.

46. **Kihara, F., Tsuji, Y., Miura, M., Ishibashi, S., and Ide, T.,** Events blocked in prereplicative phase in senescent human diploid cells, TIG-1, following serum stimulation, *Mech. Ageing Dev.,* 37, 103, 1986.
47. **Tsuji, Y., Ishibashi, S., and Ide, T.,** Induction of heat-shock proteins in young and senescent human diploid fibroblasts, *Mech. Ageing Dev.,* 36, 155, 1986.
48. **Ide, T., Ninomya-Tsuji, J., Ferrari, S., Philiponis, V., and Baserga, R.,** Expression of growth regulated genes in tsJT6T cells: a temperature sensitive mutant of the cell cycle, *Biochemistry,* 25, 7041, 1986.

Chapter 3

CHANGES IN GENE EXPRESSION DURING SENESCENCE/ IMMORTALIZATION OF MOUSE EMBRYO FIBROBLASTS

Craig L. J. Parfett and David T. Denhardt

TABLE OF CONTENTS

I. INTRODUCTION

Rodent embryo cells cultured in serum-containing medium essentially cease to proliferate when the population has accumulated 15 to 20 doublings. Cells then enter a ''crisis'' period during which little or no growth occurs and many cells die. Continued passage of the senescent culture under appropriate conditions often leads to the outgrowth of a variant population with a renewed and indefinitely extended proliferative capacity.[1] The rodent species from which the embryo cells originate largely determines the ease with which immortal populations are recovered — mouse embryo cultures almost always giving rise to cell lines if sufficient numbers of cells are passed. The aneuploid nature of many cell lines intimates that genetic alterations may be important in the underlying mechanism of immortalization; however, immortalization often precedes overt chromosomal rearrangements, and the existence of immortal diploid lines suggests that although chromosomal rearrangements may often be associated with immortalization, they may not be a necessary factor.[2] A relationship between the cells' potential for reorganization at the cytogenetic level and the probability of generating immortal lines has been noted.[3] Treatment with mutagens enhances the frequency of conversion to the immortal phenotype, lending support to a role for mutations.[4]

Certain combinations of oncogenes can also cooperate both to immortalize and to transform primary cells, although it is not clear that immortalization is a direct effect of the activities of these genes.[4] Current ideas on the multistage progression toward oncogenic transformation include immortalization as a key step. Thus, new patterns of gene expression following chromosomal rearrangements or ill-defined epigenetic changes, possibly involving known oncogenes or as yet unidentified loci, are thought to confer important new growth properties upon cells within the mortal population. These cells harboring changes important for long-term survival and growth in the *in vitro* environment would obviously be favored during subsequent passage.

The influence of culture conditions on the phenotype of the cells after selection for growth *in vitro* indicates that adaptive responses and differential gene expression likely play significant roles in establishment. Todaro and Green[1] determined many of the parameters that affected the establishment (immortalization) process; in particular, they showed that the concentration of the cells when plated and the length of time they were permitted to proliferate before passing were major determinants of the cell line that emerged. Recently it has been shown that for mouse embryo cells, the appearance of the senescent phenotype (''crisis'') and the outgrowth of aneuploid lines can be averted when the usual serum-supplemented medium is replaced with a defined medium.[5] Differences in subculture procedure also affect the frequency with which immortal variants are recovered from the senescing population.[6] It is possible that components in the serum/medium or nutritional deficiencies impact upon the cells in culture and that, as a consequence, particular changes in gene expression become necessary to maintain continued proliferation under that specific situation. In a defined set of circumstances, it is reasonable to expect to identify gene products that are associated with the immortalized cells but not produced by the low-passage diploid precursor cells with limited growth potential.

We and others have documented the expression of a growth factor-responsive gene encoding a secreted protein (known as mitogen-regulated protein [MRP] or proliferin) in many immortal murine cell lines; however, expression is not detected in the primary (or early passage) mouse embryo fibroblast populations from which such cell lines are typically derived.[7,8] MRP was first identified as a glycoprotein secreted in elevated amounts by cultured Swiss mouse 3T3 cells stimulated with serum or certain growth factors.[9] About five copies of the gene exist in the mouse haploid genome, and regulation of serum-induced expression in 3T3 cells is largely at the level of the transcripts' abundance in the cytoplasm.[7,10,11] MRP is apparently identical to the prolactin-related protein called proliferin,[12] as evidenced both by the identity of the nucleotide sequence of a partial cDNA clone of MRP with a portion of the proliferin coding sequence and by the close immunological relationship between the two proteins.[7,13] Expression in mouse tissues appears to be restricted to the embryo's portion of the midgestational placenta.[14]

This report summarizes some of our data showing that the change in MRP/proliferin expression begins during "crisis" in mouse embryo fibroblasts passed according to the 3T3 regimen[1] and that these changes occur both in senescent cells and in the immortal 3T3 line that develops. We also report changes in the expression of an endogenous murine leukemia virus (MLV) genome or related sequences. The possible functional significance of expression of these genes to the immortalized cell is discussed.

II. MATERIALS AND METHODS

CD-1 Swiss mouse embryos (14 to 16 d old) were minced and cells disaggregated with 0.125% trypsin in 0.1% KCl and 0.44% sodium citrate. They were seeded at 2.5×10^4 cells/cm^2 and grown to confluence (two to three generations) in glass roller bottles in Dulbecco's minimal essential medium supplemented with 10% fetal calf serum. To generate 3T3 lines, the cells were plated at densities of 1.5×10^4 cells/cm^2 (2.1×10^6 cells per 150-mm dish) and passed every 3.5 d. Total cytoplasmic RNA was isolated from cells and electrophoresed on 1.1% agarose gels containing 6% formaldehyde, as described previously.[15] The RNA was blotted onto Gene Screen Plus (NEN), as specified by the supplier, and the resulting blot hybridized in 10% dextran sulfate to an appropriate probe labeled with ^{32}P by nick translation.

In situ hybridization was performed essentially as described by Lawrence and Singer[16] on cells grown on plastic culture dishes. The final hybridization solution was 50% formamide, 10 mM Tris-HCl, pH 7.5, 1 mM EDTA, 600 mM NaCl, 25 µg/ml poly[dA], 5% dextran sulfate, and 7 µg/ml probe labeled with tritiated 2'-deoxythymidine monophosphate (^{3}H-dTMP) by nick translation. The probe was a plasmid fragment containing the protein coding sequence of a near full-length MRP cDNA clone linked to the shorter *ScaI-EcoRI* segment of pBR322; the control was pBR322 plasmid DNA without an insert.

III. RESULTS

The observation that cytoplasmic MRP mRNA could not be detected in low-passage mouse embryo fibroblast (MEF) cells has been repeatedly confirmed in this laboratory. In a total of 19 permanent cell lines investigated, 15 were observed to have detectable MRP mRNA in their cytoplasm (Table 1). All cell lines isolated according to the 3T3 regimen and one SV40-transformed 3T3 clone were positive, but in both cell lines isolated by the 3T12 regimen, MRP mRNA was not detected. During passage according to the 3T12 regimen, the concentration of cells upon replating is fourfold greater than for 3T3 cells. The B16F1 and JB6 cell lines are immortal derivatives of epidermal tissue.

Since cell lines such as the Swiss 3T3 line of Todaro and Green[1] have been maintained in culture for a considerable time period, it is not apparent when MRP expression might arise. To investigate this point, we assessed expression of the gene from the earliest stages in the evolution of an immortal 3T3 population. MEFs isolated from 14- to 16-day-old Swiss mouse embryos were serially passaged in 150-mm dishes according to an appropriate 3T3 regimen. At increasing passage numbers, total cytoplasmic RNA was prepared on the second day after passage, approximately 16 h after the cultures had been refed with fresh serum-supplemented medium. The individual RNA preparations were electrophoresed on an agarose gel and blotted onto nylon membranes for analysis.

Cell counts taken at the time the populations were trypsinized and replated are indicative

TABLE 1
MRP mRNA Expression Among Independent Murine Cell
Lines

Cell line designation	Presence of MRP mRNA (+ or −)
NIH 3T3	+
BALB/c 3T3	+
Swiss 3T3[1]	+
Swiss 3T3 (three lines, this laboratory)	+
Swiss 3T3 L1	+
BALB/c BNL	+
Swiss 3T6	+
C3H 10T1/2	+[37]
SV40 BALB/c 3T3	+[37]
Swiss (two clones, this report)	+
Krebs ascites carcinoma	+[37]
Ehrlich ascites carcinoma	+
BALB/c 3T12	−
Swiss 3T12	−
JB6	−
B16 F1 melanoma	−

Note: This is a listing of cell lines that have been screened for MRP/proliferin mRNA. Northern blots of cytoplasmic RNA were probed with either a partial cDNA clone encoding an antigen bound by the antiserum against authentic MRP[7] or a clone encoding the complete MRP/proliferin polypeptide. Except for those cell lines generated in this laboratory, all other lines are described in the American Type Culture Collection (ATCC) catalog.

of the proliferative ability of the cells at that passage. Typical data for MEF cells passed according to the 3T3 regimen are illustrated in the top panel of Figure 1. The initial period, during which cells maintain their high rate of proliferation, appears to last approximately four passages; with each subsequent passage, the number of cells recovered from the plate surface declines. During the first and second passages, considerable cell death occurs and quantitation is not meaningful. The data are somewhat scattered, since the number of cells attached to the plate at the time of harvest was not determined and no correction for this variable was made. However, the "crisis" period in which little growth occurred was apparent by the fifth passage and continued for four more passages. By the tenth passage, the growth rate of the culture began to recover and eventually attained a rate near that of the original MEF population.

The autoradiograms of Northern blots of total cytoplasmic RNA shown in Figure 1 illustrate the changes in expression of certain sequences that occur as the mouse embryo fibroblasts are passed. Data from two separate experiments are combined; the specific passage at which each RNA preparation was made is indicated on one of two levels on the abscissa. These combined data show some of the variability between experiments. The actin (2.0 kb) blot is a control showing that comparable amounts of RNA are present in each lane. The lower two panels show the same blot probed for MRP and for sequences related to MLV.

The MRP and MLV autoradiograms are overexposed to emphasize the absence of mature transcripts in the early-passage cultures. The 1-kb mature MRP transcript is indicated; we believe the higher molecular weight species are unspliced nuclear transcripts contaminating the cytoplasmic RNA preparations. A weak MRP signal can be seen by passages 3 to 5, betraying that some of the cells in the population entering "crisis" are already beginning to express the gene. By passages 9 and 10, considerable MRP and mRNA was detected, and *in situ* hybridization analysis revealed that many of the cells contained MRP message.

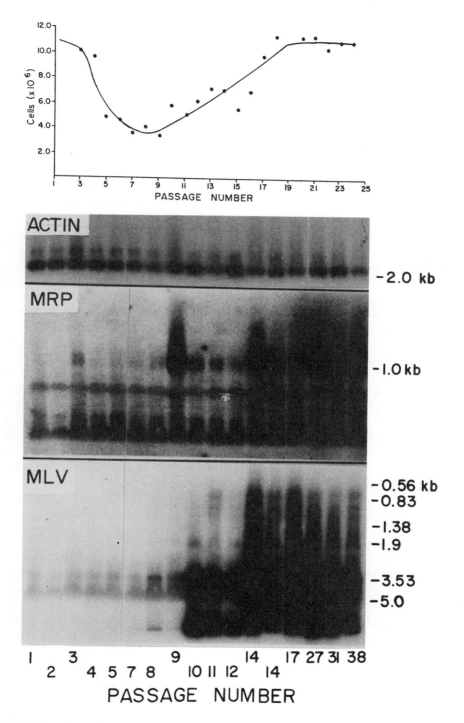

FIGURE 1. Changes in gene expression during senescence/immortalization. The number of cells recovered from a 150-mm dish 3.5 d after plating 2.1 × 10⁶ cells is plotted against the passage number in the top panel. The lower three panels show Northern blots of the total cytoplasmic RNA extracted from cells at various passages. In experiment I, which is plotted in the top panel, RNA was extracted at passages 1, 3, 9, 14, 17, 27, 31, and 38. In experiment II, RNA was extracted at passages 2, 4, 5, 7, 8, 10, 11, 12, and 14.

Heterogeneity among cells in a stimulated 3T3 population with regard to the level of expression of proliferin mRNA was also seen by Linzer and Wilder.[11]

Integrated into the genome of most mouse strains are a number of sequences related to MLV.[17,17a] In some strains of mice, expression of the endogenous viral genome occurs in a tissue-specific or developmentally regulated fashion; also, mouse embryo cells of various strains, when cultured *in vitro,* acquire the ability to express viral information. The bottom panel of Figure 1 shows that around passages 10 to 14, our mouse embryo fibroblasts (the Charles River [Montreal] Swiss CD-1 strain) begin to express substantial amounts of several species of RNA that respond to an MLV probe, specifically the *env* portion.[17b] There does not seem to be much of a correlation between the passage at which mature MRP mRNA is first detected and the passage at which the MLV *env* transcripts are first expressed.

MRP expression was first detected as the cells were entering the crisis period, when the proliferation rate of the mass population was low and very few established cells were present. Cells with clonogenic potential were detected after five serial passages during the derivation of the 3T3 line by plating a portion of the population at a tenfold lower density (2.5×10^5 cells per 150-mm dish). However, fewer than 1 in 10^4 cells formed a colony of 50 to 100 cells during a 3-week incubation, and only 2 of 21 of the largest colonies could be picked with cloning cylinders and grown to large cultures, the rest apparently having exhausted their potential to divide. Thus, the earliest detection of MRP mRNA was correlated with populations in which the cells with a senescent phenotype greatly predominated.

The extent of MRP expression among the colony-forming and non-colony-forming cell types was examined by *in situ* hybridization. The population of cells and colonies remaining on representative plates after the 3-week incubation were stimulated for 16 h with fresh medium containing 20% calf serum and then processed for *in situ* hybridization with an MRP cDNA probe labeled by nick translation. Very few cells had a hybridization signal above that obtained with the control pBR322 DNA probe. The rare MRP-positive cells were associated with the sparse population of senescent, nonproliferating cells and with tiny colonies distributed evenly over the plate surfaces. This militates against the possibility that MRP is produced exclusively by a small fraction of preexisting immortal cells that subsequently expand to establish the 3T3 line. One example of a positive cell is shown in Figure 2. In the three large colonies examined on the same plates, no MRP mRNA was detected by these procedures; however, mRNA extracted from two established populations picked from the previously expanded colonies did contain MRP mRNA at levels comparable to ''producing'' cell lines.

MRP expression is regulated by agents that control cell proliferation (e.g., platelet-derived growth factor [PDGF], fibroblast growth factor [FGF]). The proto-oncogenes and their activated derivatives are also known to possess important growth-regulatory properties; therefore, the levels of these mRNAs in the cytoplasm of low-passage MEF and 3T3 cells were examined for possible changes in the regulation of their transcripts. Probes for various proto-oncogenes, including c-*myc,* c-*ras,* c-*sis,* c-*myb,* c-*fms,* c-*fes,* c-*abl,* erbB, and p53, were hybridized to Northern blots of cytoplasmic RNA prepared from primary MEFs and 3T3 cells. Like MRP/proliferin, the mRNAs for c-*myc,* c-*fos,* c-*ras,* c-*myb,* and p53 are elevated in response to induction by serum growth factors.[18] No significant differences in the abundance or size of any of the mRNAs responding to these probes were detected; however, the signals for c-*fms,* c-*abl,* and c-*fes* were very weak or nonexistent. In similar studies by Shuin et al.[19] the only difference in the expression of a number of tested oncogenes was an 8.2-kb c-*abl* transcript that was expressed in C3H embryos but not in a 1OT1/2 line derived from them.

IV. DISCUSSION

The experiments described herein correlated the *de novo* expression of the MRP/proliferin

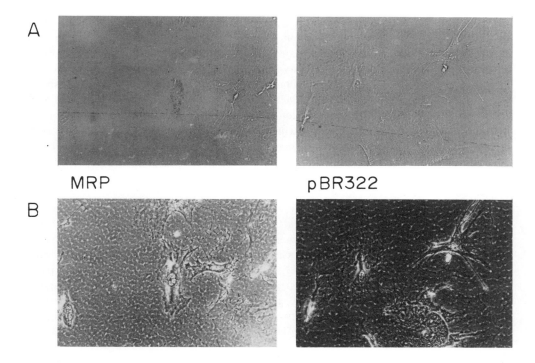

FIGURE 2.   Autoradiographs of senescent MEF cells hybridized *in situ* with ³H-labeled probe to MRP/proliferin mRNA. Shown are unstained cells beneath the developed photographic emulsion with silver grains in the upper photographs (A) and the same fields photographed with phase-contrast optics in the lower photographs (B). Cells on the left were probed for MRP; those on the right were probed with a pBR322 control. Specific activities of the nick-translated probes were approximately 1×10^8 dpm/μg and exposure time was 2 weeks at 4°C.

genes with the time at which mouse embryo fibroblasts, passaged according to a 3T3 regimen, begin to senesce in culture. As the fibroblast population evolves into an established 3T3 line, MRP expression is maintained and can be found in such lines long after their initial establishment. We think it unlikely that a preceding alteration in the expression of a known oncogene explains the accumulation of MRP mRNA after serum stimulation in 3T3 cells, although this conclusion must wait until all known oncogenes have been examined. MRP expression also does not seem to be a consequence of the expression of endogenous MLV-related sequences.

Our inability to detect MRP or its transcripts in an evolving 3T12 culture[20] implies that expression of this gene product is not always required for the immortalization process. Furthermore, the expression of MRP is not sufficient to immortalize cells, since its mRNA was detected in cells that were obviously senescent. Even though it is possible to document changes in gene expression during aging and immortalization *in vitro,* we do not know why cells often cease to grow in culture, nor do we know if there exists a consistent process of differentiation or if there is a common set of epigenetic changes necessary before all cells may become established. Questions that need to be addressed concern (1) the functions of gene products (such as MRP) whose expression is altered in cell lines as compared to their progenitor cells, and (2) the mechanisms that underlie such changes.

Other proteins have been identified that are not present in tissues or primary embryo cells but are found in cellular extracts of established cell lines.[21,22] As does MRP, other proteins increase as animal cells age in culture.[23-26] However, it remains to be seen whether any of these changes in expression are simply gratuitous or whether they actually contribute to senescence or immortalization. In this context it is interesting to note that the *ras* oncogene has been reported to be more mitogenic when microinjected into aged rat cells than when introduced into their counterparts of low passage number, suggesting that new growth-enhancing properties can arise during aging *in vitro*.[27] This effect is contrasted by the increased cell content of material inhibitory to DNA synthesis as mouse cells age in culture.[28] Senescent and quiescent human diploid fibroblasts also develop inhibitory factors suspected to be surface membrane-bound proteins,[29,30] and the appearance of a DNA synthesis inhibitory function within the poly[A]mRNA fraction of senescent human cells has been detected by microinjection into proliferation-competent cells.[31] Generation of macromolecular agents with diverse effects on cellular growth may be a consistent feature of cells aging *in vitro* and may in turn reflect the response of normal growth-regulatory pathways to the abnormal environment encountered within the culture vessel.

MRP has not been shown to be a proto-oncogene, but given its relation to proteins with known growth stimulatory activities (prolactin, placental lactogen, growth hormone), the possibility is open that it is capable of acting as an autocrine factor in certain situations. The autocrine stimulation hypothesis states that cells can gain growth autonomy by any means giving them the capability of producing and responding to a growth factor that controls processes normally limiting in that cell type.[32] Alternatively, MRP/proliferin could participate more indirectly by influencing the sensitivity of the immortal fibroblasts to another serum growth factor such as insulin-like growth factor (IGF) — similar to a mechanism which has been implicated in the activities of the prolactin family members.[33-36] Any effects of MRP/proliferin on 3T3 cell growth *in vitro* remain to be demonstrated.

V. SUMMARY

The increase in the cytoplasmic abundance of mRNA species encoding MRP and MLV proteins has been followed during the senescence and immortalization of mouse embryo cells passed according to a 3T3 regimen. Both senescent cells and immortal 3T3 cells acquire the ability to make MRP. The ability to synthesize the endogenous MLV sequences appears to correlate with the acquisition of the immortal phenotype.

ACKNOWLEDGMENTS

We thank Martha Holman, Greta MacKenzie, Cathy McGregor, Marilyn McLeod, and Beth Orphan for assistance with various aspects of this research, which was supported by grants from the National Cancer Institute and the Medical Research Council of Canada.

REFERENCES

1. **Todaro, G. T. and Green, H.,** Quantitative studies of the growth of mouse embryo cells in culture and their development into established lines, *J. Cell Biol.,* 17, 299, 1963.
2. **Cerni, C., Mougneau, E., and Cuzin, F.,** Transfer of ''immortalizing'' oncogenes into rat fibroblasts induces both high rates of sister chromatid exchange and appearance of abnormal karyotypes, *Exp. Cell Res.,* 168, 439, 1987.
3. **Macieira-Coelho, A.,** Cancer and aging, *Exp. Gerontol.,* 21, 483, 1986.
4. **Newbold, R. F.,** Multistep malignant transformation of mammalian cells by carcinogens: induction of immortality as a key event, in *Carcinogenesis,* Vol. 9, Barrett, J. C. and Tennant, R. W., Eds., Raven Press, New York, 1985, 17.
5. **Loo, D. T., Fuquay, J. I., Rawson, C. L., and Barnes, D. W.,** Extended culture of mouse embryo cells without senescence: inhibition by serum, *Science,* 236, 200, 1987.
6. **Curatolo, L., Erga, E., and Morasca, L.,** Culture conditions induce the appearance of immortalized C_3H mouse cell lines, *In Vitro,* 20, 597, 1984.

7. **Parfett, C. L. J., Hamilton, R. T., Howell, B. W., Edwards, D. R., Nilsen-Hamilton, M., and Denhardt, D. T.,** Characterization of a cDNA clone encoding murine mitogen-regulated protein: regulation of mRNA levels in mortal and immortal cell lines, *Mol. Cell. Biol.,* 5, 3289, 1985.

8. **Edwards, D. R., Parfett, C. L. J., Smith, J. H., and Denhardt, D. T.,** Evidence that post-transcriptional changes in the expression of mitogen regulated protein accompany immortalization of mouse cells, *Biochem. Biophys. Res. Commun.,* 147, 467, 1987.

9. **Nilsen-Hamilton, M., Shapiro, J. M., Massoglia, S. L., and Hamilton, R. T.,** Selective stimulation by mitogens of incorporation of ^{35}S-methionine into a family of proteins released into the medium by 3T3 cells, *Cell,* 20, 19, 1980.

10. **Wilder, E. L. and Linzer, D. I. H.,** Expression of multiple proliferin genes in mouse cells, *Mol. Cell. Biol.,* 6, 3283, 1986.

11. **Linzer, D. I. H. and Wilder, E. L.,** Control of proliferin gene expression in serum-stimulated mouse cells, *Mol. Cell. Biol.,* 7, 2080, 1987.

12. **Linzer, D. I. H. and Nathans, D.,** Nucleotide sequence of a growth-related mRNA encoding a member of the prolactin-growth hormone family, *Proc. Natl. Acad. Sci. U.S.A.,* 81, 4255, 1984.

13. **Nilsen-Hamilton, M., Hamilton, R. T., and Alvarez-Azaustre, E.,** The relationship between ''mitogen regulated protein'' (MRP) and ''proliferin'' (PLF), a member of the prolactin/growth hormone family, *Gene,* 51, 163, 1987.

14. **Linzer, D. I. H., Lee, S.-J., Ogren, L., Talamantes, F., and Nathans, D.,** Identification of proliferin mRNA and protein in mouse placenta, *Proc. Natl. Acad. Sci. U.S.A.,* 82, 4356, 1985.

15. **Edwards, D. R., Parfett, C. L. J., and Denhardt, D. T.,** Transcriptional regulation of two serum-induced RNAs in mouse fibroblasts: equivalence of one species to B2 repetitive elements, *Mol. Cell. Biol.,* 5, 3280, 1985.

16. **Lawrence, J. B. and Singer, R. H.,** Intracellular localization of messenger RNAs for cytoskeletal proteins, *Cell,* 45, 407, 1986.

17. **Coffin, J.,** Endogenous virus, in *RNA Tumor Viruses,* Weiss, R., Teich, N., Varmus, H., and Coffin, J., Eds., Cold Spring Harbor Laboratory, Cold Spring Harbor, NY, 1984, chap. 10.

17a. **Stoye, J. and Coffin, J.,** Endogenous viruses, in *RNA Tumor Viruses,* 2nd ed., Weiss, R., Teich, N., Varmus, H., and Coffin, J., Eds., Cold Spring Harbor Laboratory, Cold Spring Harbor, NY, 1985, 357.

17b. **Denhardt, D. T. et al.,** in preparation.

18. **Denhardt, D. T., Edwards, D. R., and Parfett, C. L. J.,** Gene expression during the mammalian cell cycle, *Biochim. Biophys. Acta,* 865, 83, 1986.

19. **Shuin, T., Billings, P. C., Lillehaug, J. R., Patierno, S. R., Roy-Burman, P., and Landolph, J. R.,** Enhanced expression of c-*myc* and decreased expression of c-*fos* protooncogenes in chemically and radiant-transformed C3H/10T1/2 Cl 8 mouse embryo cell lines, *Cancer Res.,* 46, 5302, 1986.

20. **Denhardt, D. T., Edwards, D. R., Parfett, C. L. J., and Smith, J.,** Regulation of gene expression in mortal and immortal murine fibroblasts — 3T3 but not 3T12 cells make MRP/proliferin, in *Molecular Mechanisms in the Regulation of Cell Behavior,* Waymout, C., Ed., Alan R. Liss, New York, 1987, 59.

21. **Macnab, J. C. M., Orr, A., and La Thangue, N. B.,** Cellular proteins expressed in herpes simplex virus transformed cells also accumulate on herpes simplex virus infection, *EMBO J.,* 4, 3223, 1985.

22. **Mejia, N. R. and MacKenzie, R. E,** NAD-dependent methylenetetrahydrofolate dehydrogenase is expressed by immortal cells, *J. Biol. Chem.,* 260, 14616, 1985.

23. **Pool, T. B.,** A comparison of protein synthesis between serum deprived and postmitotic human diploid fibroblasts, *Cell Biol. Int. Rep.,* 6, 658, 1982.

24. **Lincoln, D. W., II, Braunschweiger, K. I., Braunschweiger, W. R., and Smith, J. R.,** The two-dimensional polypeptide profile of terminally non-dividing human diploid cells, *Exp. Cell Res.,* 154, 136, 1984.

25. **Sottile, J., Hoyle, M., and Millis, A. J. T.,** Enhanced synthesis of a Mr = 55,000 dalton peptide by senescent human fibroblasts, *J. Cell. Physiol.,* 131, 210, 1987.

26. **Wang, E. and Lin, S. L.,** Disappearance of statin, a protein marker for non-proliferating and senescent cells, following serum-stimulated cell cycle entry, *Exp. Cell Res.,* 167, 135, 1986.

27. **Sullivan, N. F., Sweet, R. W., Rosenberg, M., and Feramisco, J. R.,** Microinjection of the *ras* oncogene protein into nonestablished rat embryo fibroblasts, *Cancer Res.,* 46, 6427, 1986.

28. **Due, C. and Ebbesen, P.,** Change in activity of fibroblast mitotic inhibitor from mouse embryo cells aging *in vitro* and undergoing spontaneous transformation, *Exp. Gerontol.,* 15, 315, 1980.

29. **Stein, G. H. and Atkins, L.,** Membrane-associated inhibitor of DNA synthesis in senescent human diploid fibroblasts: characterization and comparison to quiescent cell inhibitor, *Proc. Natl. Acad. Sci. U.S.A.,* 83, 9030, 1986.

30. **Pereira-Smith, O. M., Fisher, S. F., and Smith, J. R.,** Senescent and quiescent cell inhibitors of DNA synthesis, *Exp. Cell Res.,* 160, 297, 1985.

31. **Lumpkin, C. K., Jr., McClung, J., Pereira-Smith, O. M., and Smith, J. R.,** Existence of high abundance antiproliferative mRNA's in senescent human diploid fibroblasts, *Science,* 232, 393, 1986.

32. **Sporn, M. B. and Todaro, G. J.,** Autocrine secretion and malignant transformation of cells, *N. Engl. J. Med.,* 303, 878, 1980.

33. **Adams, S. O., Nissley, S. P., Handwerger, S., and Rechler, M. M.,** Developmental patterns of insulin-like growth factor-I and -II synthesis and regulation in rat fibroblasts, *Nature,* 302, 150, 1983.

34. **Schoinle, E., Zapf, J., Humbel, R. E., and Froesch, E. R.,** Insulin-like growth factor I stimulates growth in hypophysectomized rats, *Nature,* 296, 252, 1982.

35. **Zezulak, K. M. and Green, H.,** The generation of insulin-like growth factor-I-sensitive cells by growth hormone action, *Science,* 233, 551, 1986.

36. **Nicoll, C. S., Anderson, T. R., Hebert, N. J., and Russell, S. M.,** Comparative aspects of the growth-promoting actions of prolactin on its target organs: evidence for synergism with an insulin-like growth factor, in *Prolactin, Basic and Clinical Correlates,* MacLeod, R. M., Thorner, M. O., and Scapagnini, U., Eds., Liviana Press, Padova, 1985, 393.

37. **Linzer, D. I. H. and Nathans, D.,** Changes in specific mRNAs following serum stimulation of cultured mouse cells: increase in a prolactin-related mRNA, in *Cancer Cells,* Levin, A. J., Vande Woude, G. F., Topp, W. C., and Watson, J. D., Eds., Cold Spring Harbor Laboratory, Cold Spring Harbor, NY, 1984, 111.

Chapter 4

NUCLEAR RESPONSIVENESS TO INTRACELLULAR SIGNALS IN NORMAL AND SENESCENT LYMPHOCYTES

Kerin L. Fresa and Stanley Cohen

TABLE OF CONTENTS

I. INTRODUCTION

A. THE IMMUNE RESPONSE

The immune system is composed of a complex, interdigitated network of cells and soluble factors. The function of the immune system is primarily to maintain homeostasis by providing resistance against "nonself" invaders such as pathogenic microorganisms, toxins, and, perhaps, neoplastic cells. The induction of an immune response against a particular agent involves multiple interactions between macrophages (which process and present antigens to lymphocytes), helper and suppressor lymphocytes (which amplify and diminish, respectively, the magnitude of the response), and effector lymphocytes (which function, ultimately, to produce antibodies or lymphokines or to lyse virally infected, neoplastic, or allogeneic cells). These interactions are the result of either direct cell-to-cell contact or are mediated through the production of soluble mediator molecules.

Upon stimulation of lymphocyte populations with antigen or mitogen, the induction of the immune response depends on two critical processes: activation or **differentiation** of precursor cell populations, leading to the generation of cells with specialized effector functions, such as production of antibodies or lymphokines, and **proliferation** of these cell populations. When lymphocyte populations are stimulated with antigen, helper T cell populations with specificity toward that antigen become activated to produce lymphokines, including the T cell growth factor, interleukin-2 (IL-2). IL-2 then stimulates the clonal proliferation of these and other T cell populations with specificity toward that antigen. Polyclonal activation and proliferation of T lymphocytes can be achieved by stimulation of these cells with either one of the mitogenic plant lectins, phytohemagglutinin (PHA) or concanavalin-A (Con-A). Activation of the responding cell populations can be assessed in terms of production of IL-2 and other lymphokines, and proliferation of these cells is measured by uptake of tritiated thymidine (^{3}H-TdR). In fact, it is believed that mitogen-induced proliferation is dependent on the ability of the responding T lymphocytes to both produce and respond to IL-2 through the induction of IL-2 receptors on the cell surface.

B. DEFECTS IN PROLIFERATIVE CAPACITY OF LYMPHOCYTES FROM AGED INDIVIDUALS

It has been observed that T lymphocytes from aged individuals are less responsive to mitogenic stimulation, as measured by ^{3}H-TdR incorporation, than lymphocytes from young control individuals.[1-4] There are several possible defects in aged lymphocytes that may account for their relative inability to take up ^{3}H-TdR in response to mitogenic stimulation. It is possible that cells from aged individuals may be more susceptible to damage due to the presence of the isotope in the assay and that such damage could lead to cell cycle arrest; additionally, cells from aged individuals may show decreased susceptibility to the mitogenic signal itself.[5-6] Decreased susceptibility to mitogenic stimulation in lymphocytes from aged individuals may be due to defective activation of these cells to produce IL-2 or to a relative inability of these cells to respond to the proliferative signal of IL-2. Addition of exogenous IL-2 to cultures of lymphocytes from aged individuals only partially restores the proliferative response to mitogen, suggesting that while these lymphocytes may be deficient in their ability to produce IL-2, defects in the ability to respond to IL-2 may also be involved.[7-11] Indirect evidence suggests that these cells may also have a decreased number and/or affinity of IL-2 receptors at the cell surface.[10,12] Thus, although defects in both the ability to produce and respond to IL-2 are associated with the relative inability of lymphocytes from aged individuals to respond to mitogenic stimulation, the precise intracellular basis for this phenomenon remains undefined. This review will focus on our current understanding of cytoplasmic and nuclear signals involved in activation and proliferation of lymphocytes, as well as on defects in these signaling mechanisms which may be involved in the decreased proliferative responses that are observed in lymphocytes from aged donors.

II. INTRACELLULAR SIGNALING MECHANISMS INVOLVED IN PROLIFERATION OF LYMPHOCYTES

A. HISTORICAL EVIDENCE THAT NUCLEAR DNA SYNTHESIS IS UNDER CYTOPLASMIC CONTROL

A number of early studies suggest that cytoplasmic factors control nuclear activity. In most cases, the nuclei of cells in binucleate and multinucleate organisms exit G_1 and enter the S phase of the cell cycle synchronously, suggesting that extranuclear factors control DNA synthesis.[13] Evidence that cytoplasmic factors control nuclear DNA synthesis was obtained by Graham et al.,[14,15] who showed that nuclei from normally quiescent cells such as adult frog brain or liver cells can be induced to undergo DNA synthesis when injected into the cytoplasm of frog eggs. Similar results were obtained in studies by Harris and his associates.[16,17] Using Sendai virus-mediated cell fusion techniques, they demonstrated that suppression of nuclear DNA synthesis in differentiated cells could be reversed following fusion of these cells with continuously proliferating HeLa or Ehrlich ascites cells. However, heterokaryons formed by the simple fusion of the parental cells usually contained nuclei and cytoplasm from both parents, making conclusions about the origin of factors that control DNA synthesis in the heterokaryon difficult. Johnson and Harris[17] therefore employed enucleated erythrocytes as a source of quiescent, differentiated cells for Sendai virus-mediated fusion. Nucleated erythrocytes were found to be preferable for these studies, as the nuclei of these cells were completely dormant, and since Sendai virus was hemolytic at the concentrations required to induce cell fusion, the erythrocytes were lysed by the virus prior to fusion and effectively contained little or no cytoplasm. Thus, the heterokaryons that were generated in these fusions contained nuclei and cytoplasm from HeLa cells and nuclei alone from the erythrocyte ghosts. When the dormant nuclei were introduced into the HeLa cytoplasm by cell fusion, these nuclei underwent substantial enlargement and began synthesizing DNA and RNA. These results suggest that nuclear DNA synthesis may be controlled by factors present in the cytoplasm. Indeed, cytoplasmic extracts from mitogenic hormone-stimulated cells from early frog embryos and from continuously proliferating cells can initiate DNA synthesis in isolated nuclei from adult frog spleen and liver cells. Das[18] observed that stimulation of quiescent 3T3 cells with epidermal growth factor (EGF) resulted in the appearance of extractable cytoplasmic factors that were capable of inducing DNA synthesis in isolated nuclei. This activity was not detectable in cytoplasmic extracts from quiescent, contact-inhibited 3T3 cells. The ability to induce DNA synthesis in isolated nuclei appeared to be mediated by nondialyzable protein(s) susceptible to heat and trypsin digestion. These factors appeared to be distinct from internalized EGF or its receptor. The results suggest that actively dividing cells may contain cytoplasmic factors that are capable of inducing DNA synthesis in isolated nuclei, and they also suggest that growth factors, after binding at the cell surface, may induce a single cytoplasmic factor or a series of such factors that send the "message" to begin DNA synthesis to the nucleus.

In recent years, it has been established that the interaction of growth factors with specific receptors on the cell surface initiates a series of intracellular changes that ultimately lead to nuclear DNA synthesis. These changes include rearrangement and endocytosis of membrane receptors with internalization of the bound ligands, autophosphorylation of receptors, phosphorylation of other cellular proteins, ionic fluxes, alterations in intracellular enzyme activities and in cyclic nucleotide and intracellular lipid metabolism, and changes in the cellular cytoskeleton.[19]

These changes are most likely followed by changes in gene transcription, with subsequent changes in protein, nucleic acid, and carbohydrate metabolism, which prepare the cell for division.[19] These changes must also include the generation of factors that transport the cytoplasmic signals for DNA synthesis and cell division to the nucleus.

B. INTRACELLULAR CHANGES ACCOMPANYING B AND T CELL ACTIVATION AND PROLIFERATION

As mentioned earlier, IL-2 is a growth factor which is important for the proliferation and differentiation of antigen-specific T cells and is also important in the induction of proliferation following mitogenic stimulation of T cells. The binding of IL-2 to high-affinity specific receptors at the cell surface has been well characterized. It is well established for a number of nonlymphoid cell types that activation of membrane-associated protein kinases (PKs) is a critical step in regulation of cellular proliferation by growth factors and phorbol esters, as well as during viral transformation.[20-25] For example, it has been demonstrated that binding of EGF and insulin to their specific receptors at the cell surface results in the activation of receptor-associated PKs.[23,24] Several investigators have studied the phosphorylation patterns of membrane-associated and cytoplasmic proteins and cellular PK activity following interaction of IL-2 with its receptor. Gaulton and Eardley[26] have demonstrated that addition of IL-2 to Con-A- or PHA-activated murine splenic T cells or to IL-2-dependent T cell lines results in the rapid phosphorylation of a variety of membrane-associated proteins, including the IL-2 receptor itself. This autophosphorylation is similar to that observed in a variety of other systems,[23,24] most notably following the interaction of EGF with its cell membrane-associated receptor.[24] The phosphorylation of the IL-2 receptor is evident within 1 min of introducing IL-2 and peaks approximately 15 min after IL-2 addition. It was postulated that the binding of IL-2 to its receptor results in the activation of membrane-associated PKs.

This hypothesis is substantiated by the work of Farrar and Anderson,[27] who studied the intracellular mechanisms by which this ligand-receptor interaction promotes proliferation and differentiation of T lymphocytes. They examined protein kinase C (PKC) activity in the plasma membrane and cytoplasm of murine IL-2-dependent cells at various times after IL-2 stimulation. A fourfold increase in plasma membrane PKC activity was observed within 3 min after IL-2 stimulation, with a concomitant decrease in cytoplasmic PKC activity. Peak plasma membrane PKC activity was observed 10 min after IL-2 stimulation, at which time cytoplasmic PKC activity was lowest. At 20 min after IL-2 stimulation, a rapid decline in plasma membrane PKC activity could be observed; this activity returned to baseline levels by 60 min after stimulation. In order to determine if there was a direct correlation between IL-2-induced PKC plasma membrane association and proliferation of these cells, plasma membrane PKC activity and proliferation were measured in response to varying concentrations of IL-2. A clear correlation was observed between the degree of proliferation and plasma membrane PKC activity. Furthermore, it appeared that during activation, PKC was mobilized from the cytoplasm to the plasma membrane, as the total cellular pool of PKC did not differ between nonstimulated and IL-2-stimulated cells.

Several changes in lipid metabolism have also been observed in lymphocytes stimulated with antigen or mitogen. It has been postulated that lipid turnover, particularly the hydrolysis of phosphatidylinositol (PI), is one of the initial signals in receptor-mediated signal transduction.[28,33] Kozumbo et al.[34] demonstrated that activation of cloned resting T cell populations to produce lymphokines following stimulation with antigen or mitogen is associated with phospholipid hydrolysis. It appears that binding of antigen or mitogen to specific receptors activates an endogenous phospholipase C.[35-37] This enzyme catalyzes the transient hydrolysis of the membrane phospholipid, PI, and its polyphosphorylated derivatives, PI-4,5-bisphosphate (PIP$_2$) and PI-4-monophosphate (PIP) to diacylglycerol (DAG) and inositol phosphates (IP).[38-41] IP functions as a signal to mobilize Ca^{2+} from an intracellular pool, resulting in increased cytoplasmic concentrations of Ca^{2+}.[42] DAG functions to increase the affinity of PKC for Ca^{2+}, resulting in activation of the enzyme.[29,43] PKC has a broad substrate specificity, phosphorylating serine and threonine residues of many endogenous proteins, and may be involved in the phosphorylation events that are associated with lymphocyte activation. IP is rapidly degraded to inositol and is used in the resynthesis of PI.[28,44] DAG is rapidly converted to phosphatidic acid and arachidonic acid, which are consumed in the resynthesis of PI.[28] However, it is interesting to note that in the study presented by Kozumbo et al.,[34] IL-2 stimulation of these cells did not result in hydrolysis of PI to IP and DAG. Similar results were reported by Mills et al.,[45] who were unable to detect hydrolysis of PIP_2 in human T lymphocytes following IL-2 stimulation. These results appear to dispute the theory postulated by Farrar and Anderson,[27] who suggested that PI hydrolysis may be the mechanism by which IL-2 transmits signals for cell division across the cell membrane. However, since binding of IL-2 also fails to mobilize intracellular Ca^{2+} or to generate production of IP in lymphocytes,[27,45] it is possible that the translocation and activation of PKC in lymphocytes as a result of stimulation with IL-2 may proceed via a mechanism different from that following activation with antigen or mitogen. It should also be kept in mind that activation of lymphocytes to secrete lymphokines, including IL-2, may be intrinsically tied to proliferation of these cells. Therefore, activation of PKC by the inositol phosphate pathway may be involved, although indirectly, in the signaling mechanisms for cell proliferation.

Regardless of the mechanism, it appears that PKC translocation and activation are only initial events in the intracellular pathway leading to induction of DNA synthesis in lymphocytes and that there is another messenger that transports the signal for DNA synthesis to the nucleus. As is the case with T lymphocytes that are stimulated with antigen or mitogen, stimulation of B lymphocytes with anti-immunoglobulin appears to lead to receptor (surface immunoglobulin) cross-linking, PI hydrolysis, and PKC activation.[46-49] Recent reports have also described the appearance of novel cytoplasmic proteins that may be involved in signal transduction to the nucleus in B lymphocytes stimulated with anti-immunoglobulin.[50] Following binding of anti-immunoglobulin to the B lymphocyte cell membrane, a membrane-bound, trypsin-like serine protease is activated.[50] This serine protease appears to split precursor proteins in the cytoplasm into active cytoplasmic factors that are capable of inducing nonhistone protein-specific protein kinase activity in the nucleus.

Similar cytoplasmic factors have been detected in lymphocytes from the lipopolysaccharide (LPS)-responsive C3H/HeN strain of mice following stimulation with that endotoxin.[51] Microinjection of cytoplasmic extracts from LPS-stimulated C3H/HeN lymphocytes into B lymphocytes of nonresponder C3H/HeJ origin resulted in the ability of C3H/HeJ lymphocytes to proliferate in response to stimulation with LPS. These data suggest that the ability to respond to LPS stimulation is associated with a cytoplasmic mediator which is induced in responder strains of mice following stimulation with LPS. Furthermore, transfer of this factor into lymphocytes from nonresponder strains of mice confers the ability to proliferate in response to stimulation with LPS. Gel filtration of the cytoplasmic extracts from LPS-stimulated C3H/HeN lymphocytes has indicated that the factor responsible for this activity has a molecular weight of 100 kDa. These results all suggest that the intracytoplasmic factors that directly influence nuclear function may be accessible to experimental manipulation. Since all the pathways described in the previous sections reflect early, cell membrane-related events involved in signaling for proliferation, we thought it of interest to directly examine the subsequent steps involved in nuclear activation.

C. ACTIVELY DIVIDING LYMPHOCYTES CONTAIN A CYTOPLASMIC ACTIVATOR OF DNA SYNTHESIS

We wished to determine if cytoplasmic factors capable of induction of DNA synthesis in lymphocytes could be detected in proliferating cells. For this purpose, we initially studied several continuously proliferating lymphoblastoid cell lines and normal murine as well as human lymphocytes that had been stimulated with mitogen.[52] Cytoplasmic extracts were prepared from these cells and tested for their ability to induce DNA synthesis in isolated, quiescent frog spleen nuclei. This reaction was carried out in a cocktail containing 2'-deoxyadenosine 5'-triphosphate, 2'-deoxyguanosine 5'-triphosphate, 2'-deoxycytidine 5'-

triphosphate, adenosine 5′-triphosphate (ATP), phosphoenolpyruvate, pyruvate kinase, and ^{3}H-thymidine triphosphate (^{3}H-TTP). The requirement for all four deoxyribonucleotides and ATP strongly suggested that DNA synthesis, and not DNA repair, was being measured in these assays. Further evidence that DNA synthesis, not DNA repair, was being measured under similar experimental conditions was presented in Jazwinski et al.,[53] who documented replicative "forks" and "eyes" in the DNA by electron microscopy.

We showed that cytoplasmic extracts from spontaneously proliferating human T cell lymphoblastoid cells (MOLT-4 cells), murine plasmacytoma cells designated P3X63Ag8.653, and human B lymphoma-derived RPMI 8392 cells could induce high levels of DNA synthesis in isolated nuclei from adult frog spleen cells, as measured by ^{3}H-TdR incorporation into trichloroacetic acid (TCA)-precipitable material. Maximal levels of ^{3}H-TTP incorporation were observed after a 90-min incubation. Studies using various concentrations of the cytoplasmic extracts elicited a sigmoidal dose-response curve with increasing concentrations of extract. This pattern of response differs from that of several extracellular inducers of DNA synthesis, such as antigens, mitogens, and growth hormones, which typically have a bell-shaped dose-response curve, with inhibition at higher inducer concentrations. Minimal DNA synthesis was observed in nuclei incubated with extract and in nuclei incubated with cytoplasmic extracts from nonproliferating cells. Cytoplasmic extracts incubated alone also failed to incorporate ^{3}H-TTP. These results suggest that continuously proliferating lymphoblastoid cells contain a cytoplasmic factor that may be involved in the induction of nuclear DNA synthesis.

Extracts prepared from human peripheral blood lymphocytes (PBL) that were treated with PHA were also found to induce DNA synthesis in isolated nuclei. This effect was not due to a direct effect of residual mitogen on the nuclei, as no DNA synthesis could be observed in nuclei incubated directly with PHA. Extracts from untreated PBL induced little or no DNA synthesis in isolated nuclei, suggesting that the factor responsible for inducing DNA synthesis in isolated nuclei is only found in actively dividing cells. We have termed the factor that is present in the cytoplasms of actively dividing cells and which can induce DNA synthesis in isolated nuclei the activator of DNA synthesis, or ADR.

ADR activity has also been detected in cytoplasmic extracts from IL-2-responsive lymphocyte populations that have been stimulated with that factor,[54] suggesting that this factor is also involved in IL-2-dependent T cell growth. However, since the IL-2-responsive T cell blasts that were used in this study had been stimulated with PHA, it was important to determine that ADR was stimulated as a result of IL-2 treatment and that the ADR activity that we had observed under these conditions was not residual activity stimulated by PHA. Since dexamethasone inhibits PHA-induced proliferation in a dose-dependent manner and the degree of inhibition is inversely proportional to the amount of IL-2 produced by PHA-stimulated, dexamethasone-treated lymphocytes, we examined the ability of lymphocytes treated in this manner to produce ADR. Addition of dexamethasone to lymphocytes stimulated with PHA resulted in a significant decrease in ADR activity in these cells. This decrease in ADR activity could be reversed if exogenous IL-2 was added to the cultures. These results suggest that induction of ADR is not an early event in T cell activation by PHA, but rather appears to be a later event resulting either directly or indirectly from binding of IL-2 to IL-2-responsive cells.

III. CHARACTERIZATION OF ADR

A. PHYSICAL PROPERTIES OF ADR

Initial studies suggested that ADR is a protein, as cytoplasmic extracts that were treated with trypsin failed to induce DNA synthesis in isolated nuclei.[52] This effect was due neither to the soybean inhibitor used to stop the trypsinization reaction nor to the effect of the trypsin on the nuclei. ADR was found to be stable at 4°C for 24 h, and it could still be detected after freeze-thawing.[52] Furthermore, ADR activity was retained following lyophilization and reconstitution of the cytoplasmic extracts.[52] Dialysis studies have indicated that ADR has a molecular weight larger than 3500.[52] ADR was found to be inactivated by incubation at 60°C for 20 min.[52] These results taken together suggest that ADR is a relatively stable protein that is larger than 3.5 kDa.

In order to more closely approximate the size of the protein, cytoplasmic extracts containing ADR activity were subjected to Amicon® ultrafiltration. For these studies, cytoplasmic extracts of 8392 cells were serially filtered through XM50 and XM100A membranes. The filtrates and the fractions that were retained by the membranes were collected and assayed for ADR activity. Only the fraction that was retained by the XM100A membrane possessed ADR activity, suggesting that ADR has a molecular weight of greater than 100,000.[52] Similar results were obtained using MOLT-4 cytoplasmic extracts as a source of ADR. The XM100A retentate could then be enriched for ADR activity by ammonium sulfate precipitation. ADR activity precipitated under 30 to 50% ammonium sulfate saturation. Using these methodologies, we are able to preserve 100% of the ADR activity in preparations from which the total protein had been reduced by 73%.[52]

B. THE INTRACELLULAR LOCALIZATION OF ADR

We wished to determine whether ADR is only present in the cytoplasm of actively dividing cells or if it is also secreted from cells, like conventional lymphokines. For these studies, 8392 cells were cultured for 24 h under serum-free conditions. The medium from these cultures was then dialyzed against extraction buffer.[52] The dialysate was then either diluted or concentrated, so that concentrations ranging from one tenth to ten times the original starting concentration were tested for ADR activity. No ADR activity could be detected at any of the concentrations tested.[52] The lack of activity in these preparations was not due to loss of the factor through dialysis, as we had previously shown that ADR activity was not dialyzable.[52]

Previous experiments showed that ADR could induce DNA synthesis in isolated nuclei. It was of interest to determine if ADR could also stimulate DNA synthesis in intact cells. For these experiments, intact frog spleen cells were incubated with cytoplasmic extracts of 8392 cells. These extracts failed to induce DNA synthesis in intact frog spleen cells, even after 4 d of treatment.[52] Furthermore, stimulatory activity was not observed under these conditions even after the extracts were concentrated fivefold. Intact frog spleen cells were capable of proliferation after stimulation with Con-A, indicating that the failure of cytoplasmic extracts from 8392 cells to induce proliferation in frog spleen cells was not due to a defect in the proliferative capacity of the frog spleen cells. It was also possible that, while ADR activity on isolated nuclei lacks species specificity, a species-specific membrane attachment step might be necessary when intact cells are used as an indicator of ADR activity. Therefore, cytoplasmic extracts of murine origin were tested for ADR activity using intact murine lymphocytes as indicator cells. No induction of DNA synthesis was observed under these conditions. Thus, ADR is neither secreted from cells nor has it any effect on intact cells.[52] It appears, therefore, that ADR is an entirely intracellular mediator.

C. BIOCHEMICAL CHARACTERISTICS OF ADR

In order to determine if ADR possesses protease activity similar to that observed in B lymphocytes activated with anti-immunoglobulin,[50] cytoplasmic extracts of MOLT-4 cells were tested for their ability to proteolytically degrade insolubilized, iodinated fibrin in the presence and absence of an exogenous source of plasminogen.[56] Even in the absence of fibrinogen, significant degradation of fibrin was observed following 21 h of co-incubation. However, addition of plasminogen resulted in a marked, time-dependent enhancement of proteolysis. Plasminogen alone was proteolytically inactive. These results suggest that ADR-containing, cytoplasmic extracts of MOLT-4 cells possess neutral serine protease activity that degrades fibrin directly and converts plasminogen to plasmin, and they raise the possibility that ADR itself is a protease.

It was thus of interest to determine if inhibitors of protease activity would block ADR activity on isolated nuclei. For these determinations, isolated frog spleen nuclei were incubated with increasing concentrations of MOLT-4 extracts in the presence and absence of protease inhibitors.[56] Addition of aprotinin to wells containing MOLT-4 cytoplasmic extracts and isolated nuclei resulted in a significant inhibition of ADR activity. Aprotinin could also inhibit the activity of ADR derived from IL-2-activated, normal human lymphoblasts. Other protease inhibitors, specifically N-α-tosyllysine chloromethyl ketone and leupeptin, had similar effects on ADR activity. Para-aminobenzamidine also blocked ADR activity, although higher concentrations of the inhibitor were required. However, addition of soybean trypsin inhibitor (SBTI) at comparable concentrations had no effect on ADR activity.[56]

In order to rule out the possibility that aprotinin was inhibiting DNA synthesis through a direct effect on isolated nuclei, quiescent frog nuclei were preincubated with aprotinin at concentrations that would completely inhibit DNA synthesis when added to cultures containing MOLT-4 extract and isolated nuclei. The aprotinin was then washed away prior to the addition of MOLT-4 extract. Under these conditions, no significant inhibition of DNA synthesis occurred.[56] These results suggest that the aprotinin inhibits DNA synthesis by an effect on component(s) of the MOLT-4 extract and not by an effect on the nuclei.

It was also of interest to determine if insolubilized aprotinin could absorb ADR activity. Aprotinin and, as a control, SBTI were conjugated to agarose beads. Cytoplasmic extracts of MOLT-4 cells were then allowed to interact with aprotinin-agarose, SBTI-agarose, or unconjugated agarose. The majority of ADR activity could be removed from cytoplasmic extracts of MOLT-4 cells that were allowed to interact with aprotinin-agarose. After the extracts were allowed to interact with the aprotinin-conjugated agarose, the beads were collected and packed into a column. The ADR activity could be eluted from the column with 0.4 M NaCl/0.05 M sodium acetate, pH 5.0. In contrast, adsorption of cytoplasmic extracts of MOLT-4 cells with either plain or SBTI-conjugated agarose had no effect on ADR activity. As expected, no ADR activity could be eluted from control columns containing unconjugated agarose beads. Preliminary biochemical characterization of the aprotinin-agarose eluate by sodium dodecyl sulfate/polyacrylamide gel electrophoresis (SDS-PAGE) revealed a band with an approximate molecular weight of 100,000 which possessed proteolytic activity.[56] These results were consistent with earlier studies which showed that ADR activity in MOLT-4 extracts was associated with a fraction having a molecular weight of greater than 100,000.

D. QUIESCENT CELLS CONTAIN AN INHIBITOR OF ADR ACTIVITY

As mentioned earlier, cytoplasmic extracts from actively proliferating lymphoblastoid cell lines and from IL-2- or mitogen-stimulated PBL are capable of inducing nuclear DNA synthesis. This activity is associated with a protein of approximately 100 kDa that possesses serine protease activity, which we have termed ADR. Cytoplasmic extracts of resting cells, including unstimulated PBL, have little to no detectable ADR activity.[52] We therefore investigated the possibility that quiescent cells may contain a factor that inhibits ADR activity.

For these studies, cytoplasmic extracts were prepared from MOLT-4 cells or from PBL that were cultured in the presence (positive extract) or absence (negative extract) of PHA.[55] Consistent with previous studies,[52] the MOLT-4 extract and the positive extract induced high levels of DNA synthesis in isolated frog nuclei. Conversely, the negative extract failed to induce detectable DNA synthesis in isolated frog nuclei. Interestingly, the negative extracts were capable of inhibiting induction of nuclear DNA synthesis by positive extracts.[57] Nuclear

DNA synthesis in wells containing both positive and negative extracts was only 10 to 50% of that in wells containing only the positive extract. This suppression was maximal after 90 min of co-culture and was also observed when isolated nuclei were obtained from human PBL instead of from frog spleen cells.[57]

We postulated that the ability of the negative extract to suppress induction of DNA synthesis by ADR was due to the presence of an inhibitor of ADR activity. The inhibitory activity was abolished if the extracts were treated with trypsin prior to assay, suggesting that the inhibitor was a protein. The factor was stable at 4°C for 24 h and following freeze-thawing. Furthermore, the inhibitory activity was only slightly diminished by heating to 56°C for 20 min.[57] These results suggest that the inhibitor of ADR activity present in cytoplasmic extracts of quiescent lymphocytes is a relatively heat-stable protein.

To determine the approximate molecular weight of this factor, negative extracts were subjected to Amicon® ultrafiltration. The inhibitory activity was retained by an XM50 membrane, suggesting that the inhibitory factor was ≥50 kDa. Those components of the extract that were under 50 kDa were unable to inhibit ADR-induced DNA synthesis in isolated nuclei.[57]

Since the inhibitory factor was isolated from resting PBL that had been cultured *in vitro* for 3 d, we wished to determine whether this factor could also be recovered from freshly isolated PBL or if it appeared only after *in vitro* cultivation. For these experiments, cytoplasmic extracts were prepared from freshly isolated PBL and from unstimulated PBL after various times of culture. Cytoplasmic extracts of freshly isolated PBL had no detectable ADR-inhibitory activity. Inhibitory activity could be demonstrated in extracts of PBL that had been cultured *in vitro* for as little as 2 to 6 h. The inhibitory activity was maximal after 18 h of culture.[57] Failure to detect inhibitory activity in freshly isolated PBL may have been due to the fact that these cells may not be truly quiescent, as suggested by the observation that freshly isolated PBL incorporate low but significant levels of ^{3}H-TdR.

It was necessary to rule out the possibility that macrophages, and not quiescent lymphoid cells, were producing the inhibitory factor or that macrophages were secreting a factor that induced lymphoid cells to produce this factor. For these determinations, unfractionated PBL and macrophage-depleted PBL were cultured overnight.[57] Cytoplasmic extracts of these cells were prepared and tested for their ability to suppress ADR-induced nuclear DNA synthesis.[57] The extracts were equally able to inhibit ADR activity, suggesting that the factor is not produced by macrophages and that its production is not influenced by macrophages or their products.

IV. DEFECTS IN INTRACELLULAR SIGNALING MECHANISMS IN LYMPHOCYTES FROM AGED INDIVIDUALS

Several investigators, including ourselves, have shown that decreased proliferative responses observed in lymphocytes from aged individuals are associated with defects in intracellular signaling mechanisms. Proust et al.[58] have studied PKC activity and translocation, PI hydrolysis, and cytoplasmic Ca^{2+} levels in lymphocytes from aged individuals in order to determine if changes in these intracellular signaling mechanisms are associated with decreased proliferative responses. They found that basal levels of PKC were comparable in lymphocytes from young and aged individuals. However, following Con-A stimulation of these cells, induced PKC activity in lymphocytes from aged individuals was only half of that observed in lymphocytes from young control donors stimulated in the same manner. However, they were unable to detect differences in PIP_2 hydrolysis, as measured by the generation of IP_3, or in free cytoplasmic Ca^{2+} levels in Con-A stimulated lymphocytes from aged individuals when compared to Con-A stimulated young lymphocytes. These results suggest that some intracytoplasmic signaling mechanisms may be deficient in lymphocytes from aged individuals and may partially account for the decreased proliferative responses observed in lymphocytes from these individuals.

We wished to determine if changes in the ability to produce or respond to ADR are also associated with proliferative defects in aged lymphocytes. For these studies, PBL from aged (66 to 72 years) and young adult (22 to 30 years) human donors were treated with PHA for 66 h. Cytoplasmic extracts were prepared from these cells and tested for the ability to induce DNA synthesis in isolated nuclei from adult frog spleen cells.[59]

We showed that cytoplasmic extracts from PHA-stimulated lymphocytes from aged donors were as active in the induction of DNA synthesis in isolated frog nuclei as were extracts from young donors, even though the PHA-induced proliferative capacity of lymphocytes from some, but not all, aged donors was significantly less than that of lymphocytes from young donors. Dose-response analysis revealed no quantitative differences in ADR production in lymphocytes from aged donors compared to that of young donors. As expected, cytoplasmic extracts from unstimulated lymphocytes of both young and aged donors failed to induce significant DNA synthesis in isolated frog nuclei. These results suggest that the decreased proliferative capacity of lymphocytes from aged donors is not associated with an inability to produce ADR.

We then investigated the possibility that decreased proliferation in response to stimulation with mitogen in lymphocytes from aged individuals is associated with a decreased ability to respond to the proliferative signal provided by ADR. To test this hypothesis, we assayed the DNA synthetic response of nuclei isolated from lymphocytes of aged individuals to an exogenous source of ADR. We isolated nuclei from lymphocytes from young donors as well as from aged donors who showed either intact or defective ability to proliferate in response to simulation with PHA. Nuclei derived from young lymphocytes and from aged lymphocytes with intact PHA responsiveness were comparable in their ability to synthesize DNA in response to stimulation with an exogenous source of ADR. However, nuclei from aged lymphocytes with defective proliferative responses to PHA showed a relative inability to synthesize DNA in response to stimulation with ADR. There was a clear correlation between the response of intact cells to PHA stimulation and the response of isolated nuclei from these cells to ADR. These results suggest that the decreased proliferative capacity observed in lymphocytes from some aged individuals is associated with an inability of the nuclei of these cells to respond to ADR.

V. DISCUSSION

The ability of lymphocytes to proliferate in response to mitogenic stimulation depends on the ability of these cells to become activated to both secrete and respond to IL-2. It appears that a number of intracellular signaling mechanisms are set in motion following stimulation with mitogen, including PIP_2 hydrolysis with generation of IP_3 and DAG. These early events serve to increase intracellular levels of Ca^{2+} and to activate PKC, and they appear to be intrinsically tied to the signal for these cells to begin lymphokine production and secretion. Secretion and utilization of IL-2 appears to be necessary for induction of DNA synthesis. Following binding of IL-2 to its receptor, a number of different signaling mechanisms are set in motion to send the proliferative signal from the cell membrane to the nucleus. These signaling mechanisms include the rapid phosphorylation of certain membrane proteins and the activation of PKC through a mechanism that appears to be distinct from PI hydrolysis. We have shown that stimulation of IL-2-responsive (activated) T lymphocytes with IL-2 results in the appearance of a cytoplasmic factor that is capable of inducing DNA synthesis in isolated nuclei. We propose that this late-appearing protein serves as the final cytoplasmic message to the nucleus to begin DNA synthesis.

Lymphocytes from aged individuals often show decreased proliferative responses to stimulation with mitogen. This impairment appears to be the result of defects at a number of levels, including the ability to secrete and respond to IL-2. Lymphocytes from aged individuals also have lower intracellular levels of PKC following stimulation with mitogen than those found in young lymphocytes stimulated in the same manner. We have shown that while lymphocytes from aged individuals are able to produce ADR at levels comparable to those produced by lymphocytes from young individuals, nuclei from aged lymphocytes are defective in their ability to respond to ADR. Thus, it appears that the decreased proliferative capacity of aged lymphocytes is associated with defects in the intracellular signaling network at the level of the cell membrane, the cytoplasm, and the nucleus.

REFERENCES

1. **Weksler, N. E. and Hutteroth, T. H.,** Impaired lymphocyte function in aged humans, *J. Clin. Invest.,* 53, 99, 1974.
2. **Hori, Y., Perkins, E. H., and Halsall, M. K.,** Decline in phytohemagglutinin responsiveness of spleen cells from aging mice, *Proc. Soc. Exp. Biol. Med.,* 144, 48, 1973.
3. **Foad, B. S. I., Adams, L. E., Yamaguchi, Y., and Litwin, A.,** Phytomitogen responses of peripheral blood lymphocytes in young and older subjects, *Clin. Exp. Immunol.,* 17, 657, 1974.
4. **Joncourt, F., Bettens, F., Kristensen, F., and DeWeck, A. L.,** Age-related changes in mitogen responsiveness in different lymphoid organs from outbred NMRI mice, *Immunobiology,* 158, 439, 1981.
5. **Staiano-Coico, L., Darzynkiewicz, Z., Hefton, J. M., Dutkowski, R., Darlington, G., and Weksler, M. E.,** Increased sensitivity of lymphocytes from people over 65 to cell cycle arrest and chromosomal damage, *Science,* 219, 1335, 1983.
6. **Hefton, J. M., Darlington, G. J., Casazza, B. A., and Weksler, M. E.,** Immunologic studies of aging. V. Impaired proliferation of PHA responsive human lymphocytes in culture, *J. Immunol.,* 125, 1007, 1980.
7. **Gillis, S., Kozak, R., Durante, M., and Weksler, M. E.,** Immunological studies of aging: decreased production of and response to T cell growth factor by lymphocytes from aged humans, *J. Clin. Invest.,* 67, 937, 1981.
8. **Thomas, M. L. and Weigle, W. O.,** Lymphokines and aging. Interleukin 2 production and activity in aged animals, *J. Immunol.,* 127, 2101, 1981.
9. **Miller, R. A. and Stutman, O.,** Decline in aging mice of the anti-2,4,6-trinitrophenyl cytotoxic T cell response attributable in loss of Lyt-2-, interleukin 2 producing helper cell function, *Eur. J. Immunol.,* 11, 751, 1981.
10. **Gilman, S. C., Rosenberg, J. S., and Feldman, J. D.,** T lymphocytes of young and aged rats. II. Functional defects and the role of interleukin 2, *J. Immunol.,* 128, 644, 1982.
11. **Cheung, H. T., Wu, W. T., Pahlavani, M., and Richardson, A.,** Effect of age on the interleukin 2 messenger RNA level, *Fed. Proc.,* 44, 573, 1985.
12. **Chang, M.-P, Makinodan, T., Peterson, W. J., and Strehler, B. L.,** Role of T cells and adherent cells in age-related decline in murine interleukin 2 production, *J. Immunol.,* 129, 2426, 1982.
13. **Gonzales-Fernandez, A., Gimenez-Martin, G., Diez, J. L., de la Torre, C., and Lopez-Saez, J. F.,** Interphase development and beginning of mitosis in the different nuclei of polynucleate homokaryotic cells, *Chromosoma,* 36, 100, 1971.
14. **Graham, C. F.,** The regulation of DNA synthesis and mitosis in multinucleate frog eggs, *J. Cell Sci.,* 1, 363, 1966.
15. **Graham, C. F., Arms, K., and Gurdon, J. B.,** The induction of DNA synthesis by frog egg cytoplasm, *Dev. Biol.,* 14, 349, 1966.
16. **Harris, H., Watkins, J. F., Ford, C. E., and Schoefl, G. I.,** Artificial heterokaryons of animal cells from different species, *J. Cell Sci.,* 1, 1, 1966.
17. **Johnson, R. T. and Harris, H.,** DNA synthesis and mitosis in fused cells. II. HeLa-chick erythrocyte heterokaryons, *J. Cell Sci.,* 5, 625, 1969.
18. **Das, M.,** Mitogenic hormone-induced intracellular message: assay and partial characterization of an activator of DNA replication induced by epidermal growth factor, *Proc. Natl. Acad. Sci. U.S.A.,* 77, 112, 1980.
19. **Rosen, O. M.,** After insulin binds, *Science,* 237, 1452, 1987.
20. **Brugge, J. S. and Erikson, R. L.,** Identification of a transformation-specific antigen induced by an avian sarcoma virus, *Nature,* 269, 346, 1977.

21. **Witte, O. N., Dasgupta, A., and Baltimore, D.,** Abelson murine leukeumia virus protein is phosphorylated *in vitro* to form phosphotyrosine, *Nature,* 283, 826, 1980.

22. **Beemaan, K.,** Transforming proteins of some feline and avian sarcoma viruses are related structurally and functionally, *Cell,* 24, 145, 1981.

23. **Kasugh, M., Karlson, F. A., and Kahn, C. R.,** Insulin stimulates the phosphorylation of the 95,000 dalton subunit of its own receptor, *Science,* 215, 185, 1980.

24. **King, L. E., Jr., Carpenter, G., and Cohen, S.,** Characterization by electrophoresis of EGF stimulated phosphorylation using A-431 membranes, *Biochemistry,* 19, 1524, 1980.

25. **Sibley, D. R., Peters, J. R., Nambi, P., Caron, M. G., and Lefkowitz, R. J.,** Desensitization of turkey adenylate cyclase, *J. Biol. Chem.,* 259, 9742, 1984.

26. **Gaulton, G. N. and Eardley, D. D.,** Interleukin 2-dependent phosphorylation of interleukin 2 receptors and other T cell membrane proteins, *J. Immunol.,* 136, 2470, 1986.

27. **Farrar, W. L. and Anderson, W. B.,** Interleukin 2 stimulates association of protein kinase C with plasma membrane, *Nature,* 315, 233, 1985.

28. **Berridge, M. J. and Irvine, R. F.,** Inositol triphosphate, a novel second messenger in cellular signal transduction, *Nature,* 312, 315, 1984.

29. **Berridge, M. J.,** Inositol triphosphate and diacylglycerol as second messengers, *Biochem. J.,* 220, 345, 1984.

30. **Nishizuka, Y.,** Turnover of inositol phospholipids and signal transduction, *Science,* 225, 1365, 1984.

31. **Majerus, P. W., Neufeld, E. J., and Wilson, D. B.,** Production of phosphoinositide-derived messengers, *Cell,* 37, 701, 1984.

32. **Berridge, M. J., Heslop, J. P., Irvine, R. F., and Brown, K. D.,** Inositol lipids and cell proliferation, *Biochem. Soc. Trans.,* 13, 67, 1985.

33. **Majerus, P. W., Wilson, D. B., Connolly, T. M., Bross, T. E., and Neufeld, E. J.,** Phosphoinositide turnover provides a link in stimulus-response coupling, *Trends Biochem. Sci.,* 10, 169, 1985.

34. **Kozumbo, W. J., Harris, D. T., Gromkowski, S., Cerottini, J.-C., and Cerutti, P. A.,** Molecular mechanisms involved in T cell activation. II. The phosphatidylinositol signal transducing mechanism mediates antigen-induced lymphokine production but not interleukin 2 induced proliferation in cloned cytotoxic T lymphocytes, *J. Immunol.,* 138, 606, 1987.

35. **Monroe, J. G. and Cambier, J. C.,** B cell activation. IV. Induction of membrane depolarization and hyper I-A expression by phorbol diesters suggests a role for protein kinase C in murine B lymphocytes, *J. Immunol.,* 132, 1472, 1984.

36. **Ku, Y., Kishimoto, A., Takai, Y., Ogawa, Y., Kimura, S., and Nishizuka, Y.,** A new possible regulatory system in human peripheral lymphocytes. II. Possible relationship to phosphatidylinositol turnover induced by mitogens, *J. Immunol.,* 127, 1375, 1981.

37. **Allan, D. and Michell, R. H.,** Phosphatidylinositol cleavage in lymphocytes. Requirement for calcium ions at a low concentration and effects of other cations, *Biochem. J.,* 142, 599, 1974.

38. **Fisher, D. B. and Meuller, G. C.,** An early alteration in the phospholipid metabolism of lymphocytes by phytohemagglutinin, *Proc. Natl. Acad. Sci. U.S.A.,* 60, 1396, 1968.

39. **Hui, D. Y. and Harmony, J. A. K.,** Phosphatidylinositol turnover in mitogen-activated lymphocytes: suppression by low density lipoproteins, *Biochem. J.,* 192, 91, 1980.

40. **Hasegawa-Sasaki, H. and Sasaki, T.,** Rapid breakdown of phosphatidylinositol accompanied by accumulation of phosphatidic acid and diacylglycerol in rat lymphocytes stimulated by concanavalin A, *J. Biochem.,* 91, 463, 1982.

41. **Taylor, M. V., Metcalfe, J. C., Hesketh, T. R., Smith, G. A., and Moore, J. P.,** Mitogens increase phosphorylation of phosphoinositides in thymocytes, *Nature,* 312, 462, 1984.

42. **Streb, H., Irvine, R. F., Berridge, M. J., and Schultz, I.,** Release of Ca^{2+} from a nonmitochrondrial intracellular store in acinar pancreatic cells by inositol-1,4,5-triphosphate, *Nature,* 306, 67, 1983.

43. **Kishimoto, A., Takai, Y., Mori, T., Kikkawa, U., and Nishizuka, Y.,** Activation of calcium and phospholipid dependent protein kinase by diacylglycerol, its possible relationship to phosphatidylinositol turnover, *J. Biol. Chem.,* 255, 2273, 1980.

44. **Downes, C. P., Mussat, M. C., and Michell, R. H.,** The inositol triphosphate phosphomonoesterase of the human erythrocyte membrane, *Biochem. J.,* 203, 169, 1982.

45. **Mills, G. B., Stewart, D. J., Mellors, A., and Gelfand, E. W.,** Interleukin 2 does not induce phosphatidylinositol hydrolysis in activated T cells, *J. Immunol.,* 136, 3019, 1986.

46. **Coggeshall, K. M. and Cambier, J. C.,** B cell activation. VII. Membrane immunoglobulins transduce signals via activation of phosphatidylinositol hydrolysis, *J. Immunol.,* 133, 3382, 1984.

47. **Coggeshall, K. M. and Cambier, J. C.,** B cell activation. VI. Effects of exogenous diglyceride and modulators of phospholipid metabolism suggest a central role for diacylglycerol generation in transmembrane signalling by mIg, *J. Immunol.,* 134, 101, 1985.

48. **Muraguchi, A., Kehrl, J. H., Butler, J. L., and Fauci, A. S.,** Sequential requirements for cell cycle progression of resting human B cells after activation with anti-Ig, *J. Immunol.,* 132, 176, 1984.

49. **Maino, V. C., Hayman, M. J., and Crumpton, M. J.,** Relationship between enhanced turnover of phosphatidylinositol and lymphocyte activation by mitogens, *Biochem. J.,* 146, 247, 1975.

50. **Kishimoto, T., Kikutani, H., Nishizawa, Y., Sakaguchi, N., and Yamamura, Y.,** Involvement of anti-Ig-activated serine protease in the generation of cytoplasmic factor(s) that are responsible for the transmission of Ig-receptor-mediated signals, *J. Immunol.,* 123, 1504, 1979.

51. **Eda, Y., Ohara, J., and Watanabe, T.,** Restoration of LPS responsiveness of C3H/HeJ mouse lymphocytes by microinjection of cytoplasmic factor(s) from LPS-stimulated normal lymphocytes, *J. Immunol.,* 131, 1294, 1983.

52. **Gutowski, J. K. and Cohen, S.,** Induction of DNA synthesis in isolated nuclei by cytoplasmic factors from spontaneously proliferating and mitogen-activated lymphoid cells, *Cell. Immunol.,* 75, 300, 1983.

53. **Jazwinski, S. M., Wang, J. L., and Edelman, G. M.,** Initiation of replication in chromosomal DNA induced by extracts of proliferating cells, *Proc. Natl. Acad. Sci. U.S.A.,* 73, 2231, 1976.

54. **Gutowski, J. K., Mukherji, B., and Cohen, S.,** The role of cytoplasmic intermediates in IL-2-induced T cell growth, *J. Immunol.,* 133, 3068, 1984.

55. **Gillis, S., Crabtree, G. R., and Smith, K. A.,** Glucocorticoid induced inhibition of T cell growth factor production. I. The effect on mitogen-induced lymphocyte proliferation, *J. Immunol.,* 123, 1624, 1979.

56. **Wong, R. L., Gutowski, J. K., Katz, M., Goldfarb, R. H., and Cohen, S.,** Induction of DNA synthesis in isolated nuclei by cytoplasmic factors: inhibition by protease inhibitors, *Proc. Natl. Acad. Sci. U.S.A.,* 84, 241, 1987.

57. **Gutowski, J. K., West, A., and Cohen, S.,** The regulation of DNA synthesis in quiescent lymphocytes by cytoplasmic inhibitors, *Proc. Natl. Acad. Sci. U.S.A.,* 82, 5160, 1985.

58. **Proust, J. J., Filburn, C. R., Harrison, S. A., Buchholz, M. A., and Nordin, A. A.,** Age-related defect in signal transduction during lectin activation of murine T lymphocytes, *J. Immunol.,* 139, 1472, 1987.

59. **Gutowski, J. K., Innes, J., Weksler, M. E., and Cohen, S.,** Induction of DNA synthesis in isolated nuclei by cytoplasmic factors. II. Normal generation of cytoplasmic stimulatory factors by lymphocytes from aged humans with depressed proliferative responses, *J. Immunol.,* 132, 559, 1984.

60. **Gutowski, J. K., Innes, J. B., Weksler, M. E., and Cohen, S.,** Impaired nuclear responsiveness to cytoplasmic signals in lymphocytes from elderly humans with depressed proliferative responses, *J. Clin. Invest.,* 78, 40, 1986.

Chapter 5

ACTIVATION DEFECTS IN T CELLS FROM OLD MICE

Richard A. Miller, Adam Lerner, Ben Philosophe, and John Macauley

In mammals, aging brings a progressive derangement of homeostatic control systems, systems that often involve interacting cell types, and a corresponding increase in vulnerability to internal (e.g., neoplastic) and external (e.g., infectious) threats. The immune reaction is currently among the best characterized, and hence most malleable, of these defensive systems. Immunologists can now, with a moderate degree of fidelity, mimic in culture a wide range of multicellular processes (e.g., antibody production and cellular anti-viral and anti-tumor responses) that serve *in vivo* to protect the organism from these threats. Many of these protective responses have been shown to work less well in older animals. Investigators are now able to exploit recent work that has led to improved definition of the interacting cell types, of the lymphokines they use to signal one another (and to influence cells outside the immune system per se), and of the genes whose expression controls immune reactivity.

Most immune reactions are initiated by T lymphocytes, which respond to specific antigenic stimuli by clonal proliferation, by the production of lymphokines, and by direct interactions with other T cells and with antibody-producing B lymphocytes. Although the number of T cells that can respond to any single antigen is quite small, one can induce proliferation and lymphokine secretion in the majority of T cells by using plant lectins, including concanavalin A (Con-A) and phytohemagglutinin (PHA), that are thought to react with the T cell's antigen receptor (or perhaps other components of the signal-transducing pathway). Lymphocytes from old rodents and old humans do not proliferate as vigorously in response to mitogenic lectins as do T cells from younger individuals,[1,2] a defect thought to reflect age-dependent declines in protective immune responses in the intact organism itself.

Our lab, like many others, has begun to dissect the process of mitogen-induced T cell proliferation into its component steps to see at what stage or stages age-related deficits can first be demonstrated. A key development was the realization (reviewed by Cantrell and Smith[3]) that T cell entry into the DNA-synthetic S phase of the mitotic cycle requires receipt of a signal mediated by a soluble lymphokine, interleukin 2 (IL-2), acting upon a high-affinity receptor (IL-2R). Unstimulated T cells neither produce IL-2 nor express IL-2R, but activation leads to production of both within 12 to 24 h. Several reports[4-6] have shown that aging leads to a decline in the amount of IL-2 accumulated in cultures of T cells upon mitogen stimulation. A corresponding deficit in IL-2 mRNA production has also been reported.[7] These observations, together with the demonstration that high doses of IL-2 can, at least under some circumstances, lead to improved immune responsiveness of T cells from old donors,[4,6,8] suggest that diminished IL-2 production is an important component of senescent immune deficiency. Despite the ability of exogenous IL-2 to improve responsiveness of old cells, there is now good evidence that aging leads to a decline in responsiveness to IL-2 as well. In limiting dilution experiments, for example, we used a functional measure of IL-2 responsiveness by adding saturating amounts of IL-2 to cultures of Con-A-activated T cells,[9] and we found a three- to fivefold decline in the number of IL-2-reactive T cells in 18-month-old mice. In a second approach,[10] we used a fluorescent antibody (7D4) now known to be specific for the 55-kDa component of the high-affinity IL-2R[11] to show that Con-A induces fewer T cells in old mice to become IL-2R positive. We have also analyzed RNA preparations made 6 h after Con-A stimulation of spleen cells from young and old mice using a probe specific for the 55-kDa component of the IL-2R (Figure 1); an age-associated decline in RNA production sufficient to account for the deficit in IL-2R protein

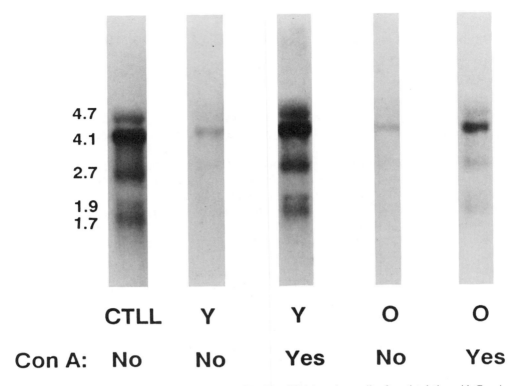

FIGURE 1.   Age-dependent decline in expression of IL-2R mRNA by spleen cells after stimulation with Con-A. Total cellular RNA was extracted with LiCl/urea and purified by sequential phenol-chloroform extraction and ethanol precipitation. Purified RNA was then subjected to electrophoresis in a gel containing 1% agarose and 6% formaldehyde and then blotted onto nitrocellulose filter paper. The filter was then probed with a murine cDNA that contained the entire 804-base coding region of the p55 component of the IL-2R. Each lane contained 20 μg RNA. Lane 1: RNA from CTLL, a murine cytotoxic T cell line that expresses p55 constitutively (positive control); lanes 2 and 3: RNA from spleen cells of 3-month-old female (C57BL/6 × CBA) F_1 mice incubated for 6 h with (lane 3) or without (lane 2) 4 μg/ml Con-A; lanes 4 and 5: RNA isolated from spleen cells of 26-month-old mice incubated for 6 h with (lane 5) or without (lane 4) Con-A stimulation. Y = young; O = old.

expression was found.[10] Definitive tests for age-related changes in the number of high-affinity binding sites and for changes in expression of the 70-kDa polypeptide that converts the low-affinity binding sites to high-affinity sites[11] still need to be carried out. It seems likely, however, that declines in production of both IL-2 and IL-2R molecules, and very probably of other gene products required for entry into S phase, underlie the age-dependent loss of T cells able to progress through even a single mitotic cycle after lectin exposure.[12,13]

Expression of IL-2 and IL-2R genes occurs 4 to 6 h after mitogen exposure, and surely depends upon still earlier lectin-induced transitions. The cellular proto-oncogene c-*myc* is thought to play a critical role in the return of resting (G$_0$) cells to active cycling, and we have indeed recently shown that there is a decline with age in the amount of c-*myc* mRNA produced after Con-A stimulation of mouse spleen cells.[14] Since Con-A increases both c-*myc* gene transcription levels and RNA accumulation in young cells to an equal extent, we expected the age-dependent loss of accumulation to reflect a decline in mitogen-induced transcription as well; however, we were surprised to find no change with age in transcript production, as measured by nuclear run-off methods. Since the mature c-*myc* RNA species decays with equal speed in T cells from old and young mice,[14] we speculate that the defect in c-*myc* expression may depend upon alterations in intranuclear processing. In contrast, other labs have shown that age-associated declines in production of RNAs for specific gene products in liver cells often do involve parallel declines in transcription rates.[15] It will be important to determine to what extent these (and other, e.g., translational) modes of altered gene expression are characteristic of aging in a wide range of developmental transitions among a variety of cells types.

The details of the process by which lectin/receptor interaction leads, 1 to 2 h later, to the expression of c-myc and other genes involved in cellular activation are not yet entirely worked out. The earliest detectable intracellular changes to occur after T cells are exposed to mitogen include[16] an increase in free calcium ion concentration (Ca_i) and the production of inositol trisphosphate (IP_3) by phospholipase C-mediated hydrolysis of phosphatidyl-inositol-bisphosphate (PIP_2; see Imboden and Stobo[17]). Each of these can be demonstrated within 5 min of activation. We recently examined the effects of aging on these very early signs of T cell activation. We found T cells from old mice to be defective in their calcium responses (but not in the production of inositol phosphates) and in responses to Con-A, and we found further that the proliferative defect can be "corrected" by mitogenic agents that do not require transduction of a signal by a specific surface receptor. These results, which are discussed in detail below, suggest that at least some of the defect in T cell proliferative response may be attributable to defects in receptor-dependent signal transduction.

In one study,[18] we used the calcium-sensitive fluorochrome indo-1 to estimate Ca_i values in T cells from young (2- to 6-month-old) and old (18- to 26-month-old) mice. We found no significant effect of age on the Ca_i levels of unstimulated cells (100 to 130 nM). There was, however, a significantly smaller increase in Ca_i levels in T cells from old mice in the first 5 min after Con-A stimulation, by which time these values have peaked and begun to decline toward a stable plateau level, which they then maintain for several hours. These results were obtained with Con-A doses that were optimal for mitogenesis (2 to 10 μg/ml). The age-related deficit in Ca^{2+} transport could be overcome by using higher Con-A doses (e.g., 30 μg/ml); these doses, however, lead to cell death rather than to proliferation and therefore seem of less relevance to physiologic conditions. Using a flow cytometric method that permitted the determination of Ca_i levels in individual indo-1-labeled T cells, we showed that our findings reflected a decline with age in the number of T cells that could respond to Con-A stimulation by increases in intracellular Ca^{2+} concentration. The defect is thus "clonal", in the sense that it seems to affect some but not all T cells, and therefore is in accord with the conclusions of limiting dilution culture studies[10] and cytofluorimetric analyses of transitions that occur later in the cell cycle.[12,13]

Increases in intracellular Ca^{2+} concentrations can be induced in resting murine T cells by antibodies to the T cell receptor (TCR) and the closely associated T3 polypeptides,[19] by antibodies specific for the Ly-6[20] and Thy-1 molecules,[21] and by antigen-laden macrophages,[22] as well as by Con-A and PHA. Relationships among these surface molecules and transmembrane signal transduction are only gradually becoming clearer. It is assumed that anti-TCR antibodies act by mimicking the physiological role of antigen. Con-A and PHA each bind to a wide variety of glycoproteins on the cell surface, and it is not yet clear which one(s) of these are able to transduce the activation signal. Early evidence[23] has suggested that PHA and Con-A activate T cells through different surface receptors. The CD2 determinant is present on nearly all T-lineage cells, including thymus cells, and is thought to play a role in some forms of T cell activation. One recent study[24] using monoclonal antibodies to the CD2 determinant to block mitogen-induced Ca^{2+} increases suggested that PHA might use the CD2 molecule to activate and that Con-A might act through some other receptor (conceivably the T cell's antigen receptor). The issue has been clouded, however, by the finding that other PHA preparations could activate cells even in the presence of anti-CD2 blocking antibodies.[25] Biochemical approaches, although incomplete, have indicated that PHA may bind to idiotype-bearing chains of the TCR complex, while Con-A may bind to the invariant T3 chains of the same complex.[26]

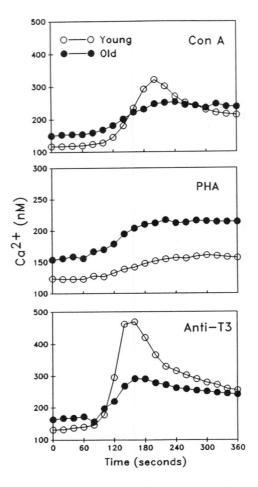

FIGURE 2.   Age-dependent decline in Ca^{2+} response to Con-A and to anti-T3 antibody, but not to PHA-P. Purified T cells from young (O—O) or from old (●—●) mice were labeled with indo-1 and suspended in a phosphate-buffered saline solution at 1.5×10^6 cells/ml at 37°C. The intracellular Ca^{2+} concentration was determined fluorimetrically every 15 s, as described by Miller et al.[18] After 60 s (to determine baseline Ca^{2+} concentrations), cells were stimulated with either Con-A at 10 μg/ml, PHA-P at 30 μg/ml, or monoclonal antibody to murine T3 at a 1:10,000 dilution of ascitic fluid, and readings were continued for another 300 s. (Rabbit antibody to mouse IgG at 5 μg/ml, which is itself nonstimulatory under these conditions, was added to the anti-T3 cuvette to provide the required cross-linkage.) Note that the scale for the PHA experiment is different from the other two panels because maximal responses to PHA are comparatively weak.

Despite these ambiguities, we thought it would be of interest to see if the process that leads to age-related loss in Ca_i responsiveness after Con-A exposure[18] would also affect responses to other activating agents. Although our survey of activating agents is not yet complete, an interesting disparity has emerged: although responses to anti-TCR antibodies, like those to Con-A, are diminished in old mice, responses to PHA seem to be preserved. Figure 2 shows a typical illustrative experiment. Con-A leads to a vigorous response in T cells from the young animal (top panel) but to a slower and smaller reaction from old T cells, as expected.[18] Anti-T3 antibody (bottom panel) also induces a greater response in

TABLE 1
Percentage of Cells Showing
Increased Ca$_i$ in Response to
Con-A and to PHA-P

Age	Mitogen	Mean	Range
Young	Con-A	74	71—77
Old	Con-A	51	47—56
Young	PHA	27	24—31
Old	PHA	33	24—41

Note: Values were determined by flow
cytometric analysis of indo-1-la-
beled cells, as described in Miller
et al.,[18] and represent the fraction
of purified T cells that exhibited
an increase in cytoplasmic Ca^{2+}
concentration during a 5-min incubation with Con A (10 μg/ml)
or PHA (30 μg/ml). N = 4 for
young mice; N = 3 for old mice.

young T cells. We have also obtained similar evidence (not shown) for age-associated defects in responses to antibody F23.1,[27] which is specific for a determinant expressed on the β-chain of about 20% of mouse T cell receptors. Since T3 and F23.1 determinants (unlike the unidentified Con-A receptors) are known to be part of the T cell's antigen-receptor complex, these results further strengthen our assumption that the age-related deficit is relevant to the *in vivo* response to foreign antigens. These reagents will also let us test (and we suspect rule out) the hypothesis that the functional decline reflects merely a loss of TCR density on cells from old mice.

As illustrated by the middle panel of Figure 2, however, we cannot detect any age-related decline in the Ca$_i$ response to optimally mitogenic doses of PHA. Similar results have now been obtained in a series of experiments involving 20 young and 12 old animals. We also used flow cytometry to count the number of T cells that could respond to PHA by an increase in CA$_i$, and again we found age-associated differences in the responses to Con-A, but not to PHA (Table 1). These experiments were initially carried out using a specific form of PHA preparation, called PHA-P, reported to activate human T cells through the CD2 molecule,[24] but have now been repeated with other PHA preparations, including those[25] whose action cannot be blocked with anti-CD2 antibodies. We do not know to what extent the PHA-responsive T cells (preserved in aging mice) are a subset of the cells that respond to Con-A. The cell sorter should allow us to isolate, from old and young mice, PHA-nonreactive cells, which can then be tested directly for the ability to respond to Con-A.

We have also measured the production of inositol phosphates from inositol-containing phospholipids as a second very early measure of lectin-induced activation. In some systems, including lectin-mediated activation of a T cell lymphoma,[17] generation of IP$_3$ by hydrolysis of PIP$_2$ is thought to mediate much of the early increase in Ca$_i$ by causing release of Ca^{2+} from the endoplasmic reticulum. In resting T cells, however, most of the lectin-initiated increase in Ca$_i$ seems to involve flow of Ca^{2+} across the plasma membrane; the change in concentration can be blocked by extracellular EGTA.[28,28a] There are suggestions that trans-membrane Ca^{2+} fluxes may also be regulated by production of either IP$_3$[29] or by inositol tetrakisphosphate (IP$_4$),[30] which can itself be produced from IP$_3$. Phospholipase-mediated hydrolysis also generates diacylglycerols that by activation of protein kinase C are likely to influence later steps in the activation sequence.

We have measured production of inositol phosphate (IP_1), inositol biphosphate (IP_2), and IP_3 by a method[31] in which cells are first labeled with ^3H-inositol, which becomes incorporated into membrane phosphatidylinositol (PI) and its phosphate esters. After brief exposure of the prelabeled cells to an activating agent (e.g., Con-A), we use ion-exchange chromatography to estimate labeled, inositol-containing components of the water-soluble extract. The procedure can provide estimates of IP_1, IP_2, and IP_3, but not IP_4. Since the more heavily phosphorylated members of this family are rapidly dephosphorylated to IP_1, which can then be used to regenerate PI, experiments that involve longer periods of mitogen exposure (e.g., 15 to 60 min) must be carried out in the presence of 5 mM LiCl to inhibit the dephosphorylation of IP_1 to PI. Short-term experiments (5 min, without LiCl) provide a measure of the initial burst of phospholipase C activity, while the LiCl experiments give a longer-term measure of cumulative flux through the PI cycle, dependent both on the rate of hydrolysis and the rate of production of phospholipase substrates from PI.

Figure 3 shows typical results for both 5-min (left) and 30-min (right) protocols. In each case, we tested an individual old mouse together with two young control mice. At the early time point, there is substantial radioactivity in all three of the IP species; by 30 min, as expected, most of the IP_3 and IP_2 had been dephosphorylated to the IP_1 species. There was no change with age in either test. Similar results have been obtained in five other experiments (three short-term and two long-term). It is interesting to note that IP_x production is also preserved in another age-sensitive system for studying agonist-induced changes in Ca^{2+} concentration, the rat parotid salivary gland.[32] Increased production of IP_x can also be induced by anti-T3 and PHA;[32a] we are currently examining the effect of age on these responses as well. Although the age-associated defect in Con-A-induced Ca_i increases clearly cannot be explained by alterations in IP_3, it remains possible that either diminished production of other products of the phospholipase reaction or diminished sensitivity to IP_3 may account for the loss of Ca^{2+} mobilization. The steps by which ligand binding induces increases in Ca_i are not yet clearly established, but at a minimum our results show that some, but not all, of the very early lectin-induced events are defective in old T cells.

To what extent might the deficit in Ca^{2+} concentration account for the poor proliferative responses seen for old T cells? If old T cells are defective only in the earliest states of the activation process (including transmembrane Ca^{2+} transport and activation of protein kinase C by diacylglycerol), but are able to respond to these intracellular signals once they are generated, then one might predict strong mitogenic responses in old T cells stimulated by agents that bypass surface receptors to induce the intracellular events directly. It has recently been established that T cells can indeed by induced to divide by a combination of a calcium ionophore and phorbol myristic acetate (PMA), which serves to activate protein kinase C;[33] levels of proliferation are often even higher than those induced by Con-A.

We tested T cells from old mice to see whether they would proliferate when exposed to PMA (1 ng/ml) and the ionophore ionomycin. We found[34] that these intracellular mediators could largely overcome the age-related proliferative defect observed in cells stimulated with Con-A. In the average experiment, the young cells responded 3.9-fold better to Con-A than did the old cells, but only 1.3-fold better in responses induced by PMA and optimal ionomycin concentrations (Table 2). These results suggest that the pathways by which cells respond to intracellular signals (i.e., protein kinase activators and increases in Ca_i concentration) may be largely preserved by aging, and that a failure to generate the intracellular signals in response to extracellular ligands may be particularly important in the aging immune system. It is important to note, however, that the signals provided by PMA and ionomycin are surely stronger and more prolonged than those ordinarily induced by even the strongest antigenic or mitogenic agents, and they may be able to obscure subtle age-related deficits in responses to intracellular messengers.

We are now examining the effects of these activating agents on expression of specific

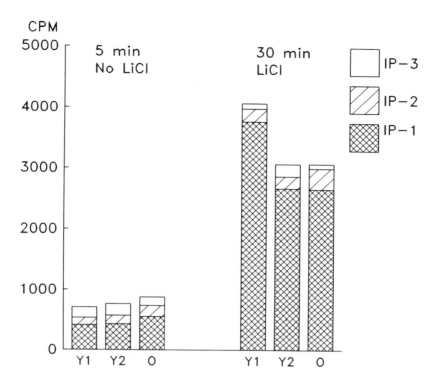

FIGURE 3. Absence of an age-dependent effect on production of IP_x species after Con-A stimulation of mouse T cells. Two representative experiments are shown, each involving two young mice (Y1 and Y2) and one old (O) mouse. The species generated from radiolabeled inositol-containing phospholipids was determined by the method of Bijsterbosch et al.[31] after stimulation of purified T cells with Con-A for 5 min without LiCl (left panel) or after 30 min in the presence of 5 mM LiCl to inhibit degradation of IP_x to PI. Background cpm = 134 in the absence of Con-A for the 30-min time point (total IP).

TABLE 2
Proliferation by Old and
Young T Cells: Augmentation
by PMA and Ionomycin[34]

Age	Con-A	PMA/ionomycin
Young	245 ± 16	270 ± 11
Old	92 ± 16	228 ± 19

Note: Values represent cpm of ^{3}H-thymidine incorporation ($\times 10^{-3}$) ± SEM (N = 6 for each age) by 10^5 cells/0.2-ml microwell for optimal doses of Con-A and ionomycin (with PMA at 1 ng/ml) during the last 6 h of 48-h culture. Each experiment included one young (2- to 5-month) and one old (20- to 24-month) mouse. The mean young/old ratio, averaged over six experiments, was 3.9 ± 1.1 (SEM) for Con-A and 1.3 ± 0.2 for PMA plus ionomycin.

gene products. Some stimuli (e.g., Con-A, PHA, high concentrations of anti-T3, PMA with ionomycin) induce "complete" activation, with expression of c-*myc*, IL-2, and IL-2R, and eventual proliferation. However, other stimuli (e.g., PMA alone or low doses of anti-T3) lead to IL-2R expression without IL-2 production, and some (e.g., ionomycin alone) induce activation up to the c-*myc* stage but do not lead to expression of either IL-2R or IL-2. We hope to exploit these differences in activation requirements to learn more about age-related alterations in responsiveness to the individual intracellular signals.

SUMMARY

T cells from old mice exhibit deficits very early in the sequence of events by which extracellular activators (e.g., plant lectins and antibodies to the T cell receptor) induce DNA synthesis and eventual mitosis. Con-A-induced increases in intracellular Ca^{2+} concentration, demonstrable within 2 min of lectin exposure, are lower in old T cells. This deficit represents a lower frequency of reactive cells within the phenotypically defined T cell set and cannot be attributed to diminished production of IP_3. Many G_1-stage changes in gene expression, including production of mRNA for c-*myc*, IL-2, and the IL-2 receptor, occur less vigorously in T cells from old mice; in at least one instance (c-*myc*), the defect seems to involve altered intranuclear processing of primary transcripts. The importance of early deficits in trans-membrane signal transductions is suggested by the ability of T cells from old mice to proliferate well when stimulated by a combination of phorbol ester and calcium ionophore that together bypass the requirement for reception of extracellular signals.

ACKNOWLEDGMENTS

We thank Ruth Pimental and Bruce Jacobson for technical assistance and Jeff Bluestone for his gift of anti-T3 hybridoma cells. This work was supported by NIH grants AG-03978 and AG-07114. R.A.M. is supported by a Scholar Award from the Leukemia Society of America and an RCDA from the National Institute on Aging. A.L. is supported by training grant AG-00115.

REFERENCES

1. **Walters, C. S. and Claman, H. N.,** Age-related changes in cell-mediated immunity in BALB/c mice, *J. Immunol.,* 115, 1438, 1975.
2. **Antel, J. P., Oger, J.-F., Dropcho, E., Richman, D. P., Kuo, H. H., and Arnason, B. G. W.,** Reduced T-lymphocyte cell reactivity as a function of human aging, *Cell. Immunol.,* 54, 184, 1980.
3. **Cantrell, D. A. and Smith, K. A.,** The Interleukin-2 T-cell system: a new cell growth model, *Science,* 224, 1312, 1984.
4. **Thoman, M. L. and Weigle, W. O.,** Lymphokines and aging: Interleukin-2 production and activity in aged animals, *J. Immunol.,* 127, 2101, 1981.
5. **Gillis, S., Kozak, R., Durante, M., and Weksler, M. E.,** Immunological studies of aging. Decreased production of and response to T cell growth factor by lymphocytes from aged humans, *J. Clin. Invest.,* 67, 937, 1981.
6. **Miller, R. A. and Stutman, O.,** Decline, in aging mice, of the anti-TNP cytotoxic T cell response attributable to loss of Lyt-2⁻, IL-2 producing helper cell function, *Eur. J. Immunol.,* 11, 751, 1981.
7. **Wu, W., Pahlavani, M., Cheung, H. T., and Richardson, A.,** The effect of aging on the expression of Interleukin 2 messenger RNA, *Cell. Immunol.,* 100, 224, 1986.
8. **Thoman, M. L. and Weigle, W. O.,** Reconstitution of *in vivo* cell-mediated lympholysis responses in aged mice with Interleukin 2, *J. Immunol.,* 134, 949, 1985.
9. **Miller, R. A.,** Age-associated decline in precursor frequency for different T cell-mediated reactions, with preservation of helper or cytotoxic effect per precursor cell, *J. Immunol.,* 132, 63, 1984.

10. **Vie, H. and Miller, R. A.,** Decline, with age, in the proportion of mouse T cells that express IL-2 receptors after mitogen stimulation, *Mech. Ageing Dev.,* 33, 313, 1986.
11. **Sharon, M., Klausner, R. D., Cullen, B. R., Chizzonite, R., and Leonard, W. J.,** Novel interleukin-2 receptor subunit detected by cross-linking under high-affinity conditions, *Science,* 234, 859, 1986.
12. **Kubbies, M., Schindler, D., Hoehn, H., and Rabinovitch, P. S.,** BrdU-Hoechst flow cytometry reveals regulation of human lymphocyte growth by donor-age-related growth fraction and transition rate, *J. Cell. Physiol.,* 125, 229, 1985.
13. **Staiano-Coico, L., Darzynkiewicz, Z., Melamed, M. R., and Weksler, M. E.,** Immunological studies of aging. IX. Impaired proliferation of T lymphocytes detected in elderly humans by flow cytometry, *J. Immunol.,* 132, 1788, 1984.
14. **Buckler, A., Vie, H., Sonenshein, G., and Miller, R. A.,** Defective T lymphocytes in old mice. Diminished production of mature c-*myc* RNA after mitogen exposure not attributable to alterations in transcription or RNA stability, *J. Immunol.,* 140, 2442, 1988.
15. **Richardson, A.,** personal communication.
16. **Weiss, A., Imboden, J., Hardy, K., Manger, B., Terhorst, C., and Stobo, J.,** The role of the T3/antigen receptor complex in T-cell activation, *Annu. Rev. Immunol.,* 4, 593, 1986.
17. **Imboden, J. B. and Stobo, J. D.,** Transmembrane signalling by the T cell antigen receptor. Perturbation of the T3-antigen receptor complex generates inositol phosphates and releases calcium ions from intracellular stores, *J. Exp. Med.,* 161, 446, 1985.
18. **Miller, R. A., Jacobson, B., Weil, G., and Simons, E. R.,** Diminished calcium influx in lectin-stimulated T cells from old mice, *J. Cell. Physiol.,* 132, 337, 1987.
19. **Oettgen, H. C., Terhorst, C., Cantley, L. C., and Rosoff, P. M.,** Stimulation of the T3-T cell receptor complex induces a membrane-potential-sensitive calcium influx, *Cell,* 40, 583, 1985.
20. **Yeh, E. T. H., Reiser, H., Daley, J., and Rock, K. L.,** Stimulation of T cells via the TAP molecule, a member in a family of activating proteins encoded in the Ly-6 locus, *J. Immunol.,* 138, 91, 1987.
21. **Kroczek, R. A., Gunter, K. C., Seligmann, B., and Shevach, E. M.,** Induction of T cell activation by monoclonal anti-Thy-1 antibodies, *J. Immunol.,* 136, 4379, 1986.
22. **Nisbet-Brown, E., Cheung, R. K., Lee, J. W. W., and Gelfand, E. W.,** Antigen-dependent increase in cytosolic free calcium in specific human T-lymphocyte clones, *Nature,* 316, 545, 1985.
23. **Greaves, M. F., Bauminger, S., and Janossy, G.,** Lymphocyte activation. III. Binding sites for phytomitogens on lymphocyte subpopulations, *Clin. Exp. Immunol.,* 10, 537, 1972.
24. **O'Flynn, K., Krensky, A. M., Beverley, P. C. L., Burakoff, S. J., and Linch, D. C.,** Phytohaemagglutinin activation of T cells through the sheep red blood cell receptor, *Nature,* 313, 686, 1985.
25. **O'Flynn, K., Russul-Saib, M., Ando, I., Wallace, D. L., Beverley, P. C. L., Boylston, A. W., and Linch, D. C.,** Different pathways of human T-cell activation revealed by PHA-P and PHA-M, *Immunology,* 57, 55, 1986.
26. **Kanellopoulos, J. M., De Petris, S., Leca, G., and Crumpton, M. J.,** The mitogenic lectin from *Phaseolus vulgaris* does not recognize the T3 antigen of human T lymphocytes, *Eur. J. Immunol.,* 15, 479, 1985.
27. **Staerz, U. D., Rammensee, H.-G., Benedetto, J. D., and Bevan, M. J.,** Characterization of a murine monoclonal antibody specific for an allotypic determinant on T cell antigen receptor, *J. Immunol.,* 134, 3994, 1985.
28. **Tsien, R. Y., Pozzan, T., and Rink, T. J.,** T-cell mitogens cause early changes in cytoplasmic free Ca^{2+} and membrane potential in lymphocytes, *Nature,* 295, 68, 1982.
28a. **Miller, R. A., Lerner, A., Philosophe, B., and Macauley, J.,** unpublished data.
29. **Kuno, M. and Gardner, P.,** Ion channels activated by inositol 1,4,5-trisphosphate in plasma membrane of human T-lymphocytes, *Nature,* 326, 301, 1987.
30. **Irvine, R. F. and Moor, R. M.,** Micro-injection of inositol 1,3,4,5,-tetrakisphosphate activates sea urchin eggs by a mechanism dependent on external Ca^{2+}, *Biochem. J.,* 240, 917, 1986.
31. **Bijsterbosch, M. K., Meade, C. J., Turner, G. A., and Klaus, G. G. B.,** B lymphocyte receptors and polyphosphoinositide degradation, *Cell,* 41, 999, 1985.
32. **Roth, G. S.,** Effects of aging on mechanisms of alpha-adrenergic and dopaminergic action, *Fed. Proc.,* 45, 60, 1986.
32a. **Lerner, A. and Miller, R. A.,** submitted.
33. **Mastro, A. M. and Smith, M. C.,** Calcium-dependent activation of lymphocytes by ionophore, A23187, and a phorbol ester tumor promoter, *J. Cell. Physiol.,* 116, 51, 1983.
34. **Miller, R. A.,** Immunodeficiency of aging: restorative effects of phorbol ester combined with calcium ionophore, *J. Immunol.,* 137, 805, 1986.

Chapter 6

AGE-RELATED DIFFERENCES IN DNA POLYMERASE ALPHA SPECIFIC ACTIVITY: POTENTIAL FOR INTERACTION IN DNA REPAIR

David L. Busbee, Victor L. Sylvia, and Geoffrey M. Curtin

TABLE OF CONTENTS

I. INTRODUCTION

A. DNA REPAIR PERSPECTIVES

Over 20 years ago, investigators reported that a repair mechanism of prokaryotic organisms was capable of removing photoproduct lesions from UV-damaged DNA.[1,2] Subsequent studies quickly revealed that this repair process involved the hydrolytic excision of nucleotides around the lesion site, that it required the expression of several different genetic loci, and that the loss of DNA repair competence resulted in increased lethality in UV-[3,4] or carcinogen-treated[5] cells. The determination that this repair mechanism involved removal of a sequence of nucleotides encompassing a photoproduct lesion from damaged DNA with subsequent resynthesis of the excised DNA resulted in this process being called DNA excision repair.[6,7] Damaged DNA is nicked by an endonuclease, or glycosylases may remove carcinogen-altered bases prior to nicking of the DNA strand by an apurinic site-specific endonuclease; a variable segment of nucleotides is then removed by exonucleolytic hydrolysis, and the excised nucleotide segment is resynthesized by DNA polymerases using the remaining single strand of DNA as a template, with rejoining of the strand ends by polynucleotide ligase. Excision repair constitutes the primary mechanism for prereplication removal of DNA lesions and for high-fidelity resynthesis of the excised nucleotide segment in radiation- or carcinogen-treated eukaryotic organisms.[8]

DNA excision repair in eukaryotes has been categorized as long patch (UV-like) or short patch (X-ray like) by the size of the nucleotide segment removed, which in turn is dependent on the DNA lesion initiating the repair processes.[9] Long-patch repair of single-stranded DNA segments, averaging about 100 nucleotides, requires the sequential function of DNA polymerases alpha and beta.[9-12] Short-patch repair may remove from 1 to 15 nucleotides, with an average of about 3 nucleotides, and utilizes polymerase beta for DNA resynthesis.[10-13] In both short- and long-patch repair, the resynthesis of excised nucleotides is independent of scheduled, mitotically associated DNA synthesis, and has been loosely called unscheduled DNA synthesis (UDS).[14-18] Chemically initiated UDS in eukaryotic cells and mutagenicity or carcinogenicity of the chemical in question are generally positively correlated, resulting in the use of UDS initiation as an evaluation mechanism to assess interaction of chemicals with DNA in higher organisms.[18-21]

B. DEFECTIVE REPAIR

Extensive examination of the effects of decreased DNA excision repair has revealed a number of genetic disorders in which DNA repair dysfunction is correlated with the occurrence of neoplastic disease. Among these are xeroderma pigmentosum (XP), ataxia telangiectasia (AT), Fanconi's anemia (FA), and Bloom's syndrome.[22-27] Although XP, AT, and FA are distinctly different in character from each other and from Bloom's syndrome, they are all autosomal recessive disorders, leading to the proposal that they have in common the absence of expression or function of one or more enzymes of DNA excision repair.[27-31] Bloom's syndrome differs in that affected individuals exhibit apparently normal UDS but show significant sister chromatid exchange, suggesting that this autosomal recessive disorder may exhibit a functional but aberrant form of excision repair.[32,33] In each of these instances, decreased DNA repair efficiency is associated with the increased occurrence of specific cancer types. Since elevated levels of the occurrence of a variety of cancer types are a characteristic of increased age, investigators have speculated that dysfunction of DNA excision repair may be age related.

II. CANCER AS A FUNCTION OF AGING

The occurrence of cancer as an apparent function of aging in humans has generated intense interest, resulting in at least two broad theories explaining age-associated carcinogenesis in man. One school of thought has proposed that increased cancer incidence in the elderly is the result of an accumulation of separate, low probability carcinogenic events occurring due to increased exposure time with increased age.[34-36] A second theory suggests that physiological changes normal to the process of aging may decrease cellular and organismic surveillance and repair mechanisms and thus may increase the incidence of cancer as a function of increased age.[37-40] Data are available to support both theories, leading one to conclude that increased age-associated occurrence of independent initiation and promotion events, coupled with decreased age-associated capacity of the organism to deal with these events, may contribute singly or in concert to the onset of cancer associated with aging.

It is generally accepted that DNA excision repair constitutes the mechanism by which cells of higher eukaryotes deal with DNA damage and that such repair is essential for the survival of cells at single, multicellular, and organismic levels. Since damage to DNA occurs naturally as a result of radiation, environmental exposure, and spontaneous oxidative events, and since the capacity to repair DNA damage is essential to the survival and/or longevity of living organisms, decreased DNA repair as a function of age provides a convenient physiological mechanism to explain increased cancer occurrence in the elderly. Investigators do not, however, agree that a decline in DNA excision repair efficiency constitutes one of the hypothetical age-related phenomena decreasing an individual's capacity to deal with mutations.

Investigators employing human fibroblasts as research models have reported that levels of UDS in cells and survival of UV-irradiated cells do not decrease as a function of increased age.[41,42] Rather, Dell'Orco and Anderson[43] reported that both UDS and cell survival were elevated in adult-derived fibroblasts if the cells were serum deprived. In contrast, Joe et al.[44] showed that serum-deprived human lymphocytes treated with a carcinogenic benzo(a)pyrene metabolite do not engage in either UDS or mitogen-stimulated DNA synthesis, and Lambert et al.[45] reported an age-related decrease in the repair of UV-initiated DNA lesions in human leukocytes. Further, the capacity of fibroblasts to be induced for the synthesis of microsomal carcinogen-activating enzymes has been shown to decrease with increased age,[46,47] as does the binding of carcinogenic hydrocarbon metabolites to cellular DNA,[48] and both DNA repair capacity and accuracy of DNA synthesis are reported to be directly correlated with increased life span of test organisms.[49-52] While early data indicating decreased accuracy of DNA synthesis with increased age are supported by recent findings showing that the fidelity of DNA resynthesis during excision repair decreases in human fibroblasts with increased donor age,[53] there are also contradictory data on fidelity of repair as a function of age, with Silber et al.[54] reporting that no age-related differences in the fidelity of excision repair-associated DNA synthesis are found in murine cells. These studies combine to show that contrasting data in aging studies are the rule rather than the exception. In a review of the association between decreased DNA repair and increased age in humans, Lehmann[55] reported the literature to be full of apparently contradictory data generated in studies of aging, DNA repair capacity, and longevity. Lehmann suggested that many studies were less than rigorous, with loosely interpreted data, and concluded that a number of these studies were designed with multiple variables that did not analyze biochemical endpoints. In an attempt to address a single age-related issue and to limit the variables involved in data interpretation, we have examined aspects of the activity of DNA polymerase alpha isolated from fibroblasts derived from human males varying in age between a 16-week fetus and a 66-year-old man.

III. DNA POLYMERASE ALPHA

A. ENZYME SIZE AND SUBUNITS

DNA polymerase alpha (deoxynucleotidyltransferase, E.C. 2.7.7.7) is the major eukaryotic enzyme adding deoxynucleotide monophosphates to the growing end of a DNA primer during either mitotically associated or excision repair-associated DNA synthesis.[56] While early studies initially indicated that catalytic activity of the enzyme could be found associated with a variety of proteins, varying in size between 70 and 1000 kDa,[57-61] analyses of immunoaffinity-purified polymerase alpha suggested the enzyme to be reasonably consistent in molecular weight among related organisms. The catalytic subunit of polymerase alpha has been recently characterized to be a large protein, between 156 and 200 kDa,[61-69] the primary structure of which is similar among species.[70] The DNA polymerase alpha holoenzyme from different species has been reported to vary somewhat in noncatalytic accessory proteins critical to the function of the enzyme.[68-73] *Drosophila melanogaster* exhibited a 182-kDa catalytic subunit and three accessory proteins of 73, 60, and 50 kDa,[69] while calf thymus polymerase contained a catalytic subunit of 148 to 180 kDa and accessory bands at 73, 59, and 48 kDa in a relative stoichiometry of 1:1:3:3.[73] Human DNA polymerase alpha has been reported to be composed of a catalytic subunit of about 185 kDa, trace amounts of 125- and 140-kDa proteins, and three additional accessory proteins, varying between 49 and 77 kDa.[68] However, intraspecies and cell cycle-dependent variations in DNA polymerase alpha chromatographic characteristics have been reported for human[74-76] and mouse[77] polymerases. These data suggesting enzyme variability are consistent with those of Lamothe and co-workers, who isolated three distinct DNA polymerase alpha forms from HeLa cells.[78] One of the enzyme forms isolated by Lamothe et al., alpha-2, exhibited a holoenzyme molecular weight of 600 kDa, with accessory proteins of 140, 69, 57, 55, and 24.5 kDa. They suggested that variations in the chromatographic methods of enzyme isolation may result in inconsistent loss of accessory proteins from the holoenzyme. Accessory activities proposed to be associated with the polymerase alpha core enzyme and which may be either accessory protein related or a function of the catalytic subunit which is regulated in some way by an accessory protein include DNA primase,[79,80] primer-template binding,[62,78,80-82] diadenosine-5′,5‴-p1-p4-tetraphosphate (Ap4A)-binding,[83-86] 5′→3′-exonuclease,[87] and 3′→5′-exonuclease.[88]

B. POSSIBLE SUBUNIT FUNCTIONS

DNA primase is copurified with and/or tightly associated with a variety of eukaryotic DNA alpha polymerases,[80,89,90] as evidenced by the ability of the polymerase holoenzyme to initiate DNA synthesis on unprimed homopolymer template.[91-93] Hubscher[92,94] found the primase activity to be inseparable from the catalytic core, while Kaguni et al.,[95] Chang et al.,[96] and Cotterill et al.[69] demonstrated the primase to be a separable protein that copurifies with the polymerase catalytic subunit. Other proposed subunits of polymerase alpha, the 3′→5′- and 5′→3′-exonucleases, have been found associated with prokaryotic polymerases, in polymerase alpha from lower eukaryotes, and in plant polymerases,[90,97-99] but they have only recently been isolated from cells of higher eukaryotes.[87,88] Grummt and associates,[104,105] Baril et al.,[102] and Rapaport and co-workers[100,101,103] have published extensively on Ap4A-binding protein as a component of DNA polymerase alpha but have not clearly delineated the relationship between the binding of Ap4A to the polymerase and the function of the enzyme. An additional intriguing aspect of DNA polymerase alpha which has not been extensively investigated is the enzyme's variable capacity to recognize and bind DNA template/primer.

A noncatalytic accessory protein of the DNA polymerase alpha holoenzyme has been determined to specify recognition for and binding to DNA template/primer by affecting the K_m of the holoenzyme for the DNA primer stem 3′-hydroxy terminus.[62,78,81,106] Lamothe et al.[78] reported that polymerase alpha-2 contains two cofactors, a 96-kDa tetrameric aggregate of 24.5-kDa subunits and a 51-kDa subunit, that are associated with enzyme binding to DNA template/primer. The capacity of DNA polymerase alpha to utilize denatured DNA as a template is significantly enhanced by these proteins, which effectively reduce the polymerase K_m for the primer stem 3'-hydroxy terminus.[81] Burke et al.[107] have shown a similar function for a 70-kDa subunit, the presence of which allows polymerase alpha to utilize primed single-stranded viral DNA as template. DNA polymerase alpha binding to DNA template/primer, presumedly a function of one or more accessory proteins, has been reported to be increased as a function of ATP hydrolysis[108] or in the presence of ATP and protein kinases,[109] suggesting that phosphoryl groups may be a determining factor regulating the capacity of DNA polymerase alpha to bind DNA template/primer. This original proposal has been partially corroborated by Krauss et al.,[110] who found that protein kinase C-associated phosphorylation of DNA polymerase alpha increases both the fidelity of synthesis and the specific activity of the enzyme. They did not, however, determine phosphorylation effects on DNA polymerase alpha binding to DNA template/primer. Sylvia et al.[111,112] and Busbee et al.[113] reported that DNA polymerase alpha isolated from mouse sarcoma or human fibroblasts may exist in either low or high activity forms, that the low activity enzyme form can be activated and ^{32}P-labeled in the presence of ^{32}P-ATP, phosphatidylinositol (PI), and phosphatidylinositol kinase (PIK) (a system known to generate ^{32}P-labeled phosphatidylinositol-4-monophosphate [PIP]), and that treatment of the polymerase with PIP alone can activate the enzyme concomitant with increasing the enzyme's binding affinity for enzymatically gapped DNA template/primer.

C. CAN DNA POLYMERASE ALPHA BE ACTIVATED?

The possibility that DNA polymerase alpha might exist in low and high activity forms was initially proposed by Fisher and Korn,[114] who suggested that a mechanism of polymerase alpha activation could logically include interaction of a low molecular weight activator molecule with the enzyme. Zierler et al.[115] have reported a cell cycle-related increase in activity in *Xenopus laevis* ova DNA polymerase alpha and speculated that the enzyme may be stockpiled in an inhibited form and activated during stages of the cell cycle when immediate increases in DNA synthesis are required. The existence of active and inactive forms of DNA polymerase alpha was further reported by Murray,[116] who demonstrated that enzyme activity in senescent fibroblasts is reduced two- to fourfold below enzyme activity in young cells. Murray further proposed that low levels of polymerase activity are correlated with the declining growth rate in senescing cells. While differences in the total activity and specific activity of DNA polymerase alpha isolated from young or old cells are not well understood, it is clear that these differences exist. DNA polymerase alpha isolated from lymphocytes of older human donors was reported by Agarwal and co-workers[117] to be thermolabile, while polymerase isolated from the cells of younger donors was more stable. Krauss and Linn[74,75] reported that polymerase alpha from cells with decreased replicative activity differs chromatographically and exhibits decreased fidelity of DNA synthesis when compared to the polymerase of actively growing cells. These data from a number of different laboratories suggest that alpha polymerases from fetal-derived vs. adult-derived cells show differences in specific activity and chromatographic characteristics and that polymerase alpha of older, slowly growing cells has a labile form, while enzyme from younger, rapidly growing cells is more stable.

IV. FETAL VS. ADULT POLYMERASES

A. CHROMATOGRAPHIC CHARACTERISTICS

In a study designed to examine possible correlations between donor age of cells and decreased cellular DNA synthesis, fibroblasts established from human subjects were examined[113] to determine if multiple types of DNA polymerase alpha could be isolated from cells dependent on donor age of the cell line. The supernatants of sonicated fetal- or adult-derived fibroblasts were subjected to DEAE-cellulose chromatography with elution using a

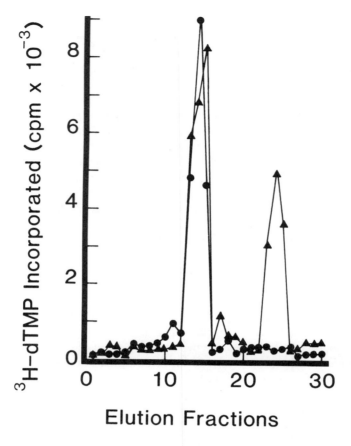

FIGURE 1. The DEAE-cellulose elution profile of DNA polymerase alpha isolated from GM6111 fibroblasts (●) or GM1717 fibroblasts (▲). GM6111 (fetal-derived) or GM1717 (adult-derived) fibroblasts were grown to a preconfluent log-phase state, harvested, sonicated in 50 mM Tris-HCl (pH 7.6) containing 0.25 M sucrose, 3 mM MgCl$_2$, 0.5 mM dithiothreitol (DDT), and 0.1 mM phenylmethylsulfonylfluoride (PMSF), and centrifuged. DNA polymerase alpha-enriched sonicate fractions from an initial DEAE-cellulose column were reapplied to DEAE-cellulose and eluted using a linear 20- to 400-mM KPO$_4$ gradient. Polymerase activity in the eluate was assessed as a function of tritiated 2′-deoxythymidine monophosphate (^{3}H-dTMP) incorporated into trichoroacetic acid (TCA)-precipitable DNA template. (From Busbee, D., Sylvia, V., Stec, J., Cernosek, Z., and Norman, J., *J. Natl. Cancer Inst.*, 79, 1231, 1987. With permission.)

linear phosphate gradient (Figure 1). Analyses of the chromatographic elution fractions from fetal-derived fibroblast sonicates showed a single DNA polymerase alpha peak (fractions 13 to 15), while adult-derived fibroblast preparations exhibited two peaks of enzyme activity, one eluting coincident with the fetal-derived enzyme peak in fractions 13 to 15 and a second peak eluting in fractions 23 to 25. When the two adult-derived fibroblast polymerase peaks were applied to an immunoaffinity column prepared using anti-DNA polymerase alpha IgG, they co-eluted in fractions 18 to 22 with one distinguishable peak (Figure 2). While the adult-derived polymerases were not separable by their affinity for monoclonal IgG, they were readily separated by their binding affinities for DNA-cellulose prepared using enzymatically activated DNA template/primer (Figure 3). DNA affinity chromatograph showed adult-derived fibroblast polymerase alpha to have low binding affinity (fractions 14 to 16)

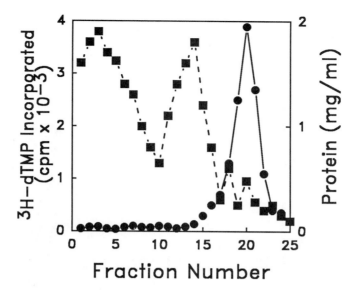

FIGURE 2. Purification of DNA polymerase alpha utilizing an immunoaffinity column. Active enzyme fractions from the DEAE-cellulose column peaks were applied to a Sepharose CL4B-IgG column. The column was extensively washed with 50 mM Tris-HCl buffer (containing 3 mM MgCl$_2$, 0.5 mM DTT, and 0.1 mM PMSF), 1 M NaCl in Tris-HCL buffer, and 3 M NaCl in Tris-HCl buffer. Polymerase alpha was eluted from the column with 3.2 M MgCl$_2$ in 50 mM Tris-HCl buffer. Polymerases A$_1$ and A$_2$ were not distinguishable using this procedure. Elution fractions were dialyzed against 50 mM Tris-HCl (pH 7.8) containing 0.5 mM DTT, 3 mM MgCl$_2$, and 0.1 mM PMSF and then assayed for enzyme activity, which was expressed as ^{3}H-dTMP incorporated into DNA template/h. DNA polymerase alpha = ●; protein = ■.

and high binding affinity (fractions 26 to 28) forms, while polymerase from fetal-derived fibroblasts showed a single peak with high binding affinity eluting in fractions 26 to 29 (Figure 3). The low DNA binding affinity polymerase in this preparation was designated A$_1$, while the enzyme with high binding affinity was designated A$_2$.

B. ACTIVE VS. INACTIVE ENZYME

The possibility that DNA polymerase alpha might exist both in an active and an inactive form capable of being activated has generated several reports suggesting that the enzyme's specific activity might be increased by treatment with ATP,[108] with ATP and protein kinases,[109,110] or with the product of phosphatidylinositol (PI) phosphorylation, PIP.[111-113,118] While data from this and other laboratories showed small increases in enzyme activity in the presence of kinases and ATP, a fairly large increase in activity (8- to 12-fold) was reported for polymerase A$_1$ in the presence of phosphoinositides.[113,118] Busbee et al.[113] reported that when A$_1$ and A$_2$ polymerases were treated with PIP and reapplied to a DNA-cellulose column, enzyme A$_2$ showed no increase in activity or affinity of binding to the column, while enzyme A$_1$ exhibited a significant increase in both specific activity (from 3,160 to 28,860 units/mg) and affinity of binding to the column (Figure 4). ^{32}P-PIP is produced by PIK phosphorylation of PI using ^{32}P-ATP as a phosphoryl donor. Sylvia et al.[112] and Busbee et al.[113] showed that radiolabel from ^{32}P-ATP could be transferred via ^{32}P-PIP to polymerase A$_1$ concomitant with an increase in enzyme activity. Polymerase A$_1$ (5 pmol) was treated with 5 pmol of ^{32}P-ATP, 100 pmol of PI, and PIK in excess, and was applied to a DNA affinity column to determine if ATP-derived radiolabel eluted with the

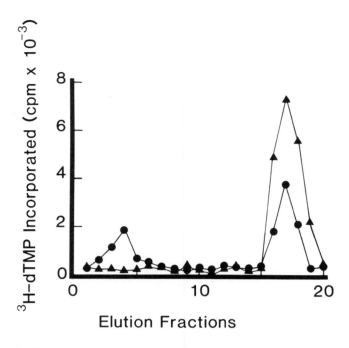

FIGURE 3.   A comparison of the DNA-binding affinity of DNA polymerase alpha from GM6111 and GM1717 fibroblasts. DNA polymerase alpha isolated through the immunoaffinity column step from GM6111 (▲) or from GM1717 (●) fibroblasts was applied to a DNA-cellulose column, extensively washed with 50 mM Tris-HCl buffer containing 3 mM MgCl$_2$, 0.5 mM DTT, and 0.1 mM PMSF, and sequentially eluted with 70 mM (fractions 1 to 10) and 370 mM (fractions 11 to 20) KCl in Tris buffer. DNA polymerase alpha activity was measured as ^{3}H-dTMP incorporated into TCA-precipitable DNA template. (From Busbee, D., Sylvia, V., Stec, J., Cernosek, Z., and Norman, J., *J. Natl. Cancer Inst.*, 79, 1231, 1987. With permission.)

activated enzyme (Figure 5). Polymerase A$_1$ was ^{32}P labeled (utilizing almost exactly 5 pmol of ^{32}P-ATP to label 5 pmol of enzyme), showed a significant increase in specific activity, and eluted in the high-affinity column fractions. This phenomenon apparently also occurs intracellularly. Joe et al.[118] showed that lipoprotein-deficient human fibroblasts did not initiate UDS upon treatment with the carcinogenic metabolite of benzo(a)pyrene, 7,8-dihydrodiol-9,10-epoxy benzo(a)pyrene (BPDE), but that lipoprotein-deficient cells supplemented with low-density lipoproteins or PI did initiate UDS (Figure 6). They further showed that ^{32}P-labeled DNA polymerase alpha could be isolated from permeabilized human fibroblasts treated with ^{32}P-ATP and induced to initiate UDS by treatment with BPDE (Figure 7) in the presence of ^{32}P-ATP and supplemental lipoproteins.

C. POSSIBLE MECHANISM OF ACTIVATION

To examine the mechanism of PIP interaction with polymerase, purified enzyme was treated with ATP, PIK, and PI which was either ^{14}C labeled on the arachidonic acid moiety or ^{3}H labeled on the myoinositol ring. ^{14}C-Labeled PIP interacted with enzyme A$_1$, effecting an immediate increase in activity and in affinity of binding to DNA. However, the enzyme was not ^{14}C labeled after the 30-min reaction was completed (Figure 8D). In the presence of PIP, ^{3}H labeled on the inositol ring, enzyme A$_1$ was ^{3}H labeled, and increased in specific activity and DNA-binding affinity (Figure 8B). This indicated that ^{3}H label from the inositol ring of PIP was retained by the activated enzyme while the diacylglycerol moiety (DAG)

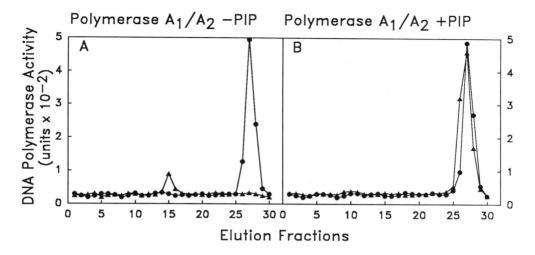

FIGURE 4. Activation of DNA polymerase alpha types A_1 and A_2 by treatment with PIP. Equal amounts of DNA polymerase alpha types A_1 (▲) and A_2 (●) protein were applied to a DNA-cellulose affinity column (left panel); an identical aliquot of enzyme was treated with PIP (100 pmol) and applied to a DNA-cellulose column (right panel). The columns were washed with 10 ml of 50 mM Tris-HCl buffer containing 3 mM MgCl$_2$, 0.5 mM DTT, and 0.1 mM PMSF and then eluted with 10 ml of 70 mM KCl (fractions 11 to 20) followed by 10 ml of 370 mM KCl (fractions 21 to 30) in 50 mM Tris-HCl buffer and collected in 1-ml fractions which were assayed for polymerase alpha activity required to incorporate 1 nmol of ^{3}H-dTMP into acid-precipitable DNA template per hour.

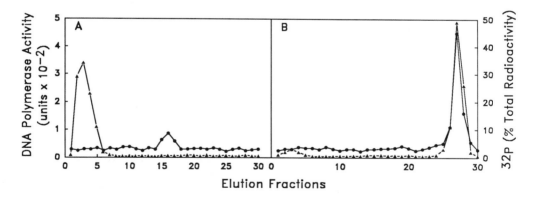

FIGURE 5. DNA polymerase alpha (●) and ^{32}P label (▲) co-elution from a DNA-cellulose affinity column correlated with the presence or absence of PIP production. DNA polymerase alpha A_1 (5 pmol), ^{32}P-ATP (5 pmol), and PI (100 pmol) were incubated at 37°C for 1 h in the absence (left panel) or presence (right panel) of 100 μl of PIK. Enzyme preparations were then applied to a DNA-cellulose affinity column, washed with 10 ml of 50 mM Tris-HCl buffer containing 3 mM MgCl$_2$, 0.5 mM DTT, and 0.1 mM PMSF (fractions 1 to 10), and sequentially eluted with 10 ml of 70 mM KCl (fractions 11 to 20) and 370 mM KCl (fractions 21 to 30) in the same Tris-HCl buffer (pH 7.8). Elution fractions were assayed for DNA polymerase alpha activity and radiometrically assessed for ^{32}P label.

was released, suggesting a phospholipase C-type hydrolysis of PIP by purified polymerase alpha. To further examine this phenomenon, DNA polymerase alpha A_1 was treated with a variety of precursors, intermediates, and end products of the PI phosphorylation cascade. Of these compounds, only PIP and its phospholipase C hydrolysis product, inositol-1,4-biphosphate (IP$_2$), were capable of activating the A_1 form of polymerase alpha (Figure 9). DNA polymerase alpha A_1, ^{3}H labeled by treatment with ATP, PIK, and ^{3}H-PI (Figure 10A), was treated with alkaline phosphatase, resulting in complete dissociation of ^{3}H label from

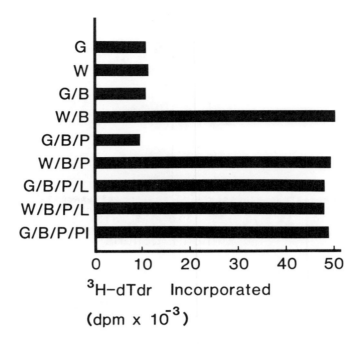

FIGURE 6.   An examination of DNA polymerase alpha activity measured as a function of ^{3}H-thymidine incorporation in BPDE-treated, lipoprotein (LP)-deficient (G) and normal (W) fibroblasts with and without lipoprotein (LP) supplementation. G and W are GM1915 and WI-38 cells, respectively. G/B and W/B show ^{3}H-thymidine incorporation into DNA of G and W cells after treatment with BPDE at 400 ng/ml. G/B/P and W/B/P show ^{3}H-thymidine incorporation into DNA of G and W cells treated with BPDE and permeabilized. G/B/P/L and W/B/P/L show ^{3}H-thymidine incorporation into DNA in G and W cells that have been treated with BPDE, permeabilized, and treated with LDL (20 μg/ml). G/B/P/PI shows G cells treated with BPDE, permeabilized, and treated with 10 μg/ml of PI. (From Joe, C. O., Sylvia, V. L., Norman, J. O., and Busbee, D. L., *Mutat. Res.*, 184, 129, 1987. With permission.)

the enzyme and in reversion of the enzyme to a form eluting with a lower affinity of binding to DNA than was noted for the original A_1 enzyme (Figure 10B). This suggested that removal of phosphoryl groups apparently renders ^{3}H-IP$_2$ unable to remain associated with the enzyme, and, since DNA affinity is decreased below that of the original A_1 enzyme, that other phosphoryl groups may also be affected by the phosphatase. Polymerase A_2 was totally inactivated by alkaline phosphatase (Figure 11), suggesting that phosphoryl groups on this known phosphoprotein must play a major role in regulation of the enzyme's activity. Whatever role phosphorylation plays in regulating the specific activity of polymerase alpha A_2, the enzyme is stable under storage conditions, while activated polymerase A_1 is not.[113] When radiolabeled, activated polymerase A_1 was held at $-80°C$ in 10% glycerol for 7 d, a decrease in amplitude of the high activity, high binding affinity enzyme peak was noted; it correlated with the appearance of a low binding affinity peak which showed about a tenfold decrease in enzyme specific activity and which was not labeled (Figure 12, lower panel). Polymerase A_2 did not exhibit this lability after 14 d at $-80°C$ (Figure 12, upper panel). These data have prompted the proposal that, since initiation of the PI phosphorylation cascade has been characterized as a receptor-mediated phenomenon in a wide variety of cells, IP$_2$ may constitute a type of second messenger capable of increasing the specific activity of existing low-activity DNA polymerase alpha in adult cells (Figure 13).

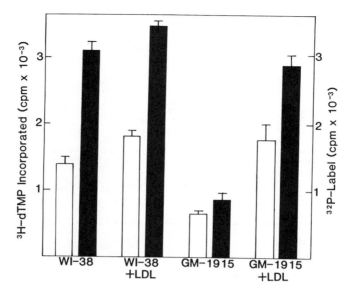

FIGURE 7. An examination of intracellular ^{32}P labeling and activation of DNA polymerase alpha in BPDE-treated human fibroblasts. GM1915 and WI-38 cells were grown to confluency, treated with 5 mM hydroxyurea (HU) for 2 h, and osmotically permeabilized. Cells were then treated with either BPDE (400 ng/ml) alone or BPDE (400 ng/ml) and LDL (20 µg/ml). DNA polymerase alpha was isolated after 1 h and assessed for *in vitro* enzymatic activity (open bars) and ^{32}P label (solid bars). Polymerase activity is expressed as ^{3}H-dTMP incorporated into acid-precipitable DNA template/primer. (From Joe, C. O., Sylvia, V. L., Norman, J. O., and Busbee, D. L., *Mutat. Res.*, 184, 124, 1987. With permission.)

While the A_1 form of DNA polymerase alpha was not found in fetal-derived cells, it comprised 12% of the fibroblast polymerase alpha protein pool isolated from cells of a 5-day-old infant, and its expression increased with age of the cell donor until it constituted 94% of the total DNA polymerase alpha protein isolated from fibroblasts derived from a 66-year-old healthy male (Table 1). Fibroblasts derived from the 66-year-old donor exhibited essentially the same generation time as did cells from 20- and 39-year-old donors, all of which grew more slowly than did fetal-derived cells.

Since these studies examined the ratios of expression of DNA polymerase alpha forms A_1 and A_2 and the capacity of polymerase A_1 and A_2 to utilized enzymatically gapped DNA template/primer without regard to the status of DNA primase in the enzyme preparations, they are appropriate to a determination of the ability of cells to engage in DNA excision repair but not to a determination of replicative DNA synthesis. When BPDE-treated fetal- and adult-derived fibroblasts were examined for ability to incorporate ^{3}H-thymidine as a function of UDS,[113] the fetal-derived cells exhibited a higher initial rate of thymidine incorporation and approximately a twofold greater total incorporation than was seen in adult-derived cells (Figure 14). The major fetal vs. adult difference in this 4-h evaluation of BPDE-initiated excision repair was the rapid decline in adult cell UDS after 1 h, while fetal cells continued UDS for up to 3 h before beginning a decline. These findings may provide a basis for evaluation of the data presented in Figure 15.[113] In a study of UV-initiated UDS in human peripheral blood lymphocytes (PBL) carried out over a period of 15 years, five persons were used as the laboratory UDS controls for PBL. When the investigators began to be concerned that the controls were declining in activity relative to activity of PBL from a constant age student donor population, they collected 10 years of control data and evaluated them to see if a pattern of UDS decline could be established. Data collected for these male subjects

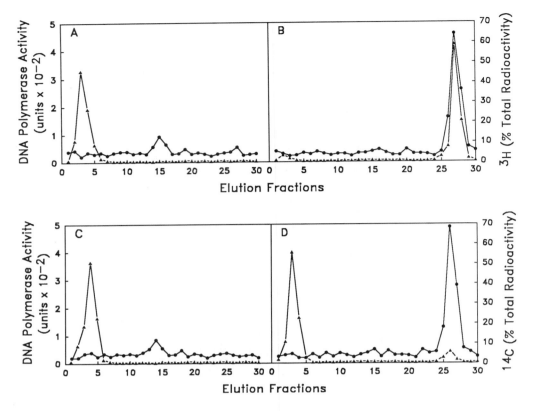

FIGURE 8. PI-([14C]-diacylglycerol) vs. PI-([3H]-inositol) interaction with DNA polymerase alpha during PIP-mediated enzyme activation. DNA polymerase alpha A_1 (50 pmol) was incubated for 30 min at 37°C in the absence (A,C) or presence (B,D) of PIK (100 μg/ml) in a reaction mixture containing 5 mM ATP and 50 pmol of either myoinositol-2-([3H])-PI (top panels) or 1-stearoyl-2-arachidonyl-([14C])-PI (bottom panels). Treated enzyme preparations were applied to a DNA-cellulose affinity column which was washed with 50 mM Tris-HCl (pH 7.8) containing 3 mM MgCl$_2$, 0.5 mM DTT, and 0.1 mM PMSF (fractions 1 to 10) and then sequentially eluted with 70 mM KCl (fractions 11 to 20) and 370 mM KCl (fractions 21 to 30) in 50 mM Tris-HCl. Elution fractions were assayed for polymerase alpha activity (●; expressed as units/mg) and for the the presence of [14C] or [3H] label (▲). [3H]-labeled inositol chromatographs with the activated enzyme (B), while [14C]-labeled DAG does not chromatograph with the activated enzyme (D).

were normalized so that UDS controls at age 34 were assigned a value of 100% for each person. As the individuals increased in age to 44 years, thymidine incorporation decreased. Evaluation of the data showed a negative slope (-1.95) with an intercept of 168 and a rather poor correlation coefficient ($r = -0.58$) between increased age and decreased thymidine incorporation as a function of UDS.

V. CONCLUDING REMARKS

At this time, there is not a consensus regarding the mechanism(s) by which aging causes an increase in cancer incidence. There is, however, no question that one inexorable result of aging is to increase the interval over which a person is exposed to mutagenic and carcinogenic events. It is also generally accepted that one or more of the physiological changes occurring over a person's life span as a natural consequence of aging may impair the ability of cells to repair DNA damage caused by spontaneous oxidative events, environmentally encountered mutagens, and radiation. To understand these mechanisms and how they contribute to the onset of cancer as a function of aging, one should, as much as possible,

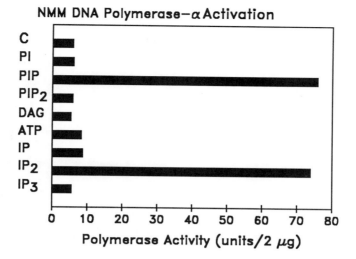

FIGURE 9. An examination of activation of DNA polymerase alpha by a series of phosphoinositides, their precursors, and their phospholipase C hydrolysis products. DNA polymerase alpha aliquots (2 μg/assay: 20 μg/ml) were treated with 1 μg/ml each of PI, PIP, phosphatidylinositol 4,5-biphosphate (PIP$_2$), inositol 1-monophosphate (IP), IP$_2$, inositol 1,4,5-trisphosphate (IP$_3$), DAG, and ATP. DNA polymerase alpha activity, determined as ^{3}H-dTMP incorporated into TCA-precipitable activated DNA template/primer, is expressed as units.

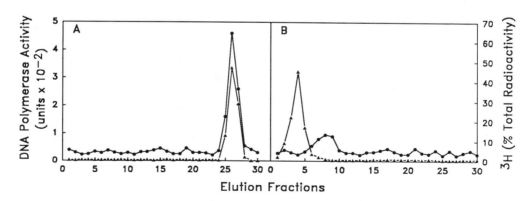

FIGURE 10. Reduction in enzyme activity and dissociation of ^{3}H-inositide from an activated preparation of polymerase A$_1$ treated with alkaline phosphatase. Activated polymerase alpha A$_1$ (1.32 μg ^{3}H-labeled enzyme) eluted from a DNA-cellulose column in the 370 mM KCl wash (upper panel, fractions 25 to 29). An identical aliquot of the ^{3}H-labeled enzyme was treated with 0.1 unit of alkaline phosphatase (type III-S) and examined for DNA-cellulose binding and polymerase activity (lower panel). DNA polymerase alpha assays (●) and radiometric determinations (▲) were completed on elution fractions to determine co-elution of ^{3}H label with the enzyme.

study the phenomena as single variables in an experimental design. Data presented here show the effects of aging in human cells on expression of DNA polymerase alpha, a major enzyme necessary for DNA synthesis occurring during excision repair. These data do not address the fidelity of DNA synthesis in human cells as a function of donor age, but are limited to considerations of (1) polymerase alpha specific activity, (2) the capacity of human DNA polymerase alpha to be activated, (3) the lability of fetal and activated adult polymerase alpha, (4) the capacity of polymerase alpha to bind DNA template/primer, and (5) the age-related expression of putative isozymes which differ in the characteristics given above.

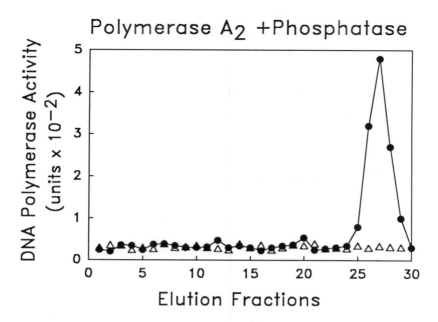

FIGURE 11.   Reduction in enzyme activity and DNA binding of active polymerase A_2 treated with alkaline phosphatase. DNA polymerase alpha A_2 eluted from a DNA-cellulose column in the 370 mM KCl wash (●, fractions 26 to 29). An identical aliquot of the activated enzyme was treated with 0.1 units of alkaline phosphatase and examined for DNA-cellulose binding and polymerase activity (△).

Diminished DNA synthesis in regenerating hepatocytes of aging animals and the inability of stimulated lymphocytes and late-passage fibroblasts to enter mitosis are indicative of the generally decreased DNA synthesis reported to be correlated with increased age in man and experimental animals.[119-124] Although several studies have suggested that this decrease is related to the diminished mitotic index of older cells and that DNA excision repair does not decrease with age,[41-43] other studies show that DNA synthesis occurring as a function of excision repair does decrease in aging rat hepatocytes and mouse skin cells[125,126] and that UV-initiated UDS in cells established from a number of different species is directly proportional to the longevity of the species.[127-129] When one considers that DNA polymerase alpha accounts for 85 to 95% of the DNA synthesis associated with both cell division and DNA excision repair,[108,116] it becomes logical to inquire whether changes in DNA polymerase alpha expression or in the expression of one of its associated and necessary proteins may constitute a physiological mechanism basic to the origin of age-associated cancer in man.

In the studies of human DNA polymerase alpha presented here, the enzyme isolated from fetal-derived cells has been characterized as stable, highly active, and having a high affinity of binding to DNA template/primer, while enzyme isolated from adult-derived fibroblasts has two forms (putative isozymes), one of which is stable, highly active, and indistinguishable from the fetal enzyme, and one of which has charge differences, a decreased affinity of binding to DNA template/primer, and approximately a tenfold lower specific activity. Highly active polymerase alpha, either fetal or adult derived, cannot be further activated by interaction with PIP or IP_2, but the low-activity enzyme form interacts readily with PIP and IP_2, resulting in a dramatic increase in affinity of enzyme binding to DNA and an approximate eight- to tenfold increase in specific activity of the enzyme. Activated polymerase alpha cannot be distinguished from the normally highly active enzyme form, with the exception that activated polymerase from adult-derived cells spontaneously reverts to an inactive form at $-80°C$, while active polymerase from either adult or fetal cells is

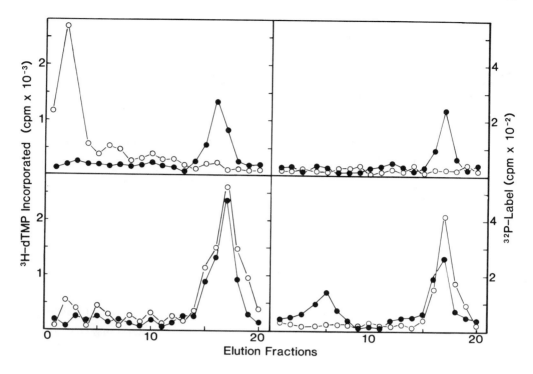

FIGURE 12.   An examination of phosphorylation, activation, and spontaneous loss of activity of DNA polymerase alpha forms A_1 (lower panel) and A_2 (upper panel) from GM1717 fibroblasts. DNA polymerase alpha fraction A_2 from GM1717 (adult-derived) fibroblasts was treated with ^{32}P-ATP, PI, and PIK and then applied to a DNA-cellulose affinity column, which was sequentially eluted with 70 and 370 mM KCl (upper left panel). Elution fractions were assessed for ^{32}P-label (○) and polymerase activity (●). After 14 d storage at $-80°$C in 10% glycerol, the enzyme was reapplied to a DNA-cellulose affinity column and eluted, showing no loss of activity (upper right panel). GM1717 DNA polymerase alpha fraction A_1 was treated identically to fraction A_2 and eluted from a DNA-cellulose affinity column immediately after phosphorylation (lower left panel) or after 7 d of storage at $-80°$C (lower right panel). (From Busbee, D., Sylvia, V., Stec, J., Cernosek, Z., and Norman, J., *J. Natl. Cancer Inst.*, 79, 1231, 1987. With permission.)

stable at $-80°$C. This is demonstrated by the ^{3}H-IP$_2$ dissociation from activated radiolabeled DNA polymerase alpha A_1, which occurs concomitantly with loss of the majority of enzyme activity and loss of the capacity to bind DNA template/primer as a function of time. While the loss of IP$_2$ occurs spontaneously, it also is seen when the enzyme is treated with alkaline phosphatase. Polymerase A_2 differs from A_1 in that it loses all activity and affinity for binding to DNA when treated with alkaline phosphatase. No evidence suggests that IP$_2$ is bound to normally active polymerase A_2, and the alkaline phosphatase elimination of A_2 activity and DNA-binding capacity is apparently the result of dephosphorylation of this known phosphoprotein.[130] These data indicate that DNA polymerase alpha A_2 from fetal- and adult-derived cells and DNA polymerase alpha A_1 from adult-derived cells are substantially different enzyme forms having the same enzymatic function, the synthesis of DNA during excision repair or preceding cell division.

Examinations of cultured cells have shown an approximate twofold increase in accumulation of mutations during the initial two thirds of the *in vitro* cell life span,[131] with a rapid accumulation of mutations during the latter part of the life span.[132] This is consistent with the report of Moreley et al.[133] that mutations do not accumulate as a linear function of time *in vivo* and with previous findings that the ability of cells to repair DNA damage *in vitro* decreases as a function of donor age.[45,46,125,126] Data presented here indicate that an isozyme of DNA polymerase alpha which has low specific activity is not expressed in fetal

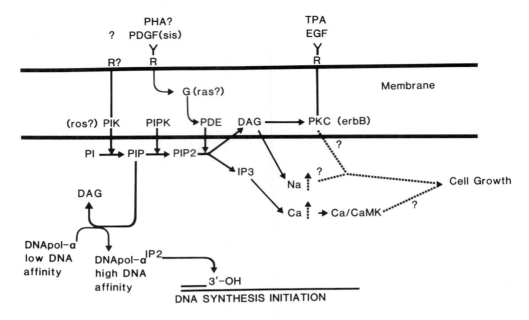

FIGURE 13.   A PI phosphorylation cascade depiction showing IP_2 as a proposed second messenger in the activation of inactive DNA polymerase alpha.

TABLE 1
Percent Total and Specific Activities of DNA Polymerase Alpha Forms A_1 and A_2 in Normal Human Fibroblasts

Cell type	Polymerase protein (% of total)		Specific activity (units/mg)	
	A_1	A_2	A_1	A_2
GM6111 (16-week fetal)	0	100	—	27,710
GM3468A (5 d)	12	88	2,730	38,520
GM0500B (10 year)	22	78	2,220	35,270
GM3440A (20 year)	26	74	2,640	27,450
GM1717 (39 year)	35	65	3,160	27,720
GM3529 (66 year)	94	6	1,620	26,870

Note: Human skin fibroblast DNA polymerase alpha from various age male donors was purified using DEAE-cellulose, immunoaffinity, and DNA-cellulose procedures. One unit of activity is defined as that amount capable of catalyzing the incorporation of 1 nmol of dTMP into activated DNA template per hour at 37°C.

cells but is increasingly expressed as a function of cell donor age, and that both lymphocytes and fibroblasts exhibit decreased UDS as a function of increased donor age. The applicability of data generated by *in vitro* activation of DNA polymerase alpha isolated from aging cell donors to the question of intracellular BPDE initiation of UDS has been partially addressed by Joe et al,[118] who reported that polymerase alpha is activated in arrested, permeabilized human fibroblasts treated with BPDE concomitant with transfer of ^{32}P label from ^{32}P-ATP to the enzyme and that intracellular activation and phosphorylation of polymerase A_1 do not

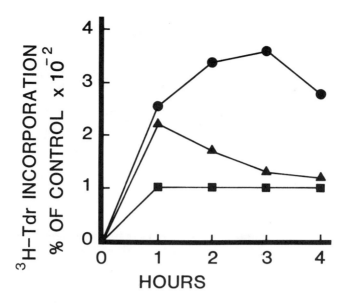

FIGURE 14. A comparison of ³H-thymidine incorporated into the DNA of human fibroblasts, GM6111 (●), GM1717 (▲), and control (■), as a function of unscheduled DNA synthesis. Fibroblasts of the same passage number were grown under identical conditions. Cells were harvested, plated at 10^6 cells/plate, treated with 3 mM hydroxyurea for 3 h to inhibit scheduled DNA synthesis, and further treated with 7,8-dihydrodiol-9,10-epoxybenzo(a)pyrene (anti) to initiate unscheduled DNA synthesis. ³H-thymidine (1 μCi/ml) was added to the separate cultures, the cells were allowed to repair for the indicated intervals, and ³H-thymidine incorporation into TCA-precipitable DNA was determined as a measure of unscheduled DNA synthesis. (From Busbee, D., Sylvia, V., Stec, J., Cernosek, Z., and Norman, J., *J. Natl. Cancer Inst.*, 79, 1231, 1987. With permission.)

occur in the absence of either low-density lipoproteins or PI. These findings are consistent with those of an earlier report[44] indicating that serum-deprived lymphocytes treated with BPDE did not initiate UDS and that BPDE-treated human low-density lipoprotein (LDL) uptake-deficient cells initiated, but did not complete, DNA excision repair, resulting in fragmentation of cellular DNA in the absence of resynthesis of excised oligonucleotides. In concert, these data suggest both a requirement and a potential mechanism for activation of DNA synthesis occurring in carcinogen-treated cells engaged in excision repair, and may explain, at least in part, why human cells show a decline in UDS as a function of increased donor age.

The data suggest that the increasing ratio of low activity to high activity forms of DNA polymerase alpha in human cells may be correlated with an age-related decrease in specific activity of the total intracellular DNA polymerase alpha pool, and indicate that the increased low to high activity polymerase alpha ratio may be correlated with a decrease in UDS as a function of age. Current data further suggest that activation of the PI phosphorylation cascade and, ultimately, the known second messengers, diacylglycerol and inositol-1,4,5-trisphosphate, and the proposed second messenger, inositol-4,5-bisphosphate, along with the increased requirement for polymerase alpha A_1 activation as this enzyme concentration increases with increased age, may introduce an additional complex set of physiological parameters which one must consider in attempting to correlate increased age with increased cancer incidence in man.

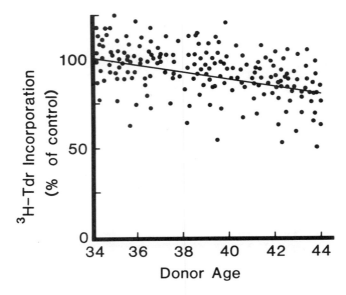

FIGURE 15.   An assessment of ³H-thymidine incorporation into DNA of non-mitogen-stimulated human peripheral blood lymphocytes treated with UV light at 260 nm. Lymphocytes isolated by Ficoll-Hypaque density-gradient centrifugation were treated with UV at 3.5 J/m² in the presence of 3 m*M* hydroxyurea and 1 μCi/ml ³H-thymidine. ³H-thymidine incorporation into TCA-precipitable DNA was determined as a measure of UV-stimulated DNA synthesis. Linear regression analysis was completed using Sigma Plot for microcomputers. Slope = −1.95, intercept = 168, r = −0.58. (From Busbee, D., Sylvia, V., Stec, J., Cernosek, Z., and Norman, J., *J. Natl. Cancer Inst.,* 79, 1231, 1987. With permission.)

REFERENCES

1. **Setlow, R. B. and Carrier, W. L.,** The disappearance of thymine dimers from DNA: an error-correcting mechanism, *Proc. Natl. Acad. Sci. U.S.A.,* 51, 226, 1964.
2. **Boyce, R. P. and Howard-Flanders, P.,** Release of ultraviolet light-induced thymine dimers from DNA in *E. coli* K-12, *Proc. Natl. Acad. Sci. U.S.A.,* 51, 293, 1964.
3. **Boyce, R. P. and Howard-Flanders, P.,** Genetic control of DNA breakdown and repair in *E. coli* K-12 treated with mitomycin C or ultraviolet light, *Z. Vererbungsl.,* 95, 433, 1964.
4. **Hill, R. F. and Feiner, R. R.,** Further studies of ultraviolet-sensitive mutants of *Eschericia coli* strain B, *J. Gen. Microbiol.,* 35, 105, 1964.
5. **Lawley, P. D. and Brookes, P.,** Cytotoxicity of alkylating agents toward sensitive and resistant strains of *Escherichia coli* in relation to extent and mode of alkylation of cellular macromolecules and repair of alkylation lesions in deoxyribonucleic acids, *Biochem. J.,* 109, 345, 1968.
6. **Hanawalt, P. C. and Haynes, R. H.,** Repair replication of DNA in bacteria: irrelevance of base defects, *Biochem. Biophys. Res. Commun.,* 19, 462, 1965.
7. **Roberts, J. J., Crathorn, A. R., and Bent, T. P.,** Repair of alkylated DNA in mammalian cells, *Nature,* 218, 970, 1968.
8. **Regan, J. D. and Setlow, R. B.,** Two forms of repair in the DNA of human cells damaged by chemical carcinogens and mutagens, *Cancer Res.,* 34, 3318, 1974.
9. **Hanawalt, P. C., Cooper, P. K., Ganesan, A. K., and Smith, C. A.,** DNA repair in bacteria and mammalian cells, *Annu. Rev. Biochem.,* 48, 783, 1979.
10. **Cleaver, J. E.,** Structure of repaired sites in human DNA synthesized in the presence of inhibitors of DNA polymerases alpha and beta in human fibroblasts, *Biochim. Biophys. Acta,* 739, 301, 1983.

11. **Keyse, S. M. and Tyrell, R. M.,** Excision repair in permeable arrested human skin fibroblasts damaged by UV (254 nm) radiation: evidence that alpha and beta polymerases act sequentially at the repolymerization step, *Mutat. Res.,* 146, 109, 1985.

12. **Miller, M. R. and Chinault, D. N.,** The roles of DNA polymerases alpha, beta, and gamma in DNA repair synthesis induced in hamster and human cells by different DNA damaging agents, *J. Biol. Chem.,* 257(17), 10204, 1982.

13. **Norman, J. O., Joe, C. O., and Busbee, D. L.,** Inhibition of DNA polymerase alpha activity by methyl methanesulfonate, *Mutat. Res.,* 159, 83, 1986.

14. **Rasmussen, R. E. and Painter, R. B.,** Evidence for repair of ultraviolet damaged deoxyribonucleic acid in cultured mammalian cells, *Nature,* 203, 1360, 1964.

15. **Rasmussen, R. E. and Painter, R. B.,** Radiation stimulated DNA synthesis in cultured mammalian cells, *J. Cell Biol.,* 29, 11, 1966.

16. **Painter, R. B. and Cleaver, J. E.,** Repair replication in HeLa cells after large doses of x-irradiation, *Nature,* 216, 369, 1967.

17. **Stich, H. F., San, R. H. C., Miller, J. A., and Miller, E. A.,** Various levels of DNA repair synthesis in xeroderma pigmentosum cells exposed to the carcinogens *N*-hydroxy and *N*-acetoxy-2-acetylaminofluorene, *Nature,* 238, 79, 1972.

18. **Stich, H. F. and San, R. H. C.,** DNA repair synthesis and survival of repair deficient human cells as a relevant, rapid, and economic assay for environmental carcinogens, in *Recent Topics in Chemical Carcinogenesis,* Odashima, S., Takayama, S., and Sato, H., Eds., University Park Press, Baltimore, 1975, 3.

19. **Ames, B. N., Lee, F. D., and Durston, W. E.,** An improved bacterial test for the detection and classification of mutagens and carcinogens, *Proc. Natl. Acad. Sci. U.S.A.,* 70, 782, 1973.

20. **Brusick, D. J.,** Mammalian DNA repair assays, in *Carcinogenesis and Mutagenesis Testing,* Douglas, J. F., Ed., Humana Press, Clifton, NJ, 1980, 251.

21. **Jones, E., Richold, M., May, J. H., and Saje, A.,** The assessment of the mutagenic potential of vehicle engine exhaust in the Ames Salmonella assay using a direct exposure method, *Mutat. Res.,* 155, 35, 1985.

22. **Cleaver, J. E.,** Defective repair replication of DNA in xeroderma pigmentosum, *Nature,* 218, 652, 1968.

23. **Hanawalt, P. C. and Sarasin, A.,** Cancer-prone hereditary diseases with DNA processing abnormalities, *Trends Genet.,* 2, 124, 1986.

24. **Levin, S. and Perlov, S.,** Ataxia telangiectasia in Israel, with observations on its relevance to malignant disease, *Isr. J. Med. Sci.,* 7, 1535, 1971.

25. **German, J., Archibald, R., and Bloom, D.,** Chromosomal breakage in a rare and probably genetically determined syndrome of man, *Science,* 148, 506, 1965.

26. **Maher, V. M. and McCormick, J. J.,** DNA repair and carcinogenesis, in *Chemical Carcinogens and DNA,* Grover, P. L., Ed., CRC Press, Boca Raton, FL, 1979, 133.

27. **Arlett, C. F. and Lehmann, A. R.,** Human disorders showing increased sensitivity to the induction of genetic damage, *Annu. Rev. Genet.,* 12, 95, 1978.

28. **Paterson, M. C., Smith, B. P., Lohman, P. M. H., Anderson, A. K., and Fishman, L.,** Defective excision repair of gamma ray damaged DNA in human (ataxia telangiectasia) fibroblasts, *Nature,* 260, 444, 1975.

29. **Poon, P. K., O'Brien, R. L., and Parker, J. W.,** Defective DNA repair in Fanconi's anemia, *Nature,* 250, 223, 1975.

30. **Swift, M., Sholman, L., Perry, M., and Chase, C.,** Malignant neoplasms in the families of patients with ataxia telangiectasia, *Cancer Res.,* 36, 209, 1976.

31. **Cleaver, J. E., Bootsma, D., and Freidberg, E.,** Human diseases with genetically altered DNA repair processes, *Genetics,* Suppl. 79, 215, 1975.

32. **German, J.,** Bloom's Syndrome. II. The prototype of human genetic disorders predisposing to chromosomal instability and cancer, in *Chromosomes and Cancer,* German, J., Ed., John Wiley & Sons, New York, 1974, 601.

33. **Tice, R., Rary, J. M., and Bender, M. A.,** An investigation of DNA repair potential in Bloom's Syndrome, *J. Supramol. Struct.,* Suppl. 2, 82, 1978.

34. **Peto, R., Roe, F. J. C., Lee, P. N., Levy, L., and Clack, J.,** Cancer and ageing in mice and men, *Br. J. Cancer,* 32, 411, 1975.

35. **Peto, R., Parish, S. E., and Gray, R. G.,** There is no such thing as ageing, and cancer is not related to it, in *Age-Related Factors in Carcinogenesis,* No. 58, Likhachev, A., Anisimov, V., and Montesano, R., Eds., IARC, Lyon, 1985, 43.

36. **Magee, P. N.,** Carcinogenesis and ageing, *Adv. Exp. Med. Biol.,* 97, 133, 1978.

37. **Burnet, F. M.,** *Immunology, Aging and Cancer. Medical Aspects of Mutation and Selection,* W. H. Freeman, San Francisco, 1976.

38. **Dilman, V. M.,** Ageing, metabolic immunodepression and carcinogenesis, *Mech. Ageing Dev.,* 8, 153, 1978.

39. **Dilman, V. M.,** *The Law of Deviation to Homeostasis and Diseases of Aging,* John Wright PSG, Boston, 1981.

40. **Dix, D., Cohen, P., and Flannery, J.,** On the role of aging in cancer incidence, *J. Theor. Biol.,* 83, 163, 1980.
41. **Hall, J. D., Almy, R. E., and Scherer, K. L.,** DNA repair in cultured fibroblasts does not decline with donor age, *Exp. Cell Res.,* 139, 351, 1982.
42. **Hennis, H. L., Braid, H. L., and Vincent, R. A.,** Unscheduled DNA synthesis in cells of different shape in fibroblast cultures from donors of various ages, *Mech. Ageing Dev.,* 16, 355, 1981.
43. **Dell'Orco, R. T. and Anderson, L. E.,** Unscheduled DNA synthesis in human diploid cells of different ages, *Cell Biol. Int. Rep.,* 5, 359, 1981.
44. **Joe, C. O., Rankin, P. W., and Busbee, D. L.,** Human lymphocytes treated with $(+/-)$-r-7,t-8-dihydroxy-t-9,10-epoxy-7,8,9,10-tetrahydrobenzo(a)pyrene require low density lipoproteins for DNA excision repair, *Mutat. Res.,* 131, 37, 1984.
45. **Lambert, B., Ringbold, U., and Skoog, L.,** Age-related decrease of ultraviolet light-induced DNA repair synthesis in human peripheral leukocytes, *Cancer Res.,* 39, 2792, 1979.
46. **Pashko, L. L. and Schwartz, A. G.,** Inverse correlation between species life span and specific cytochrome P-448 content of cultured fibroblasts, *J. Gerontol.,* 37, 38, 1982.
47. **Schwartz, A. G.,** Correlation between species lifespan and capacity to activate 7,12-dimethyl-benz(a)anthracene to a form mutagenic to a mammalian cell, *Exp. Cell Res.,* 94, 445, 1975.
48. **Schwartz, A. G. and Moore, C. J.,** Inverse correlation between species lifespan and capacity of cultured fibroblasts to bind 7,12-dimethylbenz(a)anthracene to DNA, *Exp. Cell Res.,* 109, 448, 1977.
49. **Hart, R. W. and Setlow, R. B.,** Correlation between deoxyribonucleic acid excision-repair and life-span in a number of mammalian species, *Proc. Natl. Acad. Sci. U.S.A.,* 71, 2169, 1974.
50. **Medcalf, A. S. C. and Lawley, P. D.,** Time course of O^6-methylguanine removal from DNA of N-methyl-N-nitrosourea-treated human fibroblasts, *Nature,* 289, 796, 1981.
51. **Harris, G., Lawley, P. D., and Olsen, I.,** Mode of action of methylating carcinogens: comparative studies of murine and human cells, *Carcinogenesis,* 2, 403, 1981.
52. **Pegg, A. E., Roberfroid, M., von Bahr, C., Foote, R. S., Mitra, S., Bresil, H., Likhachev, A., and Montesano, R.,** Removal of O^6-methylguanine from DNA by a human liver fraction, *Proc. Natl. Acad. Sci. U.S.A.,* 79, 5162, 1982.
53. **Linn, S., Kairis, M., and Holliday, R.,** Decreased fidelity of DNA polymerase activity isolated from ageing human fibroblasts, *Proc. Natl. Acad. Sci. U.S.A.,* 73, 2818, 1976.
54. **Silber, J. R., Fry, M., Martin, G. M., and Loeb, L. A.,** Fidelity of DNA polymerases isolated from regenerating liver chromatin of aging *Mus musculus, J. Biol. Chem.,* 260(2), 1304, 1985.
55. **Lehmann, A. R.,** Ageing, DNA repair of radiation damage and carcinogenesis: fact and fiction, in *Age-Related Factors in Carcinogenesis,* No. 58, Likhachev, A., Anisimov, A., and Montesano, R., Eds., IARC, Lyon, 1985, 203.
56. **Fry, M. and Loeb, L. A.,** *Animal Cell DNA Polymerases,* CRC Press, Boca Raton, FL, 1986.
57. **Weissbach, A.,** Eukaryotic DNA polymerases, *Annu. Rev. Biochem.,* 46, 25, 1977.
58. **Banks, G. R., Boezi, J. A., and Lehman, I. R.,** A high molecular weight DNA polymerase from *Drosophila melanogaster* embryos. Purification, structure and partial characterization, *J. Biol. Chem.,* 254, 9886, 1979.
59. **Craig, R. K. and Keir, H. M.,** Deoxyribonucleic acid polymerases of BHK/21-C13 cells. Partial purification and characterization of the enzymes, *Biochem. J.,* 145, 215, 1975.
60. **Mechali, M., Abadidebat, J., and de Recondo, A. M.,** Eukaryotic DNA polymerase α. Structural analysis of the enzyme from regenerating rat liver, *J. Biol. Chem.,* 255, 2114, 1980.
61. **Mechali, M. and de Recondo, A. M.,** Structural analysis of eukaryotic DNA polymerase-α, in *New Approaches in Eukaryotic DNA Replication,* de Recondo, A. M., Ed., Plenum Press, New York, 1983, 57.
62. **Pritchard, C. G. and DePamphilis, M. L.,** Preparation of DNA polymerase $\alpha \cdot C_1C_2$ by reconstituting DNA polymerase α with its specific stimulting cofactors, C_1C_2, *J. Biol. Chem.,* 258, 9801, 1983.
63. **Holmes, A. M. and Johnston, I. R.,** Molecular asymmetry of rat liver cytoplasmic DNA polymerase, *FEBS Lett.,* 29, 1, 1973.
64. **Karawya, E., Swack, J., Albert, W., Fedorko, J., Minna, J. D., and Wilson, S. H.,** Identification of a higher molecular weight DNA polymerase-α catalytic polypeptide in monkey cells by monoclonal antibody, *Proc. Natl. Acad. Sci. U.S.A.,* 81, 7777, 1984.
65. **Albert, W., Grummt, F., Hubscher, U., and Wilson, S. H.,** Structural homology among calf thymus α-polymerase polypeptides, *Nucleic Acids Res.,* 10, 935, 1982.
66. **Wahl, A., Kowalski, S., Harwell, L., Lord, E., and Bambara, R.,** Immunoaffinity purification and properties of a high molecular weight calf thymus DNA α polymerase, *Biochemistry,* 23, 1895, 1984.
67. **Karawya, E. M. and Wilson, S. H.,** Studies on catalytic subunits of mouse myeloma α-polymerase, *J. Biol. Chem.,* 257, 13129, 1982.

68. **Wang, T. S.-F., Hu, S.-Z., and Korn, D.,** DNA primase from KB cells. Characterization of a primase activity tightly associated with immunoaffinity purified DNA polymerase-α, *J. Biol. Chem.,* 259, 1854, 1984.

69. **Cotterill, S. M., Reyland, M. E., Loeb, L. A., and Lehman, I. R.,** Enzymatic activities associated with the 182 kDa polymerase subunit of the DNA polymerase-primase of *Drosophila melanogaster,* in *Eukaryotic Replication,* Cold Spring Harbor Laboratory, Cold Spring Harbor, NY, 1987, 147.

70. **Mechali, M. and de Recondo, A. M.,** Eukaryotic DNA polymerase-α tryptic map analysis and comparison between the rat liver and *Drosophila* enzymes, *Biochem. Int.,* 4, 465, 1982.

71. **Sauer, B. and Lehman, I. R.,** Immunological comparison of purified DNA polymerase-α from embryos of *Drosophila melanogaster* with forms of the enzyme present *in vivo, J. Biol. Chem.,* 257, 12394, 1982.

72. **Kaguni, L. S., Rossignol, J.-M., Connaway, R. C., and Lehman, I. R.,** Isolation of an intact DNA polymerase-primase from embryos of *Drosophila melanogaster, Proc. Natl. Acad. Sci. U.S.A.,* 80, 2221, 1983.

73. **Grosse, F. and Nasheuer, H.-P.,** Structure and properties of the immunoaffinity purified DNA polymerase alpha-primase complex, in *Eukaryotic Replication,* Cold Spring Harbor Laboratory, Cold Spring Harbor, NY, 1987, 151.

74. **Krauss, S. W. and Linn, S.,** Changes in DNA polymerases α, β, and γ during the replicative life span of cultured human fibroblasts, *Biochemistry,* 21, 1002, 1982.

75. **Krauss, S. W. and Linn, S.,** Studies of DNA polymerase alpha and beta from cultured human cells in various replicative states, *J. Cell. Physiol.,* 126, 99, 1986.

76. **Bhattachaarya, P., Simet, I., and Basu, M.,** Differential inhibition of multiple forms of DNA polymerase-α from IMR-32 human neuroblastoma cells, *Proc. Natl. Acad. Sci. U.S.A.,* 78, 2683, 1981.

77. **Enomoto, T., Tanuma, S.-I., and Yamada, M.-A.,** Purification and characterization of two forms of DNA polymerase α from HeLa cell nuclei, *Biochemistry,* 22, 1134, 1983.

78. **Lamothe, P., Baril, B., Chi, A., Lee, L., and Baril, E.,** Accessory proteins for DNA polymerase α activity with single-strand DNA templates, *Proc. Natl. Acad. Sci. U.S.A.,* 78, 4723, 1981.

79. **Kornberg, A.,** *DNA Replication, 1982 Supplement,* W. H. Freeman, San Francisco, 1982.

80. **De Pamphilis, M. L. and Wasserman, P. M.,** Replication of eukaryotic chromosomes: a close-up of the replication fork, *Annu. Rev. Biochem.,* 49, 627, 1980.

81. **Pritchard, C. G., Weaver, D. T., Baril, E. F., and DePamphilis, M. L.,** DNA polymerase α cofactors C_1C_2 function as primer recognition proteins, *J. Biol. Chem.,* 258, 9810, 1983.

82. **Novak, B. and Baril, E. F.,** HeLa DNA polymerase activity *in vitro:* specific stimulation by a non-enzyme protein factor, *Nucleic Acids Res.,* 5, 221, 1978.

83. **Rapaport, E. and Zamecnik, P. C.,** Presence of diadenosine $5',5'''-P^1,P^4$-tetraphosphate (Ap_4A) in mammalian cells in levels varying widely with proliferative activity of the tissue: a possible positive "pleiotropic activator", *Proc. Natl. Acad. Sci. U.S.A.,* 73, 3984, 1976.

84. **Zamecnik, P. C., Stephenson, M. L., Janeway, C. M., and Randerath, K.,** Enzymatic synthesis of diadenosine tetraphosphate and diadenosine triphosphate with a purified lysyl-sRNA synthetase, *Biochem. Biophys. Res. Commun.,* 24, 91, 1966.

85. **Randerath, K., Janeway, C. M., Stephenson, M. L., and Zamecnik, P. C.,** Isolation and characterization of dinucleotide tetra- and tri-phosphate in the presence of lysyl-sRNA synthetase, *Biochem. Biophys. Res. Commun.,* 24, 98, 1966.

86. **Baril, E. F., Malkas, L. H., Hickey, R., Li, C., Vishwanatha, J. K., and Coughlin, S.,** Multiprotein form of DNA polymerase alpha from human cells: interaction with other proteins in DNA replication, in *Eukaryotic Replication,* Cold Spring Harbor Laboratory, Cold Spring Harbor, NY, 1987, 148.

87. **Goulian, M., Carton, C., DeGrandpre, L., Heard, C., Olinger, B., and Richards, S.,** Discontinuous DNA synthesis by purified proteins from mammalian cells, in *Eukaryotic Replication,* Cold Spring Harbor Laboratory, Cold Spring Harbor, NY, 1987, 150.

88. **Braithwaite, A. W. and Jenkins, J. R.,** Isolation of mammalian DNA sequences that can function as origins of DNA replication, in *Eukaryotic Replication,* Cold Spring Harbor Laboratory, Cold Spring Harbor, NY, 1987, 87.

89. **Kornberg, A.,** *DNA Replication,* W. H. Freeman, San Francisco, 1980.

90. **Wintersberger, E.,** Yeast DNA polymerases: antigenic relationship, use of DNA primer and associated exonuclease activity, *Eur. J. Biochem.,* 84, 167, 1978.

91. **de Recondo, A. M., Lepesant, J. A., Fichot, O., Grassat, L., Rossignol, J. M., and Cazillis, M.,** Synthetic template specificity of a deoxyribonucleic acid polymerase from regenerating rat liver, *J. Biol. Chem.,* 248, 131, 1973.

92. **Hubscher, U.,** DNA polymerase holoenzymes, *Trends Biochem. Sci.,* 9, 390, 1984.

93. **Kaftory, A. and Fry, M.,** Highly efficient copying of a single stranded DNA by eukaryotic cell chromatin, *Nucleic Acids Res.,* 5, 2679, 1978.

94. **Hubscher, U.,** The mammalian primase is part of a high molecular weight DNA polymerase α polypeptide, *EMBO J.,* 2, 133, 1983.

95. **Kaguni, L. S., DiFrancesco, R. A., and Lehman, I. R.,** The DNA polymerase-primase from *Drosophila melanogaster* embryos. Rate and fidelity of polymerization on single stranded DNA templates, *J. Biol. Chem.,* 259, 9314, 1984.

96. **Chang, L. M. S., Rafter, E., Augl, C., and Bollum, F. J.,** Purification of a DNA polymerase-DNA primase complex from calf thymus glands, *J. Biol. Chem.,* 259, 14679, 1984.

97. **McLennan, A. G. and Keir, H. M.,** Deoxyribonucleic acid polymerase of *Euglena gracilis.* Primer-template utilization of, and enzyme activities associated with, the two deoxyribonucleic acid polymerases of high molecular weight, *Biochem. J.,* 151, 239, 1975.

98. **Creran, M. and Pearlman, R. E.,** Deoxyribonucleic acid polymerase from *Tetrahymena pyriformis.* Purification and properties of the major activity in exponentially growing cells, *J. Biol. Chem.,* 249, 3123, 1974.

99. **Banks, G. R. and Yarranton, G. T.,** A DNA polymerase from *Ustilago maydis.* II. Properties of the associated deoxyribonuclease activity, *Eur. J. Biochem.,* 62, 143, 1976.

100. **Rapaport, E. and Feldman, L.,** Adenosine(5')tetraphosphate(5')adenosine-binding protein of calf thymus, *Eur. J. Biochem.,* 138, 111, 1984.

101. **Rapaport, E., Zamecnik, P. C., and Baril, E. F.,** HeLa Cell DNA polymerase-α is tightly associated with tryptophanyl-tRNA synthetase and diadenosine 5',5'''-P^1,P^4-tetraphosphate binding activities, *Proc. Natl. Acad. Sci. U.S.A.,* 78, 838, 1981.

102. **Baril, E., Bonin, P., Burstein, D., Mara, K., and Zamecnik, P.,** Resolution of the diadenosine 5',5'''-P^1,P^4-tetraphosphate binding subunit from a multiprotein form of HeLa cell DNA polymerase-α, *Proc. Natl. Acad. Sci. U.S.A.,* 80, 4931, 1983.

103. **Rapaport, E., Zamecnik, P. C., and Baril, E. F.,** Association of diadenosine 5',5'''-P^1,P^4-tetraphosphate binding protein with HeLa cell DNA polymerase-α, *J. Biol. Chem.,* 256, 12148, 1981.

104. **Grummt, F.,** Diadenosine tetraphosphate triggers *in vitro* DNA replication, *Cold Spring Harbor Symp. Quant. Biol.,* 43, 649, 1978.

105. **Weinman-Dorsch, C., Hedl, A., Grummt, I., Albert, W., Ferdinand, F.-J., Friis, R. R., Pierron, G., Moll, W., and Grummt, F.,** Drastic rise of intracellular adenosine(5')tetraphospho(5')adenosine correlates with onset of DNA synthesis in eukaryotic cells, *Eur. J. Biochem.,* 138, 179, 1984.

106. **Kaguni, L. S., Rossignol, J.-M., Conway, R. C., Banks, G. R., and Lehman, I. R.,** Association of DNA primase with the beta/gamma subunits of DNA polymerase α from *Drosophilia melanogaster* embryos, *J. Biol. Chem.,* 258, 9037, 1983.

107. **Burke, J. F., Plummer, J., Huberman, J. A., and Evans, M. J.,** Restriction fragment primed φX174 single stranded DNA as template for polymerases α and β. Detection and partial purification of a DNA polymerase α stimulating factor, *Biochim. Biophys. Acta,* 609, 205, 1980.

108. **Lawton, K. G., Wierowski, J. V., Schechter, S., Hilf, R., and Bambara, R. A.,** Analysis of the mechanism of ATP stimulation of calf thymus DNA alpha polymerase, *Biochemistry,* 23, 4294, 1984.

109. **Danse, J. M., Egly, J. M., and Kemp, J.,** *In vitro* activation of DNA polymerase alpha by a protein kinase in chick embryo, *FEBS Lett.,* 184, 84, 1981.

110. **Krauss, S. W., Mochly-Rosen, D., Koshland, D. E., and Linn, S.,** Exposure of HeLa DNA polymerase alpha to protein kinase C affects its catalytic properties, *J. Biol. Chem.,* 262(8), 3432, 1987.

111. **Sylvia, V. L., Norman, J. O., Curtin, G. M., and Busbee, D. L.,** Monoclonal antibody that blocks phosphoinositide-dependent activation of mouse tumor DNA polymerase alpha, *Biochem. Biophys. Res. Commun.,* 141, 60, 1986.

112. **Sylvia, V. L., Norman, J. O., Curtin, G. M., Ragsdale-Robinson, S., and Busbee, D. L.,** Phosphatidylinositol-4-monophosphate activates DNA polymerase alpha, in *Membrane Proteins,* Proc. Bio-Rad Membrane Protein Symp., Goheen, S., Ed., San Diego, 1986.

113. **Busbee, D., Sylvia, V., Stec., J. Cernosek, Z., and Norman, J.,** Lability of DNA polymerase alpha correlated with decreased DNA synthesis and increased age in human cells, *J. Natl. Cancer Inst.,* 79(6), 1231, 1987.

114. **Fisher, P. A. and Korn, D.,** Enzymological characterization of KB cell DNA polymerase alpha, *J. Biol. Chem.,* 254, 11033, 1979.

115. **Zierler, M. K., Marini, N. J., Stowes, D. J., and Benbow, R. M.,** Stockpiling of DNA polymerases during oogenesis and embryogenesis in the frog, *Xenopus laevis, J. Biol. Chem.,* 260, 974, 1985.

116. **Murray, V.,** Properties of DNA polymerases from young and ageing human fibroblasts, *Mech. Ageing Dev.,* 16, 328, 1981.

117. **Agarwal, S. S., Tuffner, M., and Loeb, L. A.,** DNA replication in human lymphocytes during aging, *J. Cell. Physiol.,* 106, 235, 1978.

118. **Joe, C. O., Sylvia, V. L., Norman, J. O., and Busbee, D. L.,** Repair of benzo(a)pyrene-initiated DNA damage in human cells requires activation of DNA polymerase alpha, *Mutat. Res.,* 184, 129, 1987.

119. **Gelfand, M. C. and Steinberg, A. D.,** Mechanism of allograft rejection in New Zealand mice. I. Cell synergy and its age-dependent loss, *J. Immunol.,* 110, 1652, 1981.

120. **Siskind, G. W. and Weksler, M. E.**, The effect of aging on the immune response, *Annu. Rev. Gerontol. Geriatr.*, 3, 3, 1982.

121. **Nagel, J. E., Chrest, F. J., and Adler, W. H.**, Human B-cell function in normal individuals of various ages. I. *In vitro* enumeration of pokeweed induced peripheral blood lymphocyte immunoglobulin synthesizing cells and the comparison of the result with numbers of peripheral B- and T-cells, mitogen responses, and levels of serum immunoglobulins, *Clin. Exp. Immunol.*, 44, 646, 1981.

122. **Fry, M., Silber, J. M., Loeb, L. A., and Martin, G. M.**, Delayed and reduced cell replication and diminishing levels of DNA polymerase alpha in regenerating liver of aging mice, *J. Cell. Physiol.*, 181, 225, 1984.

123. **Ballard, F. J. and Read, L. C.**, Changes in protein synthesis and breakdown rates and responsiveness to growth factors with ageing in human lung fibroblasts, *Mech. Ageing Dev.*, 30, 11, 1985.

124. **Bruce, S. A., Deamond, S. F., and Ts'o, P. O. P.**, *In vitro* senescence of Syrian hamster mesenchymal cells of fetal to aged origin. Inverse relationship between *in vivo* donor age and *in vitro* proliferative capacity, *Mech. Ageing Dev.*, 34, 151, 1986.

125. **Kennah, H. E., Coetzee, M. L., and Ove, P.**, A comparison of DNA repair synthesis in primary hepatocytes from young and old rats, *Mech. Ageing Dev.*, 29, 283, 1985.

126. **Ishikawa, T. and Sakurai, J.**, *In vivo* studies on age-dependency of DNA repair with age in mouse skin, *Cancer Res.*, 46, 1344, 1986.

127. **Hart, R. W. and Setlow, R. B.**, Correlation between deoxyribonucleic acid excision-repair and lifespan in a number of mammalian species, *Proc. Natl. Acad. Sci. U.S.A.*, 71, 2169, 1974.

128. **Kato, H., Harada, M., Tsuchiya, K., and Moriwaki, K.**, Absence of correlation between DNA repair in ultraviolet irradiated mammalian cells and lifespan of the donor species, *Jpn. J. Genet.*, 55, 99, 1980.

129. **Hall, K. Y., Hart, R. W., Benirschke, A. K., and Walford, R. L.**, Correlation between ultraviolet-induced DNA repair in primate lymphocytes and fibroblasts and species maximum achievable life span, *Mech. Ageing Dev.*, 24, 163, 1984.

130. **Wong, S. W., Paborsky, L. R., Fisher, P. A., Wang, T. S.-F., and Korn, D.**, Structural and enzymological characterization of immunoaffinity-purified DNA polymerase alpha: DNA primase complex from KB cells, *J. Biol. Chem.*, 261, 7958, 1986.

131. **Gupta, R. S.**, Senescence of cultured human diploid fibroblasts. Are mutations responsible?, *J. Cell. Physiol.*, 103, 209, 1980.

132. **Fulder, S. J. and Holliday, R.**, A rapid rise in cell variants during the senescence of populations of human fibroblasts, *Cell*, 6, 67, 1975.

133. **Moreley, A. A., Cox, S., and Holliday, R.**, Human lymphocytes resistant to 6-thioguanine increase with age, *Mech. Ageing Dev.*, 19, 21, 1982.

Chapter 7

THE PROLIFERATING CELL NUCLEAR ANTIGEN, PCNA, A CELL GROWTH-REGULATED DNA REPLICATION FACTOR

Michael B. Mathews

TABLE OF CONTENTS

I. INTRODUCTION

Although a eukaryotic cell is reckoned to contain several thousand different polypeptides, surprisingly often a protein known for one property and named accordingly turns out to be identical to "another" protein, defined by quite different properties and going by a different, equally apposite name. This leads to an unfortunate multiplicity of aliases, as well as to some fascinating insights. A recent case in point is provided by a relatively abundant nuclear protein that has been "discovered" independently at least three times and has excited interest in fields as diverse as immunology, cell cycle control, and DNA replication. The existence of the so-called proliferating cell nuclear antigen (PCNA) was first revealed by the staining of tissue slices with antibodies from autoimmune patients. The same protein was recognized a few years later during studies of cell cycle-regulated proteins, and it was named cyclin. It was also purified as a cofactor for an *in vitro* DNA synthesis reaction, and it then acquired its most descriptive styling as the DNA polymerase-δ auxiliary protein.

This article will employ the term PCNA to denote this protein, mainly on the ground of historical precedence, but also because the name "cyclin" is used for other proteins and reagents, and "DNA polymerase-δ auxiliary protein" is a cumbersome term (no disrespect intended to those who work with these entities). It is one of a number of proteins which are regulated in sympathy with cell growth rate, and it was given attention early by virtue of its abundance and unusual properties and by sheer good fortune.

II. PCNA AND CYCLIN

The initial sightings of these entities were made during survey work of differing kinds, and they led to the accumulation of two large but separate bodies of observations which exhibited a marked degree of correspondence.

A. DISCOVERY OF PCNA

During a survey of sera from patients with systemic lupus erythematosus, Miyachi et al.[1] discovered a new autoantibody specificity. It gave rise to a novel precipitin line in a double-diffusion test (Figure 1) and to nuclear staining by indirect immunofluorescence (Figure 2). Unlike most autoantibodies from lupus patients (reviewed by Tan[2] and Bernstein et al.[3]), this one reacted in a tissue-specific manner.[1] Extracts of rabbit thymus, mouse thymus, and Ehrlich ascites tumor cells were positive by immunodiffusion, whereas liver and kidney extracts were negative. Correspondingly, the antigen was detected by immunofluorescence in about half of the cells of several continuous lines, such as Hep-2 and Ehrlich ascites cells, but was not visualized in renal tubular or glomerular cells or in hepatic parenchymal cells. It was, however, seen in some cells in the interstitial regions of these tissues, where dividing cells are present. Indeed, in the thymus, areas containing dividing cells (such as the thymic cortex) were positive, whereas quiescent areas (such as the thymic medulla) were negative. An interesting correlation emerged: 20 to 50% of the nuclei reacted with antibody in tissues whose extracts reacted positively by immunodiffusion, whereas only a few nuclei reacted in tissues whose extracts were negative.

The association with some aspect of cell proliferation was cemented by observations made with lymphocytes. Normal peripheral blood lymphocytes were negative by immunofluorescence, but some lymphocytes in the germinal center of lymph node follicles and spleen were positive. Most conclusively, nuclear fluorescence appeared when resting lymphocytes were stimulated to reenter the cell cycle by mitogens such as phytohemagglutinin or other lectins.

Based on these findings, E. M. Tan's group coined the term "proliferating cell nuclear antigen" for the as yet unknown cellular target of the autoantibodies. Their observations

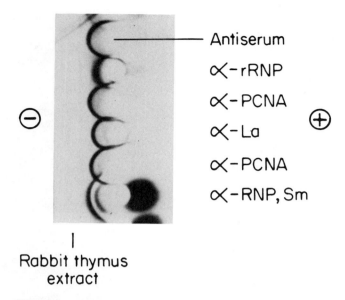

Antiserum

α-rRNP

α-PCNA

α-La

α-PCNA

α-RNP, Sm

Rabbit thymus
extract

FIGURE 1. Precipitin line formation with anti-PCNA antibody. Coun-terimmunoelectrophoresis (CIE), a sensitive form of double diffusion in which antigen and antibody are driven together in an electric field, reveals precipitin lines for PCNA and some other extractable nuclear antigens against which autoantibodies occur in patients.[96] Individual patients' sera were placed in the wells at the right, and a concentrated tissue extract serving as source of many antigens was placed in the well at the left. rRNP = ribosomal ribonucleoprotein. (Courtesy of Dr. C. C. Bunn.)

were confirmed and refined by a number of laboratories — with, for example, the demonstration that liver regeneration induced in rats by partial hepatectomy was accompanied by the appearance of PCNA immunofluorescence[4,5] — and were immediately seen to hint at the possibility that PCNA plays some regulatory role in DNA replication. Consistent with such a function, which would be expected to be conserved through evolution, the antigen is found in all mammalian species investigated (rat, mouse, hamster, and human) and even in the roots of flowing plants (cited in Nakane et al.[5]).

The first report on PCNA noted a speckled, nucleoplasmic distribution of staining, but later work revealed shifting patterns of nuclear staining that correlate with progression through the cell cycle.[6] After release from growth arrest by the addition of fresh medium, WiL-2 cells initially display speckled nucleoplasmic staining. Nucleolar staining becomes evident later in the first growth phase, G_1, and early in the DNA synthetic (S) phase. Nucleoplasmic speckles are often observed, and they strengthen during S phase at the expense of the nucleolar staining, which gradually disappears. During mitosis (M phase), fluorescence concentrates in areas devoid of condensed chromosomes. Similar results are obtained with WiL-2 cells synchronized by double-thymidine block and with mitogen-stimulated peripheral blood lymphocytes. Resting lymphocytes display no fluorescence, in contrast to WiL-2 cells, which are positive even in interphase; however, lymphocytes become positive for nucleolar fluorescence immediately after stimulation. Diffuse nucleoplasmic staining appears during S phase and gives way to a speckled pattern later in S phase. Such shifting patterns, discussed in more detail in Section V, were the first indication of the antigen's intimate relationship with DNA synthesis.

B. WHAT IS CYCLIN?

The term cyclin has, unfortunately, been applied to two completely different proteins, as well as to a proprietary antibiotic treatment used for ridding tissue cultures of mycoplasma

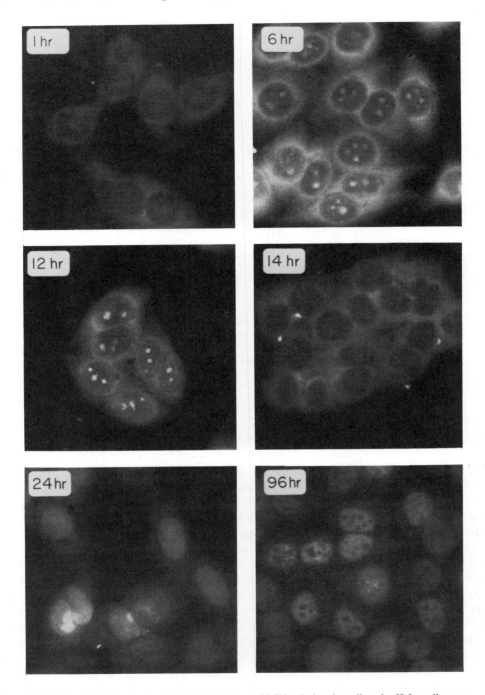

FIGURE 2.   Variations in the intracellular location of PCNA during the cell cycle. HeLa cells were synchronized by mitotic shake off, replated, and allowed to grow for the periods of time shown. DNA synthesis was first detected at about 6 h, reached a maximum at about 14 h, and then declined. Cells were fixed in methanol:acetone and were stained with human anti-PCNA antibody and fluorescein-labeled goat anti-human immunoglobulin. (Modified from Sadaie, M. R. and Mathews, M. B., *Exp. Cell Res.*, 163, 423, 1986. With permission.)

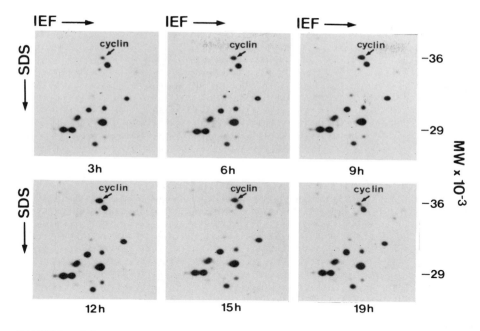

FIGURE 3. Cell cycle dependence of cyclin (PCNA) synthesis. HeLa cells were synchronized by the mitotic shake-off procedure, replated, and labeled at intervals for 1 h with ^{35}S-methionine. Equal amounts of radioactivity were analyzed by 2-D gel electrophoresis.[23] Only the relevant portion of the gel is shown. The isoelectric focusing (IEF) dimension was run with the acidic side to the right and the SDS dimension from top to bottom, as indicated by the molecular weight markers. Mid-S phase was 12 to 15 h after plating. (Reprinted by permission from *Nature*, Vol. 326, p. 515, Copyright 1987, Macmillan Magazines Ltd.)

(here spelled cycline). The cyclin protein which is *not* PCNA is found in fertilized sea urchin and clam eggs and behaves almost like a reflection of PCNA in that it disappears during M phase.[7] It also seems to play a role in oocyte maturation.[8]

In the present context, cyclin is a polypeptide first detected in 1980 by Bravo and Celis[9,10] while surveying changes in cellular proteins that could be attributed to oncogenic growth or transformation control. Their analytical method relied on the electrophoretic separation of labeled cell proteins in two-dimensional polyacrylamide gels. (An isoelectric focusing separation, based on net charge, is followed by an SDS gel separation, based on molecular weight and conducted at right angles to the first gel.) Such gels are capable of resolving several thousand polypeptides in a mammalian cell extract, each appearing as a spot. Early work resolved fewer polypeptides (about 700), of which very few changed significantly in their synthesis during the cell cycle or in their level when normal cells were compared with their virus-transformed counterparts. One exception, which stood out in part because of its location toward the acidic edge of the pattern of spots (Figures 3 and 4), was a polypeptide with an apparent molecular weight of 35 to 36 kDa and an isoelectric point of 4.8 to 4.9. The spot, numbered IEF49 in humans and in various other ways in catalogues of polypeptides from other species, was named "cyclin" when it seemed likely that it represented a single ubiquitous protein with cell cycle-related properties.[11,12]

The protein then became the subject of an exhaustive series of investigations by these workers and their colleagues; these studies have been reviewed on several occasions.[12-16] Similar studies of PCNA have been completed by Franza and Garrels,[17,18] and other laboratories have also reported on the behavior of proteins that in all probability are also PCNA.[19-21] The main conclusions of this large body of work can be distilled as follows.

Cyclin (PCNA) is found in a large number of tissues and cell types from a variety of

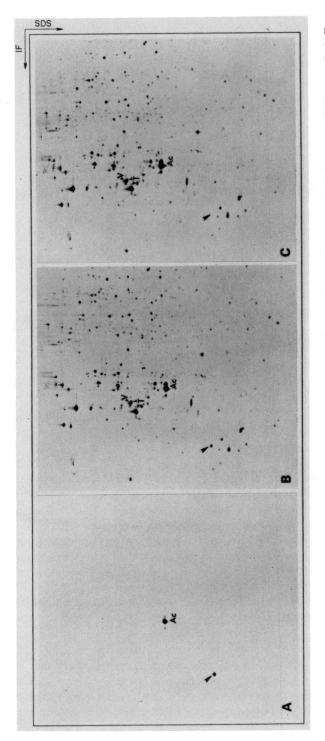

FIGURE 4.   Antibodies against PCNA react with cyclin (PCNA). An ^{35}S-methionine-labeled extract of HeLa cells was treated with human anti-PCNA autoantibody,[22] and the immune complexes were collected with *Staphylococcus aureus* protein A and examined by 2-D gel electrophoresis (A). The PCNA spot was marked with an arrowhead. Untreated cell extract (B) and antibody-treated extract (C) were analyzed in the same way; the cyclin (PCNA) spot is missing from the latter. Note that the acidic side of the isoelectric focusing dimension (IF) is on the left in these gels. Abbreviations: Ac, actin; V, vimentin; T, tubulin. (Reprinted by permission from *Nature*, Vol. 309, p. 374, Copyright 1984, Macmillan Magazines Ltd.)

vertebrates, and it is almost entirely nuclear. After enucleation of HeLa cells with cytochalasin B, it is found nearly exclusively in karyoplasts, while only traces are present in cytoplasts.[11] When cells are disrupted by homogenization, the protein is largely nuclear but not exclusively so, presumably because some leakage occurs.[22] PCNA evidently also leaches from nuclei isolated by detergent treatment of cells.[22]

The synthesis and possibly the accumulation of cyclin (PCNA) are regulated through the cell cycle and correlate temporally with DNA replication.[9,23] Cell cycle-dependent synthesis is illustrated particularly clearly in HeLa cells synchronized by mitotic shake off (Figure 3). Noncycling cells such as peripheral blood lymphocytes, senescent human skin fibroblasts, and adult mouse liver cells make little of the protein, whereas cycling cells in mouse intestine or early-passage human skin fibroblasts make it in five- to sixfold greater amounts.[11] Cyclin (PCNA) synthesis in quiescent mouse 3T3 cells is increased by treatment with serum of other growth factors which stimulate DNA replication.[24] Conversely, X-irradiation of HeLa cells[11] decreases the fraction of cells synthesizing DNA by a factor of 4 and the synthesis of cyclin (PCNA) by a factor of 6. Similarly, when epidermal growth factor (EGF) paradoxically inhibits the proliferation of A431 human carcinoma cells, cyclin (PCNA) synthesis decreases concomitantly.[25] EGF-resistant cells do not show this response. These observations, which are considered more fully in Section VI, obviously run parallel to phenomena seen by immunofluorescence with anti-PCNA serum.

Further evidence linking cyclin (PCNA) to cell proliferation has come from studies of transformed cell proteins. Various lines of transformed cells produce more of the protein than their nontransformed patients. The magnitude of the increase ranges from 1.7-fold for SV40-transformed 3T3 cells[10] to 9.3-fold for human diploid lung fibroblasts (WI-38) transformed by the same virus.[26] For a given parental cell line, the magnitude of the increase can also vary depending on the transforming agent used,[17,27] and it may be that elevated cyclin (PCNA) synthesis is associated with cell proliferation rather than with transformation per se (as discussed in Section VI.C.1).

In summary, the synthesis of cyclin (PCNA) reflects the DNA synthetic status of the cell reasonably closely. It is barely detectable or undetectable in quiescent cells and comprises on the order of 0.1% of the total protein synthesized in most dividing cells.[14] (The MOLT-4 line of T-cell leukemia cells is especially rich in this protein: it comprises over 0.6% of the total label incorporated.) With these data in hand, it would be hard to resist speculating that cyclin "could be a central component of the pathway(s) that regulate cell proliferation and that its activity may be associated with events related to DNA replication",[14] an idea remarkably similar to that which occurred to the PCNA workers. As will be seen, its association with replication is now clear, but its regulatory role is more problematic.

C. PCNA = CYCLIN

In retrospect, it is not surprising that PCNA and cyclin (PCNA) are one and the same. The two bodies of information describing these proteins emerged from different disciplines, clinical immunology and cell biology, and did not coalesce until 1984, when Bernstein et al.[3] coupled a survey of autoimmune patients with a study of the extractable nuclear antigens (ENAs) that react with their antibodies. Antibodies against PCNA, found in a minority of lupus patients, recognized a polypeptide with an apparent molecular weight of ca. 35,000 by immunoprecipitation.[3] When the immunoprecipitated polypeptide was examined by two-dimensional gel electrophoresis, it gave a spot precisely coincident with the cyclin spot, suggesting that the two entities might be identical (Figure 4). The unlikely possibility that two different proteins occupied the very same position in the two-dimensional array of polypeptides was eliminated by showing that the cyclin spot, and only this spot, was absent from an extract that had been treated with anti-PCNA antibody. Furthermore, PCNA (isolated by immunoprecipitation) and cyclin (isolated from two-dimensional gels) gave rise to very similar peptides when digested with V8 protease, indicating that they are composed of the same peptides.[22]

These data unambiguously established the identity of cyclin with PCNA. Corroborative evidence can be drawn from a variety of observations: for example, partially purified PCNA exhibits a similar molecular weight and isoelectric point to cyclin,[28] and murine monoclonal antibodies prepared against PCNA recognize the cyclin spot in a two-dimensional immunoblot.[29] Their identity, now sealed by the sequence data and by their joint identification with the DNA polymerase-δ auxiliary protein (Sections III and IV.B), offers a satisfying rationale for the correspondence between the biological properties of these two entities. Nevertheless, some puzzling observations remain to be explained. In quiescent cells, PCNA is generally undetectable by immunofluorescence, yet the protein may be present and may be made at a detectable rate. Is this a paradox, or is it merely a misleading observation due, perhaps, to difficulty in detecting fluorescence signals of low intensity? Is this finding related to the observation that PCNA labeling falls less than the thymidine labeling index as cells senesce?[30] Solutions to these conundrums are discussed in Section V.

III. STRUCTURE OF PCNA AND ITS mRNA

Partial sequence of the PCNA protein was obtained by direct methods of polypeptide analysis, and its full sequence was deduced through cDNA cloning. The sequence of amino acids at the N-terminus of PCNA, whether defined on the basis of its immunological reactivity with autoantibodies, its two-dimensional gel position, or its DNA synthesis activity (see Section IV.B), has been determined by chemical microsequencing.[23,31,32] The human and rabbit proteins are identical for their first 26 residues, correspond to the bovine protein for at least 10 residues, and only differ from the rat sequence (predicted from the cDNA) at one position, residue number 7.

The N-terminal amino acid sequence was exploited by Matsumoto et al.[33] to clone PCNA cDNA from a rat thymocyte library. They screened the library by hybridization with a 32-fold redundant oligonucleotide corresponding to the first five amino acid residues of the protein. A different strategy was adopted by Almendral and colleagues[34] to clone the human cDNA. These workers first prepared an antibody in rabbit against PCNA purified from MOLT-4 cells by two-dimensional gel electrophoresis. They used the resulting anti-PCNA antibody to screen an expression library from the same cells, and they then employed the initial clone as a probe to isolate a longer cDNA.

The cDNAs contain the entire coding sequence of 783 coding nucleotides (excluding the stop codon) and appear to be close to full length. Their nucleotide sequences are quite widely divergent, particularly in the untranslated regions. The human clone contains 118 untranslated nucleotides at its 5' end and 330 at its 3' end for a total of 1231 nucleotides (excluding the polyA tail). This approaches the experimentally determined size of 1.3 kb for the human mRNA. The rat clone has 62 nucleotides upstream and 315 nucleotides downstream for a total of 1160, which compares to 1.1 and 0.98 kb for the two forms of rat mRNA observed. The shorter mRNA may result from the use of one or both of the two additional upstream polyadenylation signals that are found in the rat cDNA. Nothing is yet known of gene structure, although Southern blots are consistent with the existence of a single major gene in both rat and human.

From their cDNA sequences, both the rat and human proteins are predicted to contain 261 amino acids and to have a molecular weight of about 29,000. The predicted size is considerably less than values estimated by SDS-polyacrylamide gel electrophoresis (SDS-PAGE), which range from 33,000 to 37,000 and are generally between 35,000 and 36,000. The discrepancy may stem from the predominance of acidic residues that give rise to the protein's acidic charge and can reduce the binding of SDS. Since the product of cell-free translation of PCNA mRNA has a gel mobility indistinguishable from that of the natural protein,[35] it is unlikely that post-translational modifications are responsible for the anomalous gel mobility. Modifications that would result in altered intrinsic charge have not been detected by labeling studies, although they cannot be completely excluded; such modifications include acetylation, glycosylation, phosphorylation, and sialylation.[36] The weak phosphate labeling observed in one-dimensional analyses of immunoprecipitates containing PCNA[3] is probably attributable to a contaminating phosphoprotein in the precipitate and is not responsible for the acidic "satellite" spot seen in some two-dimensional gel separations.[22]

In overall amino acid composition, the predictions are in good agreement with the experimentally determined composition of rabbit PCNA,[31] except that direct analysis has given low values for serine and lysine. PCNA is well conserved evolutionarily at the sequence level as well as in its immunoreactivity, physiological behavior, and other properties. Between human and rat PCNA, there are only four amino acid changes (a frequency of 1.5%), and two of these are conservative. The changes are as follows: Val → Ile at position 7, Ser → Gly at position 33, Thr → Ser at position 196, and Ser → Pro at position 223. At the nucleic acid level, however, comparison of the rat and human sequences reveals a considerable number of nucleotide changes, amounting to about 10% of the coding nucleotides. As might be expected, the overwhelming majority of the nucleotide changes are in the third positions of their codons.

What can be learned from the amino acid sequence? A region between residues 66 and 80 resembles the helix-turn-helix motif that is a feature of many DNA-binding proteins.[34] In a number of transcription-regulating proteins, α-helical regions are thought to interact with the major groove of the DNA helix. Despite the firm attachment of a fraction of PCNA to nuclei, from which it can be liberated by digestion with DNAse,[35] PCNA does not bind to affinity matrices containing either single- or double-stranded DNA.[32,37] These findings suggest that in the cell it probably does not interact directly with DNA but may do so indirectly, possibly via interactions with other proteins. Weak sequence similarities have also been noted with two viral proteins that are connected with cell cycle regulation.[33] The first is the product of the adenovirus-5 E1A gene's 12 S mRNA, a polypeptide of 243 amino acids that does not seem to bind DNA but does play a role in cell transformation and immortalization (see Section VI.C.2). The second is a herpesvirus-1 protein known as ICP8 (infected-cell polypeptide 8). This is a major HSV-1-encoded DNA-binding protein required for viral DNA synthesis. Its molecular weight is 130 kDa, and the N-terminal one fifth of the protein exhibits similarities to the PCNA sequence. Interestingly, ICP8 moves between different intranuclear locations in a fashion reminiscent of PCNA (see Section V below). Prior to replication, it occupies discrete, punctate sites throughout the nucleus that are revealed by immunofluorescence.[38] During active viral DNA synthesis, most of it migrates to the less well-defined areas where viral DNA replication takes place, but some remains in the prereplicative sites and can be clearly visualized when the replicating molecules and accompanying diffuse fluorescence are released by DNAse. Provocative though these similarities are, their significance remains to be established, and it is doubtful whether PCNA's function would have been arrived at with any speed by following such leads.

IV. FUNCTION OF PCNA

In the event, the discovery of PCNA's role in DNA synthesis came from a quite different direction. Before discussing this, it will be useful to consider the salient features of the mammalian DNA polymerases and, particularly, to introduce DNA polymerase-δ and its auxiliary protein. The subject of DNA replication itself has been reviewed recently.[39]

A. DNA POLYMERASE-δ AUXILIARY PROTEIN

Four distinct DNA polymerase activities have been detected in extracts of mammalian cells (for reviews, see Fry and Loeb[40] and Kelly and Stillman[41]). Their chief properties are summarized in Table 1.

Until recently, DNA polymerase-α was generally assumed to be the sole polymerase involved in replicative DNA synthesis, partly because of its sensitivity to the inhibitor aphidicolin. The sensitivity of the enzyme *in vitro* matches the sensitivity of cellular and SV40 viral DNA synthesis to the drug *in vivo*. This evidence can no longer be taken as unequivocal proof, since polymerase-δ is at least as sensitive to aphidicolin as polymerase-α; however, the involvement of polymerase-α is nonetheless well founded. Its removal by immunodepletion from extracts capable of SV40 DNA replication *in vitro* leads to a block in replication activity, which is restored by addition of the DNA polymerase-α-primase complex from an appropriate source.[42] Antibodies to polymerase-α also block cellular DNA synthesis in isolated nuclei[43] or when injected into cells.[44] Furthermore, cells with a temperature-sensitive form of DNA polymerase-α fail to replicate their DNA at the nonpermissive temperature.[45]

On the other hand, as pointed out recently by Downey and co-workers,[46] DNA polymerase-α lacks some of the attributes expected of the replicative enzyme. For example, polymerase-α is only moderately processive and is incapable of strand-displacement synthesis. Mounting evidence suggests that the missing features may be supplied by DNA polymerase-δ, which could function jointly with polymerase-α to achieve coordinate synthesis at the replication fork. DNA polymerase-δ is the most recently discovered and the least studied of the mammalian polymerases known to date, but it is now attracting a great deal of attention.[46-64] It is distinguished from polymerase-α, which it resembles in its sensitivity to aphidicolin, by a number of physical and chromatographic properties, by its lesser sensitivity toward modified nucleotides, and by its template specificity. Antibodies against the two polymerases have been prepared, and each type selectively blocks the activity of its target polymerase. Significantly, polymerase-δ lacks primase activity, but it possesses a 3′-to-5′ exonuclease activity which may contribute to the fidelity of DNA replication by providing a proofreading function[55] and may also lend itself to a role in DNA repair.[56,57]

Most relevant here is the distinction that DNA polymerase-δ, but not polymerase-α, is stimulated by an auxiliary protein now known to be PCNA.[37] This protein greatly enhances the ability of polymerase-δ to utilize templates that contain long single-stranded regions. These include bacteriophage M13 DNA templates annealed to a short primer and synthetic substrates such as poly(dA) annealed to a small amount of oligo(dT) primer (e.g., at an A:T ratio of 20:1). When the poly(dA) template is progressively saturated with oligo(dT) primer, the stimulation of polymerase-δ activity by the auxiliary protein declines. Similarly, gapped calf thymus DNA and synthetic templates that are predominantly double stranded are much less dependent on the auxiliary protein, although their dependence becomes more apparent when the assay is conducted at higher Mg^{2+} ion concentrations. No DNA-binding or nuclease activities seem to be intrinsic to the auxiliary protein, but it increases the DNA-binding activity of polymerase-δ.

In this and other respects, polymerase-δ and its auxiliary protein resemble the *E. coli* replicative DNA polymerase III core enzyme and its β-subunit. These similarities, which extend as far as the physical properties of the β-subunit and the polymerase-δ auxiliary protein (both are dimers of 75,000 daltons), lend further support to the idea that polymerase-δ plays a role in DNA replication *in vivo*. It was these physical properties which initially suggested that the auxiliary protein and PCNA might be one and the same.

B. PCNA IS A DNA REPLICATION FACTOR

The first direct evidence that PCNA plays a part in DNA synthesis came from the fractionation of a cell-free system capable of replicating SV40 DNA.[32] The system is derived from human 293 cells and when supplemented with SV40 large T (tumor) antigen catalyzes

TABLE 1
Some Properties of Eukaryotic DNA Polymerases[a]

	DNA polymerase			
	α	β	γ	δ
Relative activity[b]	70—90%	10—15%	1—10%	10—25%[c]
Intracellular localization	Nucleus	Nucleus	Mitochondria and nucleus	?
Native size (kDa)	200—600	40	180—315	122—290
Subunits	Several	1	Several	1—2
3'→5' exonuclease activity	([d])	No	No	Yes
RNA polymerase (primase) activity	Yes	No	No	No
Preferred primer-template	Gapped DNA	Gapped DNA	Oligo(dT):poly(dA)	Depends on PCNA
Processivity	Quasi-processive	Distributive/ slightly processive	Highly processive	Stimulated by PCNA
Aphidicolin sensitivity	Yes	No	No	Yes
Possible functions	Replication, repair	Repair	Replication (mitochondria)	Replication, repair

[a] From Fry and Loeb[40] and polymerase-δ works cited in text.
[b] In dividing cells.
[c] Of the total (α + δ).
[d] Cryptic activity reported in *Drosophila* enzyme.[98]

the faithful, origin-dependent replication of the template. The replication system has been separated into three chromatographic fractions, all of which are essential for replication activity. Two of them contain topoisomerases, the DNA polymerase α/primase complex, and an additional DNA-binding protein, among other factors not yet fully character-ized.[58-60] The replication factor present in the third fraction has been purified to apparent homogeneity and turns out to be identical to PCNA (purified on the basis of its autoantibody reactivity) in its chromatographic behavior on several column matrices as well as in its subunit molecular weight (Figure 5A).[32] PCNA (defined immunologically) can substitute for the replication factor, each sediments as a dimer, and they have the same N-terminal amino acid sequence (which closely resembles that previously published for rabbit thymus PCNA.[31]). Furthermore, the replication activity of an unfractionated system is eliminated by depletion with antibodies to PCNA, and activity is substantially restored by adding back the purified replication factor. (The fact that restoration is not complete may suggest that an additional replication factor is removed in the depletion; possibly it is complexed with PCNA.) Finally, the characteristic speckled nuclear immunofluorescence pattern given by anti-PCNA antibodies is quenched by the replication factor.[32]

These data prove that PCNA functions in replication, at least *in vitro,* and they also link the original immunocytological description of PCNA to its biochemical action. The connection to DNA polymerase-δ was made when its auxiliary protein was purified and characterized.[37] The identity of the polymerase-δ auxiliary protein with PCNA was estab-lished by physicochemical and immunological tests, such as N-terminal sequencing, gel mobility, and antibody reactivity.[23,61] What is more, the two are interchangeable, both in the SV40 replication system and in the poly(dA):oligo(dT)-templated assay for polymerase-δ function (Figure 5).[61] Though the present results fall somewhat short of a formal proof that PCNA is involved in cellular DNA replication, little reasonable doubt remains. What doubts persist are largely dispelled by the immunocytological findings described above, the detailed chronology and subcellular localization of PCNA immunofluorescence to be dis-cussed below (Section V.A), and the growing evidence of polymerase-δ's involvement in replication.[56,62,63]

C. THE ROLE OF PCNA IN DNA SYNTHESIS

PCNA's role in the process of DNA synthesis has been studied by examining the kinetics and products of reactions performed in its absence. Normal SV40 replication reactions exhibit a lag period of about 10 min before any polymerization of nucleotides can be detected. This lag is interpreted as the time required for formation of a "presynthesis complex" at the origin, as it can be overcome by preincubating the constituents together before adding nucleotides. By leaving out different fractions from the preincubation reaction, Prelich et al.[32] showed that PCNA is *not* such a component, so it must act at a later stage of DNA synthesis than formation of the initiation complex, presumably during elongation.

On a synthetic template, poly(dA):oligo(dT), polymerase-δ produces DNA chains less than 30 nucleotides long in the absence of PCNA, but chains over 200 nucleotides long are made in its presence (Figures 5C,D).[61] Similarly, in the absence of PCNA, SV40 DNA replication in a human cell extract produces chains only 100 to 500 nucleotides long.[60,64] Surprisingly, these progeny DNAs hybridize all around the template molecule but almost exclusively to the noncoding strand (the one with the same polarity as mRNA). Thus, PCNA is required for (continuous) leading strand synthesis but not for (discontinuous) lagging strand synthesis.[64] This observation is consistent with the action of PCNA in model systems, in which it increases the processivity of DNA polymerase-δ.[37] It also fits with the association of polymerase-α with a primase activity in a complex which could well be responsible for the synthesis of the discontinuous (Okazaki) fragments on the opposite strand. These ideas are summarized in the speculative scheme depicted in Figure 6, which integrates both polymerases into the action at the replication fork.

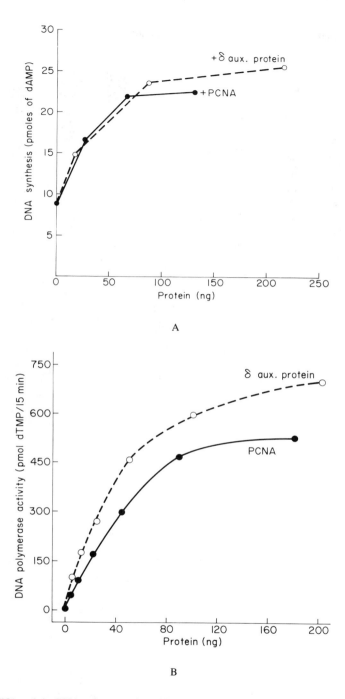

A

B

FIGURE 5. PCNA and the DNA polymerase-δ auxiliary factor are interchangeable and allow synthesis of long DNA chains. The dependence of the fractionated SV40 replication system (A) and of the polymerase-δ reaction (B) on human PCNA or the calf auxiliary protein are shown graphically. In the polymerase-δ reaction, ³H-TMP is condensed onto oligo(dT) primer in the presence of a poly(dA) template. Gels C and D, which differ in their polyacrylamide concentration, demonstrate the synthesis of short chains in the absence of PCNA (lanes 1) and longer chains in the presence of auxiliary protein (lanes 2) or PCNA (lanes 3). Lanes 4 contain marker fragments. (Reprinted by permission from *Nature,* Vol. 326, p. 517, Copyright 1987, Macmillan Magazines Ltd.)

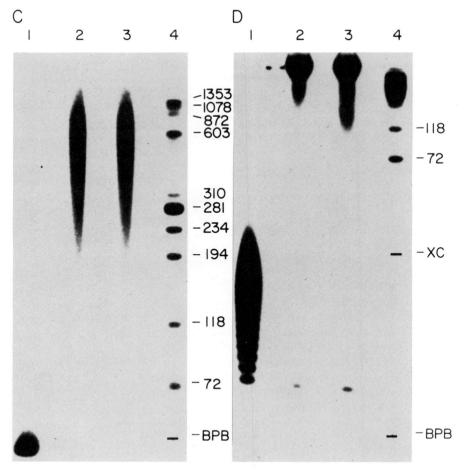

FIGURE 5 C, D.

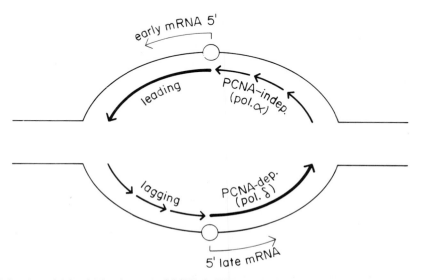

FIGURE 6.   A model for the involvement of PCNA in DNA replication. Schematically illustrated is an SV40 replication "eye", showing leading (continuous) strand synthesis mediated by DNA polymerase-δ and dependent on PCNA (thick arrows). This DNA is synthesized in the same sense as mRNA. Discontinuous (lagging) strand synthesis on the opposite strand is believed to be mediated by DNA polymerase-α (broken, thinner arrows).

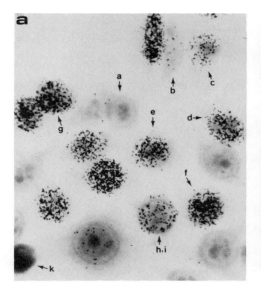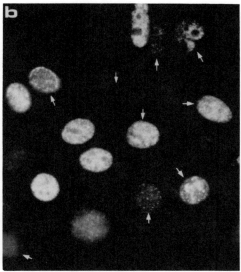

FIGURE 7. S-phase nuclei stain with anti-PCNA antibody. Asynchronous AMA cells were pulse labeled with ³H-thymidine for 30 min and fixed with methanol. Following detection of PCNA by indirect immunofluorescence (b), thymidine incorporation was visualized by emulsion autoradiography (a). The two photographs represent the same field of cells. Small letters indicate different patterns of PCNA distribution, as in Figure 9. (From Celis, J.E. and Celis, A., *Proc. Natl. Acad. Sci. U.S.A.*, 82, 3262, 1985. With permission.)

V. INTRANUCLEAR LOCALIZATION OF PCNA

Early work with autoantibodies directed against PCNA demonstrated that this antigen gives rise to nuclear staining that varies through the cell cycle in its pattern as well as in its prevalence. Studies of the protein's biosynthesis initially seemed to correlate well with the information derived by immunofluorescence, and they led to the striking visualization of PCNA localizing at replication sites but also brought to light the puzzling fact that PCNA can exist in forms that may go undetected by immunofluorescence.

A. PCNA IS ASSOCIATED WITH REPLICATION COMPLEXES *IN VIVO*

Through the use of emulsion autoradiography to reveal the sites of ³H-thymidine incorporation in mixed cell populations, it has been shown that those nuclei which stain with anti-PCNA antibody are the ones which are actively synthesizing DNA.[65,66] This has been demonstrated in mouse 3T3 cells and in human amnion cells (Figure 7), and similar results have been obtained in other cell types by cell sorting techniques.[67,68] The silver grains representing the positions of nuclei which have incorporated ³H-thymidine correlate well with immunofluorescence in their intensity and in their distribution. This is true even as asynchronous heterokaryons containing nuclei at different stages of the cell cycle within the same cytoplasm.[69]

Careful examination of individual nuclei shows that the distribution of silver grains sometimes matches quite closely the distribution of PCNA staining within the nucleus (Figure 8).[70] This is especially striking when the nucleolar regions are either intensely labeled or devoid of label. The patterns are particularly clear when the DNA is pulse labeled with BrdU instead of ³H-thymidine and is detected with a second antibody,[71] a method which gives better resolution than autoradiography. Such topographical concordance implies that PCNA is located at sites of active DNA synthesis, but the exceptions in which no such concordance is apparent are not easily explained.

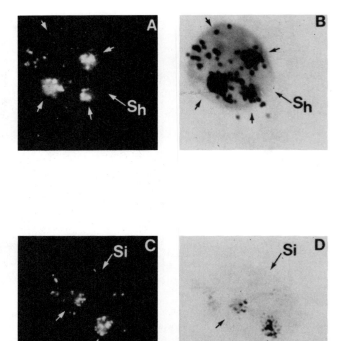

FIGURE 8.   Co-localization of PCNA with sites of DNA synthesis. Monkey kidney BSC-1 cells were labeled with ^{3}H-thymidine and examined by indirect immunofluorescence (A, C) and autoradiography (B, D), as in Figure 7. S_h and S_i refer to different patterns of PCNA and ^{3}H-thymidine distribution during S phase. (From Madsen, P. and Celis, J. E., *FEBS Lett.*, 193, 5, 1985. With permission.)

Cyclin distribution

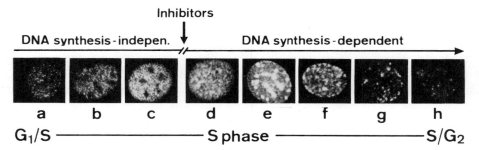

FIGURE 9.   Putative sequence of changes in PCNA distribution during S phase. The sequence was deduced from immunofluorescence studies of serum-stimulated 3T3 cells in the presence or absence of hydroxyurea or aphidicolin and after release from hydroxyurea blockage. (From Bravo, R. and Macdonald-Bravo, H., *EMBO J.*, 4, 655, 1985. With permission.)

In transformed human amnion (AMA) cells and mouse 3T3 cells synchronized by mitotic shake off, the pattern of PCNA staining shifts as the cells progress through S phase (Figure 9). The staining patterns have been arranged into a putative sequence, beginning with weak nuclear staining in G_1.[65,66] Early in S phase, granular nucleoplasmic immunofluorescence appears, then strengthens and becomes punctate. At the time of peak DNA synthesis, the staining moves to the nucleoli before again becoming punctate and subsequently declining. By the end of S phase, it has disappeared. Similar sequences are seen in several other cell lines, e.g., MOLT-4,[72] and correlate with the observation that nucleolar replication occurs relatively late in S phase in these cells.

This sequence of shifting PCNA patterns is not invariable, however. In African green monkey kidney cells (BSC-1), intense nucleolar staining is maintained from mid-S phase until it fades out late in S phase.[70] Furthermore, nucleolar staining, rather than nucleolar exclusion, has been observed as the earliest pattern (Section II.B): the nucleolar immuno-fluorescence begins late in G_1, preceding the onset of DNA synthesis in WiL-2 and HeLa cells and in mitogen-stimulated lymphocytes.[6,35] The discrepancy does not seem to be due to contaminating antibodies in the human sera, as a monoclonal anti-PCNA antibody gives similar results to those obtained with a patient's antibody,[73] but it is likely a consequence of differences between cell lines. The fixation conditions should also be considered, as distinctly different patterns are obtained with different protocols[71,74] (see below, Section V.B). Additional, distinct fluorescence patterns have also been reported on occasion; these include staining of foci near the nuclear rim and a bright punctate pattern that may characterize cells that have withdrawn from the cycle.[70] Using formaldehyde fixation, Nakane et al.[5] have described the migration of PCNA from the cytoplasm to the nuclei and then into the Golgi region; it is not clear why the protein should be associated with the secretory apparatus.

The changes in fluorescence patterns are dependent on DNA synthesis but do not result simply from an increase in PCNA synthesis. As discussed below (Section VI.A.4), hy-droxyurea and aphidicolin inhibit DNA synthesis without affecting the increase in PCNA synthesis that occurs at the G_1/S border (though PCNA accumulation has not been measured directly). In the presence of these drugs, the early stages of PCNA immunofluorescence are seen, but the cells do not progress beyond the stage of speckled nucleoplasmic staining (Figure 9).[65] When the drug is washed away, DNA synthesis resumes and the cells progress through the later stages of fluorescence. At the end of S phase, the fluorescence disappears, even though PCNA synthesized during the block by hydroxyurea is still present at undiminished levels. Thus, PCNA fluorescence can be detected in the absence of detectable DNA synthesis, and it disappears even though the protein is still present. Observations such as these suggest that PCNA is detected by the immunofluorescence technique only when it is part of a replicative complex. Perhaps these complexes can form on the DNA template even when replication is blocked by drugs. We will return to these issues in the next section.

A second line of evidence pointing to a physical association between PCNA and cellular DNA comes from studies of its release from HeLa cell nuclei. Immunoblotting has shown that a fraction of PCNA is tightly bound to nuclear components and is only released at high ionic strength (0.6 M NaCl) or by digestion with DNAse, but not by digestion with RNAse.[35] PCNA liberation by DNAse is rapid and is maximal when about a third of the DNA has been solubilized. Since a substantial fraction of PCNA is released during nuclear isolation (about 50% by the nonionic detergent procedure) but the remaining PCNA is tightly associated with the nuclei in a DNAse-sensitive form, these results imply the existence of different populations of PCNA.

B. TWO POPULATIONS OF PCNA

Despite the impressive concordance between PCNA synthesis and immunofluorescence, a number of anomalies suggest that the relationship may not be as simple as expected. First, PCNA is present in cells and is synthesized (albeit more slowly) at stages of the cell cycle (G_1, G_2, and M) when it is barely detectable by immunofluorescence. Second, immunofluorescence appears in the G_1 or G_2 phase after human amnion cells have been irradiated with X-rays to induce DNA damage and repair, even if new protein synthesis is prevented by treatment with cycloheximide (cited in Celis and Madsen[75]). Third, enucleated amnion cells are reported to synthesize PCNA, although they do not fluoresce (cited in Celis et al.[15] and Celis and Celis[66]). Fourth, digestion of fixed, detergent-treated BSC-1 cell nuclei with DNAse abolishes their immunofluorescence without releasing all the PCNA.[70]

In view of the likely existence of different populations of PCNA mentioned above, it seems that some of the PCNA is not detected by immunofluorescence even though it can react with antibody. Several possible explanations for such a circumstance can be entertained:

1.   Only newly synthesized PCNA can react with antibody.
2.   The reactive epitope(s) are only accessible to antibody when PCNA is in some special conformation or is part of some special structure (complexed with DNA, for example).
3.   The epitopes can be masked by combination with other components which sequester PCNA when it is not participating in DNA replication.
4.   Bright fluorescence is seen only when the antigen accumulates in foci at clusters of replication sites.
5.   Some PCNA is lost or denatured during sample preparation for immunofluorescence.

We do not yet have enough information to decide unequivocally among these possibilities, but some pertinent comments can be made. The first hypothesis, that only newly made PCNA gives rise to fluorescence, is incompatible with the observation (admittedly not extensively documented) that inhibition of protein synthesis does not prevent the increase in PCNA fluorescence that follows DNA damage in non-S-phase cells.[75] The second hypothesis, which supposes a requirement for a specific ligand or conformation, also seems unlikely to be correct, since the pure protein is recognized by antibody even after denaturation and transfer to a solid support in an immunoblot. It is, however, still conceivable that a fraction of the antigen molecules renature and adopt the correct, immunoreactive conformation after blotting. The fact that two monoclonal anti-PCNA antibodies exhibit essentially the same staining characteristics as the autoantibodies even though they recognize different regions and forms of the molecule, such as denatured conformations (see Section VII.B), strengthens the case against this hypothesis. It also argues against the third hypothesis, which assumes that inactive PCNA is shielded from antibody by other proteins, but does not eliminate the idea entirely. On the other hand, no such ligand has been detected to date, and extractable PCNA is fully immunoprecipitable.[22] Little can be said about the notion that fluorescence is accentuated by the concentration of antigen in discrete spots (hypothesis 4). The failure of DNA digestion to release all PCNA despite abolishing fluorescence[70] could support hypothesis 4, or indeed hypothesis 2, but this observation poses a dilemma in its own right. Since the antibodies can detect PCNA in a variety of conformations, including denatured conformations, it is far from clear how the residual PCNA reportedly present after DNAse digestion could escape detection and whether a great deal of significance should be attached to it. Therefore, it remains possible that PCNA fluorescence is enhanced when it is actively participating in replication complexes or associated with DNA.

The final proposition has been investigated in detail recently.[71] By immunoblotting, growth-arrested mouse 3T3 cells have been found to contain 30 to 40% as much PCNA 24 to 48 h after entering G_0 as growing cells. This reflects the long half-life of PCNA (ca. 20 h) and a continuing, though reduced, rate of production of the protein (10% of that of growing cells). The presence of PCNA in growth-arrested cells, while undetectable by immunofluorescence after methanol fixation, has been seen in the nuclei of cells fixed with formaldehyde and permeabilized by Triton® X-100. Under such fixation conditions, resting cells gave a homogeneous nuclear fluorescence pattern, sparing the nucleoli, upon which were superimposed intensely stained areas in the nuclei of growing cells during DNA synthesis. Only these intensely stained areas were seen if the growing cells were extracted with Triton® X-100 before formaldehyde fixation, a treatment which abolished the staining of growth-arrested cells. Based on immunoblotting and radiolabeling experiments, detergent removes 70 to 80% of the PCNA from cells in S phase, but all of it from quiescent cells. Methanol extraction does not extract PCNA, however.

These data suggest the existence of two populations of PCNA molecules: those associated with replication, which give intense fluorescence, are seen only in S phase, and survive several fixation protocols; and those not associated with replication, which are present throughout the cell cycle and are widespread in the nucleoplasm, but are loosely bound and removed by detergent treatment prior to fixation. For some reason, only the tightly bound population retains the ability to react with antibody after methanol fixation. The distribution of these molecules resembles the distribution of replication centers. It seems that newly synthesized PCNA enters a pool of "free" molecules which support DNA replication by forming a transient but stable association with DNA, presumably at the replication point.

It is not clear whether the half-lives of the two populations of PCNA are the same (see Section VI.A.1) or whether molecules which have participated in replication rejoin the pool and can be recruited from the nonreplicative into the replicative population, although the redistribution after X-irradiation would be consistent with such an event. On the basis of the two-population idea and assuming a half-life of about 20 h, cells entering G_0 would initially contain a full complement of PCNA but would be negative by immunofluorescence; as time went on, they would gradually lose their PCNA by attrition, remaining negative by immunofluorescence. Stimulation by serum would then generate a burst of PCNA synthesis to restore its concentration to a functional level; it is not known whether 100% of the full normal complement is required or whether replication can occur in the presence of lower concentrations. During a growth cycle of ca. 20 h the cells would need to produce 150% of the normal complement in each cell cycle, 50% to restore losses due to turnover and 100% to supply the two daughter cells. Conceivably, the supply of these restorative and replicative fractions could be achieved by transcription from different promoters or by regulation operating through different transcriptional elements. Maintenance of the steady-state level might be constitutive in cycling cells (though blocked in quiescent cells) and might operate during all phases of the cell cycle, while the generation of additional PCNA could operate selectively late in G_1 and during S phase and could involve more sophisticated growth-regulated pathways.

VI. REGULATION OF PCNA SYNTHESIS

PCNA synthesis is coordinated with DNA synthesis, but how closely are the two linked? And at what level is PCNA synthesis controlled? In any discussion of these questions, it is important to draw a careful distinction between the "normal" cell cycle in continuously proliferating cells and the stimulation of proliferation in quiescent cells by addition of serum or growth factors. The latter involves a transition from G_0 to G_1 and then to S phase, and it may be accompanied by biochemical actions of the stimulatory factors that are irrelevant to the cell cycle. To understand the mechanisms whereby PCNA regulation is achieved, it is also important to bear in mind that the synthesis of a protein, measured by a brief labeling with radioactive amino acid, need not relate in a simple way to its cellular concentration or to its ability to be detected by immunofluorescence (as we have just seen), although these parameters are all clearly interconnected. Most of the data currently available come from studies of tissue culture cell lines which proliferate indefinitely, and it is possible that regulatory processes may differ in cells of limited life span.

A. STIMULATION OF QUIESCENT CELLS
1. Synthesis, Turnover, and Level of PCNA

PCNA is not detected in mature, resting cells or tissues by immunodiffusion[1] or by immunoblotting or silver staining (cited in Bravo and Macdonald-Bravo[71] and Celis et al.[76]).

Similarly, in cultures of mouse 3T3 cells rendered quiescent by serum starvation, immunoblotting reveals depressed PCNA levels, although estimates of the level of residual PCNA vary from 30 of 40% of normal down to barely detectable.[60,71] The discrepancy may be attributable to the different periods of serum deprivation to which the cells were subjected in the two studies (24 to 48 h in one case and unspecified in the other), allowing for varying degrees of PCNA decay, or possibly to the different methods used in making cell extracts.

Stimulation of resting 3T3 cells with serum leads to the accumulation of PCNA to high levels, beginning at about 8 h after refeeding in one study.[60] The increased PCNA levels are accompanied by elevated synthesis of PCNA, measured in a pulse label, beginning about 12 h after refeeding in another study.[65] (The temporal discrepancy here is probably an insignificant consequence of differing experimental protocols.) This increased synthesis is presumably responsible for the rise in PCNA concentration, since the stability of PCNA does not seem to decline abruptly with the transition from the growing to the quiescent state.[71] Consistent with this notion is the low or undetectable rate of synthesis of PCNA in a variety of normal cell types (i.e., those with a finite life span).[14,77]

Although the possibility has been raised that PCNA turnover may accelerate after the completion of S phase,[19] based on present information the idea of regulation at the level of protein stability does not seem very likely. PCNA constitutes the same fraction of cell protein in pulse-labeled cells as in long-term-labeled cells, and the fraction does not decline in a chase.[18] The protein's half-life is similar to that of cell protein as a whole[18] and has been estimated at 10 to 15 h[19] and ca. 20 h.[71] It has not been exhaustively studied, however, and this aspect of PCNA metabolism may well benefit from further investigation.

2. Stimuli for PCNA Synthesis

The increased PCNA synthesis in quiescent 3T3 cells brought about by serum is concentration dependent and requires an exposure of 4 to 12 h. Some purified growth factors, such as fibroblast growth factor (FGF) and platelet-derived growth factor (PDGF), can substitute for serum, but platelet-poor plasma, which lacks growth factors, is much less effective than serum in stimulating PCNA synthesis.[24] The stimulation parallels increasing DNA synthesis, but it precedes it by 2 to 4 h. EGF is only weakly active, and growth factors such as insulin, hydrocortisone, and $PGF_{2\alpha}$, which do not induce DNA synthesis, also fail to stimulate PCNA synthesis, but the combination of EGF with insulin is effective in stimulating the synthesis of both DNA and PCNA.[13] These experiments show that PCNA synthesis is activated at a relatively late stage in the preparation for DNA synthesis, much later than *fos* or *myc* activation, for example, by a variety of stimuli that also induce DNA synthesis. A similar conclusion can be drawn from studies of mouse T-helper cells of the L2 line, which are stimulated to reenter the cell cycle upon activation by the lymphokine interleukin-2. Prior to embarking on S phase, the cells begin to synthesize a 36-kDa protein which is almost certainly PCNA at dramatically elevated rates.[20] For the first few hours, this response is sensitive to quinine, an agent which can block voltage-gated potassium ion channels, but later the response is unaffected by the drug. Thus, events that may involve ion fluxes follow activation of the proliferation pathway and are prerequisites for the later events, which include PCNA and DNA synthesis. In its turn, PCNA synthesis invariably precedes DNA replication, but is it a prerequisite for DNA synthesis? What happens if DNA synthesis is blocked? Are the two processes obligatorily coupled, or can one take place in the absence of the other? Before dealing with these questions, we turn to a consideration of the level at which regulation is effected.

3. PCNA mRNA Accumulation

With the availability of PCNA cDNA clones, it became possible to probe cytoplasmic RNA for PCNA mRNA concentration. In one experiment, PCNA mRNA was undetectable

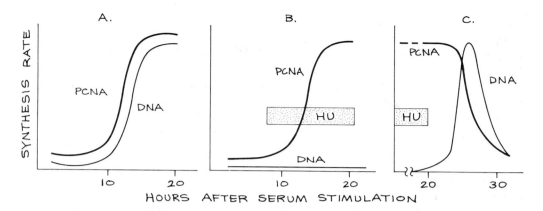

FIGURE 10. Relationship between PCNA synthesis and DNA synthesis. When quiescent cells are stimulated with serum, PCNA synthesis is induced and precedes DNA synthesis (A). Hydroxyurea (HU) blocks DNA synthesis without affecting PCNA synthesis (B). When the drug is removed, DNA synthesis resumes and PCNA synthesis declines (C). (Redrawn from Bravo, R. and Macdonald-Bravo, H., *EMBO J.*, 4, 655, 1985. With permission.)

in resting lymphocytes but became clearly visible on a Northern blot by 48 h after mitogen stimulation, at which time DNA synthesis was occurring. By 72 h, when DNA replication was declining and some cells were entering M phase, the mRNA levels had fallen greatly.[33] Similarly, in quiescent 3T3 cells, PCNA mRNA levels increased after serum stimulation. The mRNA became detectable at 10 to 12 h after stimulation, reached a peak at about 18 h, and then declined. These results indicate that PCNA synthesis is controlled by the level of its mRNA and quite possibly at the level of transcription, although neither of these points has been formally established.

4. PCNA Synthesis can be Dissociated from DNA Synthesis

Hydroxyurea inhibits DNA replication by blocking the activity of ribonucleotide reductase and stops the progression of cells through the G_1/S boundary.[78] Neither the synthesis of PCNA[65] nor the accumulation of its mRNA[34] is affected by hydroxyurea, although DNA synthesis is inhibited by >90% (Figure 10). Aphidicolin, an inhibitor of DNA polymerases-α and -δ, also has no effect on PCNA synthesis after serum stimulation of quiescent 3T3 cells or, indeed, in cultures of randomly growing cells.[79] Thus, DNA synthesis per se is not required for PCNA synthesis, consistent with the observation that the onset of DNA synthesis follows the rise in PCNA synthesis. Whether PCNA synthesis is obligatory for DNA replication is a question that has not been addressed directly *in vivo*, but it seems likely that the answer will be in the affirmative, since PCNA is almost certainly required for DNA synthesis *in vitro*.

As pointed out in an earlier section, hydroxyurea- or aphidicolin-treated 3T3 cells display only an early pattern of PCNA distribution. The fluorescence is granular, occupying the nucleoplasm excluding the nucleoli.[65] After removal of hydroxyurea, DNA synthesis resumes and the cells progress through S phase, displaying a sequence of redistribution patterns similar to those seen in normal cycling cells. Addition of aphidicolin blocks this progression, as expected, but α-amanitin does not, suggesting that further mRNA synthesis is no longer required at this stage and that all the essential components have already accumulated.[13]

Interestingly, serum-stimulated quiescent cells maintain a high level of PCNA synthesis for at least 20 h in the presence of hydroxyurea, but if the drug is washed out, DNA synthesis begins within 2 h and PCNA synthesis starts to fall after 4 h[65] (Figure 10). This implies that PCNA synthesis may be terminated by (an) event(s), such as DNA synthesis or its completion, that occur in S phase. How the accumulation and turnover of the protein and mRNA are regulated during S phase is not known. Nevertheless, it is clear that regulation of PCNA differs from that of histones. Synthesis of the histone proteins and their mRNAs is tightly coupled to DNA synthesis (reviewed by Stein et al.[80]); they are not synthesized if DNA synthesis is blocked, and in serum-stimulated 3T3 cells, histone H3 mRNA accumulation follows that of PCNA mRNA by 2 to 4 h (although they reach maximal levels at approximately the same time).[34] The regulation of PCNA synthesis more closely resembles that of enzymes concerned less directly with DNA synthesis, such as thymidine kinase, thymidylate synthase, and dihydrofolate reductase (reviewed by Johnson[81]), but a distinction between them will be drawn below (Section VI.C.2).

B. REGULATION IN THE CELL CYCLE
1. Protein Synthesis and Levels

It is well established that PCNA synthesis is maximal during S phase in cycling tissue culture cells. In one early experiment, HeLa cells were synchronized by mitotic shake off and labeled for 30-min periods at the beginning of each phase of the cell cycle. The relative rates of PCNA synthesis were 1.3 in G_1, 2.7 in S, 1.0 in G_2 and 1.9 in M; as a result, in a longer-term labeling, about 3.7 times as much radioactive label accumulated into PCNA in S phase as in either G_1 or G_2.[9] Increased synthesis during S phase has been observed in HeLa cells[9,34,36] and in AMA cells.[16,75,76] Statements to the contrary notwithstanding, there is little published evidence relating to PCNA *levels* during the cell cycle. One experiment addressed this issue directly by comparing the PCNA content of HeLa cell cytoplasmic extracts prepared from G_1 cells that had been synchronized by elutriation and were subsequently allowed to proceed through the cell cycle. Immunoblotting of these fractions revealed little change in the amount of PCNA (as a fraction of total cell protein), from which it was tentatively concluded that PCNA concentrations do not fluctuate greatly in continuously growing cells.[60] Therefore, it is unlikely that PCNA regulates DNA synthesis during the normal cell cycle, at least in HeLa cells.

This observation raises a conceptual problem. If PCNA synthesis increases in S phase, why doesn't the protein accumulate to high levels during this phase? One might propose that the synthesis of other cell proteins increases proportionally. But this seems not to be the case, as evidenced by gels loaded with equal amounts of radioactivity[23] or by quantitation of a representative selection of other proteins.[9] Alternatively, it might be supposed that the stability of PCNA falls during S phase, which seems unlikely. A third possibility takes note of the obvious fact that PCNA content only needs to double in each cell cycle and suggests that this relatively small increase may be hard to detect. The technical difficulties may have been exacerbated by the tight association of a population of PCNA with the nucleus, and it may be that in the experiment discussed,[60] a fraction of the PCNA within the cell, especially the S-phase cell, escaped measurement by resisting extraction into the soluble fraction. This third possibility would seem the most reasonable.

2. Regulatory Mechanisms

The means by which increased PCNA synthesis is achieved during the normal S phase is not known. We do not know what the signals are or the level at which they operate. However, there are indications that the situation may be a complex one. Most hybrids between a highly tumorigenic mouse cell line and a nontumorigenic Chinese hamster ovary cell line are capable of expressing both murine and hamster PCNAs, which are distinguishable in a two-dimensional gel.[82] Individual hybrid lines express the two proteins in differing ratios. This result implies that two PCNA genes resident in the same nucleus may be regulated differently, depending perhaps on their origin, sequence, or context in the cell. On the other hand, different nuclei resident in the same cytoplasm of a heterokaryon can asynchronously synthesize DNA and pass through the sequence of PCNA staining patterns, suggesting that

TABLE 2
PCNA Synthesis in Transformed Rat Cell Lines

Transforming agent	Relative PCNA synthesis	Doubling time (h)	Growth in soft agar	Tumorigenicity in nude mice
None (REF52 parent)	1	23	−	−
Simian virus 40	2.29	17	−	−
Kirsten murine sarcoma virus	1.48	15	+	+ +
Adenovirus 5	4.59	17	−	+

Data from Franza, B. R. Jr. and Garrels, J. I., in *Cancer Cells: The Transformed Phenotype*, No. 1, Levine, J. et al., Eds., Cold Spring Harbor Laboratory, Cold Spring Harbor, NY, 1984, 137. With permission.

the concentration of PCNA (or of any other cytoplasmic consituent) is not the factor (or the only factor) determining progression through S phase.[69] This finding implies that neither PCNA distribution nor DNA synthesis is determined exclusively by the cytoplasm in which the nucleus is embedded: intranuclear events or components must also be involved in determining DNA replication and PCNA distribution in the asynchronous heterokaryon.

C. TRANSFORMATION AND TUMORS
1. Relationship with PCNA Synthesis

Regardless of their site of origin in the animal, transformed cells and tumors generally synthesize PCNA at high, though variable, levels (as pointed out in Section II.B). PCNA immunofluorescence is observed in a variety of tumors, and the frequency of positive nuclei as well as the concentration of immunoreactive PCNA correlates with the growth rate of colon carcinoma cell lines.[4]

A high rate of PCNA synthesis would be expected simply from the proliferative activity of such cells, but early studies suggested that there might be a more specific connection with transformation. Reinvestigation of this issue has not yet led to a definitive resolution of the question. Franza and Garrels[17] have compared an immortalized rat embryo fibroblast line, REF52, with three of its virally transformed derivatives. The transformed cells all synthesize PCNA more rapidly than the parental line, and they all proliferate faster, too. However, there is no correlation between their PCNA synthetic rates and their doubling times (Table 2), suggesting that some transformed cells overproduce the protein. Similarly, in a long-term labeling experiment, SV40-transformed REF52 cells were found to contain fivefold more PCNA than their proliferating parental counterparts, a large difference that is unlikely to reflect the relatively slight discrepancy in their doubling times (18 to 19 h compared to 24 h).[22] Bravo and Graf,[27] on the other hand, compared primary quail cells with their retrovirus-transformed derivatives and observed elevated PCNA synthesis only in those cultures in which more cells were making DNA (Table 3). In this study, PCNA production was roughly proportional to DNA synthesis, implying that the two are coupled and that transformation has no specific effect on PCNA synthesis. Either cell lines are idiosyncratic in this regard or possibly the source of the oncogenes determines the outcome; perhaps it is significant that cells carrying oncogenes from the DNA tumor viruses adenovirus-5 and SV40 exhibited runaway PCNA synthesis, whereas those with retroviral oncogenes did not.[18] One might speculate that the transcriptional activator functions associated with the former may have something to do with this (see Section VI.A.2 below).

In the mouse/hamster hybrid lines mentioned in Section VI.B.2, tumorigenicity and

TABLE 3
PCNA Synthesis in Transformed Quail Embryo Fibroblasts

Transforming agent	Relative PCNA synthesis	³H-thymidine labeling index (%)	Tumorigenicity in nude mice
None (primary QEFs)	1	21	−
Rous sarcoma virus (*src*)	0.95	19	+
E26 virus (*myb/ets*)	1.1	25	−
MC29 virus (*myc*)	2.4	71	−
CMII virus (*myc*)	2.5	60	−

Data from Bravo, R. and Graf, T., *Exp. Cell Res.*, 156, 450, 1985. With permission.

high PCNA levels appeared to co-segregate.[82] The two species of PCNA could be separated from one another in two-dimensional gels. Interestingly, the hamster protein, derived from the nontumorigenic fusion partner, was much less evident than its murine counterpart in tumorigenic hybrids, and it was even undetectable in one tumorigenic line. These data suggest a relationship between tumorigenicity and PCNA synthesis, but no such correlation emerges from the data of Tables 2 and 3. Nevertheless, it has been suggested that anti-PCNA antibody may find some application in clinical evaluation of patients' blood or tissue samples. Some cells in lymph node smears from lymphoma patients stain with anti-PCNA antibody, as do some cells in bone marrow smears from leukemic patients.[83] In the latter disease, the antibody detects increased numbers of blast cells and of other cell types (myelocytes and metamyelocytes) with less intense staining patterns during blast crisis.[84] Neoplastic cells from hemopoietic lineages such as these stain strongly with anti-PCNA antibody, whereas most adenomas and adenocarcinomas stain weakly,[5] presumably because they are proliferating less rapidly. Conceivably, it may be possible to exploit the staining patterns of cells in different phases of the cell cycle to stage the disease or to characterize proliferating cell types.

2. Regulation by Adenovirus E1A Oncogene

Primary cultures of baby rat kidney (BRK) cells are susceptible to transformation by adenovirus, a process that involves the two viral oncogenes, E1A and E1B. One of the early steps is the induction by E1A gene products of cellular DNA synthesis and PCNA synthesis in these normally quiescent cells. Thus, this system affords the opportunity to examine the control of PCNA synthesis by identified nuclear-acting regulators, as opposed to the extracellular factors discussed above in Section VI.A.

PCNA synthesis is induced by adenovirus infection of BRK cells with a delay of 7 to 9 h following closely the appearance of E1A proteins.[77] As in serum stimulation, induction of PCNA synthesis precedes DNA synthesis by several hours and is not affected by the presence of hydroxyurea. It is accompanied by elevation of PCNA mRNA levels, assayed by translation in a cell-free system,[85] consistent with (but not proving) a transcriptional effect. The induction is dependent on the E1A gene, but it does not require the product of the gene's 13 S mRNA, which possesses a strong transcriptional activation function (reviewed by Moran and Mathews[86]). PCNA synthesis and mRNA are induced as well by an adenovirus variant which expresses only the 12 S E1A mRNA as by wild-type virus which can express all E1A products. This contrasts with activation of other viral early genes and of some cellular genes, such as that encoding thymidylate synthase, which are induced much more efficiently by the 13 S product than by the 12 S product. Thus, despite their similar responses to serum stimulation (Section VI.C.4), the expression of thymidylate synthase seems to correlate with the transcriptional activation function of the E1A 13 S product, while PCNA synthesis can be induced by the 12 S E1A product. The 12 S product is capable of immortalizing cells and of collaborating with other oncogenes, such as *ras* or adenovirus E1B, to transform them. It exhibits two functions related to these activities, an ability to induce cellular DNA synthesis and an ability to facilitate further progression through the cell cycle, as well as a transcriptional repressor activity,[87-89] and to a remarkable extent the various functions seem to be embodied in different regions, or domains, of the E1A proteins.[86] Although an appreciation of the relationships between these functions and the structure of the E1A proteins is growing rapidly, how this viral oncogene brings about changes in cell activity, such as a stimulation of PCNA synthesis or of cellular DNA synthesis, is not yet fully understood at the molecular level.

VII. ANTIBODIES AND AUTOIMMUNITY

A. ANTIBODIES AGAINST PCNA

PCNA was first recognized as an antigen reactive with certain autoimmune sera. Subsequently, the human autoantibodies have proven invaluable aids to the study of the antigen, even though they leave something to be desired as reagents. Being derived from patients' sera, they are rarely monospecific[6,35] and are not always readily available. Accordingly, other types of anti-PCNA antibodies have been developed.

Two murine hybridomas have been established which produce anti-PCNA antibodies.[29] The mice were immunized with purified rabbit thymus PCNA, and the hybridoma supernatants were screened by an enzyme-linked immunosorbent assay (ELISA) test against the same protein. Both monoclonal antibodies cross-react with human PCNA, which they can detect by immunoblotting and by immunofluorescence. However, the two monoclonals display some differences which suggest that they may recognize different epitopes. Each antibody reacts with two additional proteins in the immunoblot, but these proteins are not the same for the two antibodies, and one of the antibodies gives a much stronger immunofluorescence signal than the other. Cross-blocking tests suggest that the two antibodies have closely related, though distinct, specificities.

An anti-peptide serum has been prepared by injecting rabbits with a synthetic 13-residue peptide conjugated to keyhole limpet hemocyanin. The peptide corresponds to a sequence close to PCNA's N-terminus, between residues 11 and 23. It reacts with rabbit thymus PCNA by ELISA and by immunoblotting, as well as with human PCNA (and weakly with some additional proteins) by immunoblotting.[90]

A polyclonal serum against human PCNA was made in rabbits. The animals were injected with PCNA that had been purified by two-dimensional gel electrophoresis and eluted from the gel pieces. The antibody detects PCNA by immunoblotting and immunoprecipitation, and it cross-reacts with the calf thymus protein.[23]

Although spontaneously autoimmune mice make antibodies against several of the antigens that human autoantibodies are directed against, the production of anti-PCNA antibodies in such mice has not been reported. Whether this represents a difference between mice and men or merely reflects the low incidence of anti-PCNA antibody in autoimmunity is an undecided issue.

B. ANTIGENIC SITES ON PCNA

An understanding of the antibody-binding sites (epitopes) on the PCNA molecule could illuminate its function in DNA synthesis as well as its role in autoimmunity. Using a polymerase-δ assay for PCNA, Tan et al.[91] found that a human autoantibody reacts with PCNA in its native state, thereby blocking its ability to enhance polymerase activity. Similar results were obtained whether the antibodies were used to precipitate the antigen prior to the reaction or were preincubated with the antigen and then added directly to the polymerization reaction in a neutralization test. On the other hand, neither of the two murine monoclonal antibodies nor the rabbit antibody directed against the peptide located near PCNA's N-terminus blocked enzymatic activity, although all of them could react with the denatured protein on an immunoblot. Seemingly, the epitopes recognized by these three experimentally produced antibodies are not exposed in the native protein, although those recognized by the human antibodies are exposed.

To localize the reactive epitopes on the protein, PCNA was subjected to partial digestion and the resultant polypeptides were probed by immunoblotting.[90] The anti-peptide antibody afforded a landmark in these studies, providing a convenient means to identify the N-terminal digestion products. Two large fragments, one (15 kDa) containing the N-terminus and the other (17 kDa) seemingly from the C-terminal region, were both reactive with autoantibodies. The murine monoclonal that was tested failed to react with either fragment but did react with a 20-kDa fragment that included the N-terminus, so its epitope presumably maps in between the 15- and 17-kDa fragments. Based on their differential reactivities, the autoantisera have been classified into three groups: type A react with the 17-kDa fragment, type C with the 15-kDa fragment, and type B with both fragments. Whether type B comprises more than one class of antibody molecule is not known, nor is it clear how many epitopes exist within these peptides.

C. CLINICAL ASPECTS OF ANTI-PCNA AUTOANTIBODIES

Autoantibodies directed against such nuclear components as DNA, histones, and ribonucleoprotein particles are common in connective tissue diseases (reviewed in Tan[2] and in Bernstein et al.[3]). Antibodies against PCNA are relatively rare, occurring in 2 to 4% of lupus patients.[1,3,92] They are rarely, if ever, found in sera from patients with other connective tissue disorders, such as sicca syndrome, polymyositis, scleroderma, and rheumatoid arthritis. It is reported that the antibody rapidly disappears from circulation after corticosteroid therapy, a circumstance that may contribute to the low frequency with which the antibody is detected in patients.[92] Patients with anti-PCNA antibody do not seem to display any particular unusual demographic, clinical, or serological features that demarcate them as a group from other lupus patients.[1,92]

As with most autoimmune disorders, the etiology (cause) of lupus is unknown, apart from the drug-induced variety, which is not characterized by the presence of anti-PCNA antibodies. It is uncertain whether anti-PCNA antibodies play any role in lupus pathogenesis, although some other autoantibodies, such as anti-DNA antibodies, clearly do participate in the disease process. Why some people develop the antibodies while others do not remains a subject of active clinical investigation.

VIII. EPILOGUE

To put a name on something, or even three names, is not to describe it. What has been discussed here is the beginning of studies aimed at answering questions about the function of PCNA in DNA synthesis and its regulation in the cell cycle. This work is off to a good start, and most of the issues should be amenable to current technology given the tools now available. Taking a broader view, PCNA may serve as a paradigm for proteins that participate in replication or the cell cycle. Several polypeptides with regulatory behavior similiar to that exhibited by PCNA have been detected by gel electrophoresis,[18,20] together with other polypeptides displaying alternative modes of cell cycle regulation,[18,76,93,94] and cDNA libraries have been constructed to search for cell cycle phase-specific products.[95,96] We may have to wait longer, however, to understand the part (if any) that PCNA plays in the complicated disorders of the immune system.

ACKNOWLEDGMENTS

I thank Robert Bernstein for his seminal contributions to PCNA studies in my laboratory; my colleagues Bob Franza, Jim Garrels, Betty Moran, Gil Morris, Greg Prelich, and Bruce Stillman for many discussions and for helpful comments on this review; Barbara Weinkauff for tireless and skillful word processing; Rodrigo Bravo and Julio Celis for copies of some of their published data and permission to reproduce them; and the National Cancer Institute, Leukemia Society of America, American Cancer Society, and North Atlantic Treaty Organization for supporting work on this subject in the author's laboratory. This review is based on literature that came to my attention by January 1988.

ADDENDUM

This update takes note of progress made in the year that has elapsed since the review was written.

We have completed a comprehensive reexamination of the regulation of PCNA during the cell cycle.[99] Logarithmically growing HeLa cell cultures were separated into populations of cells enriched for various phases of the cell cycle by the technique of centrifugal elutriation, to allow examination of the synthesis, stability, and level of PCNA. In agreement with earlier work,[9] PCNA synthesis increased during S phase, but only by two- to threefold. The increased rate of PCNA synthesis reflected a similar change in the level of its mRNA. Since the protein's stability did not vary significantly during the cell cycle, this increased synthesis was sufficient to account for the approximately twofold increase in PCNA level (see Section VI.B.1). The fraction of PCNA that is tightly bound to nuclear components fluctuated more dramatically, ranging from undetectably low levels to 35% of the total pool. This fraction reached a peak in S phase, consistent with the protein's involvement in replication complexes. The total amount of PCNA was estimated at 0.63 to 1.14×10^6 copies per cell, depending on the phase of the cell cycle, but always exceeded the estimated number of replicons in the genome. These results reinforce the conclusion[60] that PCNA is unlikely to regulate DNA replication, at least in HeLa cells (Section VI.B.1).

Experiments with antisense oligonucleotides[100] point in the opposite direction, however. Oligodeoxynucleotides complementary to the 5' portion of the translated region of human PCNA mRNA were added to cultures of growing mouse 3T3 cells. Within 18 h, PCNA immunofluorescence was reduced, DNA synthesis was blocked, and the cells accumulated at or near the G_1/S boundary. Control oligonucleotides of the same sense as the mRNA did not have this effect. Although neither PCNA level nor synthesis was measured, these results suggest that PCNA synthesis is required for DNA replication and cell cycle progression in 3T3 cells. It remains to be seen whether there are cell type-specific differences between HeLa and 3T3 cells in their short-term requirement for PCNA synthesis, or nonspecific mechanisms whereby oligonucleotides can interfere with the growth of 3T3 cells, that can explain the discrepancy between the conclusions drawn from the anti-sense and the elutriation data.

There is no doubt that PCNA synthesis is required in the long term for cellular DNA synthesis (see Section IV.B). Autoantibody to PCNA introduced by microinjection into *Xenopus* oocytes reduced the replication of plasmid DNA by 35 to 67% under certain conditions,[101] and this effect was reversed by coinjection of PCNA protein. Chromosomal DNA replication was inhibited to a lesser degree. Consistent with their behavior *in vitro*[91] (see Section VII.B), monoclonal antibodies directed against PCNA were ineffective in this assay. Inhibitory effects of the autoantibody have also been reported in isolated nuclei.[102] At the same time, it must be noted that one study failed to discern a requirement for PCNA in a purified DNA replication system.[103] This study suggested that the role of PCNA is an indirect one and that it may serve to overcome the action of an inhibitory protein found in less defined replication systems.

The most compelling evidence of PCNA's replicative role *in vivo* comes from studies of the yeast *Saccharomyces cerevisiae*. Calf thymus PCNA stimulates the activity of a yeast DNA polymerase known as polymerase III[104] in much the same way as with mammalian DNA polymerase δ (Section IV.C). Moreover, a yeast analog of mammalian PCNA has been purified.[105] The yeast analog, yPCNA, appears to be a trimer or tetramer of identical 26,000-Da subunits. Yeast DNA polymerase III shares many other features with DNA polymerase-δ (see References 104 and 106) and has recently been identified as the product of the essential *CDC2* gene.[106] The yeast equivalent of DNA polymerase-δ, known as DNA polymerase I, is also essential, lending further support to the notion that replication involves the concerted action of two DNA polymerases (Figure 6). Intriguingly, a PCNA-like molecule and an enzyme resembling DNA polymerase-δ are both encoded in the genome of an insect baculovirus.[107]

PCNA synthesis is dramatically induced when cells of a cloned T lymphocyte line are stimulated to proliferate by interleukin 2[108] (see Section VI.A.2 and Reference 20). This response appears to be somewhat unusual in that PCNA synthesis reaches a maximal rate during G_1 instead of S phase. In a temperature-sensitive hamster cell line, ts13, which arrests in the G_1 phase of the cell cycle at the restrictive temperature, PCNA mRNA synthesis is not inducible by serum.[109] This observation confirms that PCNA belongs to the "late" class of growth-regulated genes (see Section VI.A.2). Consistent with this assignment, the induction of PCNA mRNA by serum addition is blocked by cycloheximide, indicating a requirement for protein synthesis in the induction. In both these respects, PCNA resembles thymidine kinase, another "late" gene, but the two genes differ in their responsivity to growth factors.[109] At this point, more detailed understanding of PCNA regulation presumably awaits the analysis of its promoter.

REFERENCES

1. **Miyachi, K., Fritzler, M. J., and Tan, E. M.,** Autoantibody to a nuclear antigen in proliferating cells, *J. Immunol.,* 121, 2228, 1978.
2. **Tan, E. M.,** Autoantibodies to nuclear antigens (ANA): their immunobiology and medicine, *Adv. Immunol.,* 33, 167, 1982.
3. **Bernstein, R. M., Bunn, C. C., Hughes, G. R. V., Francoeur, A. M., and Mathews, M. B.,** Cellular protein and RNA antigens in autoimmune disease, *Mol. Biol. Med.,* 2, 105, 1984.
4. **Chan, P.-K., Frakes, R., Tan, E. M., Brattain, M. G., Smetana, K., and Busch, H.,** Indirect immunofluorescence studies of proliferating cell nuclear antigen in nucleoli of human tumor and normal tissues, *Cancer Res.,* 43, 3770, 1983.
5. **Nakane, P. K., Moriuchi, T., Koji, T., Mitsuyoshi, S., Izumi, S., Matsumoto, K., Daidoji, H., and Takasaki, Y.,** Proliferating cell nuclear antigen (PCNA/cyclin), in *Recent Advances in Immunopathology,* Hamashima, Y., Ed., 1987, 179.
6. **Takasaki, Y., Deng, J.-S., and Tan, E. M.,** A nuclear antigen associated with cell proliferation and blast transformation, *J. Exp. Med.,* 154, 1899, 1981.
7. **Evans, T., Rosenthal, E. T., Youngblom, J., Distel, D., and Hunt, T.,** Cyclin: a protein specified by maternal mRNA in sea urchin eggs that is destroyed at each cleavage division, *Cell,* 33, 389, 1983.
8. **Pines, J. and Hunt, T.,** Molecular cloning and characterization of the mRNA for cyclin from sea urchin eggs, *EMBO J.,* 6, 2987, 1987.
9. **Bravo, R. and Celis, J. E.,** A search for differential polypeptide synthesis throughout the cell cycle of HeLa cells, *J. Cell Biol.,* 84, 795, 1980.
10. **Bravo, R. and Celis, J. E.,** Gene expression in normal and virally transformed mouse 3T3 and hamster BHK21 cells, *Exp. Cell Res.,* 127, 249, 1980.

11. **Bravo, R., Fey, S. J., Bellatin, J., Larsen, P. M., Arevalo, J., and Celis, J. E.,** Identification of a nuclear and of a cytoplasmic polypeptide whose relative proportions are sensitive to changes in the rate of cell proliferation, *Exp. Cell Res.,* 136, 311, 1981.

12. **Bravo, R., Fey, S. J., Bellatin, J., Larsen, P. M., and Celis, J. E.,** Identification of a nuclear polypeptide ("Cyclin") whose relative proportion is sensitive to changes in the rate of cell proliferation and to transformation, in *Embryonic Development,* Part A, Alan R. Liss, New York, 1982, 235.

13. **Bravo, R.,** Synthesis of the nuclear protein cyclin (PCNA) and its relationship with DNA replication, *Exp. Cell Res.,* 163, 287, 1986.

14. **Celis, J. E., Bravo, R., Larsen, P. M., and Fey, S. J.,** Cyclin: a nuclear protein whose level correlates directly with the proliferative state of normal as well as transformed cells, *Leuk. Res.,* 8, 143, 1984.

15. **Celis, J. E., Madsen, P., Nielsen, S., and Celis, A.,** Nuclear patterns of cyclin (PCNA) antigen distribution subdivide S-phase in cultured cells — some applications of PCNA antibodies, *Leuk. Res.,* 10, 237, 1986.

16. **Celis, J. E., Madsen, P., Celis, A., Nielsen, H. V., and Gesser, B.,** Cyclin (PCNA, auxiliary protein of DNA polymerase δ) is a central component of the pathway(s) leading to DNA replication and cell division, *FEBS Lett.,* 220, 1, 1987.

17. **Franza, B. R., Jr. and Garrels, J. I.,** Transformation-sensitive proteins of REF52 cells detected by computer-analyzed two-dimensional gel electrophoresis, in *Cancer Cells: The Transformed Phenotype,* No. 1, Levine, J. et al., Eds., Cold Spring Harbor Laboratory, Cold Spring Harbor, NY, 1984, 137.

18. **Garrels, J. I. and Franza, B. R., Jr.,** Transformation-sensitive and growth-related changes of protein synthesis in REF52 cells: a two-dimensional gel analysis of SV40-, adenovirus-, and Kirsten murine sarcoma virus-transformed rat cells using the REF52 protein data base, *J. Biol. Chem.,* submitted.

19. **O'Farrell, M. K.,** Metabolic turnover of proliferation-related nuclear proteins in serum-stimulated Swiss mouse 3T3 cells, *FEBS Lett.,* 204, 233, 1986.

20. **Sabath, D. E., Monos, D. S., Lee, S. C., Deutsch, C., and Prystowsky, M. B.,** Cloned T-cell proliferation and synthesis of specific proteins are inhibited by quinine, *Proc. Natl. Acad. Sci. U.S.A.,* 83, 4739, 1986.

21. **Feuerstein, N. and Mond, J. J.,** "Numatrin", a nuclear matrix protein associated with induction of proliferation in B lymphocytes, *J. Biol. Chem.,* 262, 11389, 1987.

22. **Mathews, M. B., Bernstein, R. M., Franza, B. R., Jr., and Garrels, J. I.,** Identity of the Proliferating Cell Nuclear Antigen and Cyclin, *Nature,* 309, 374, 1984.

23. **Bravo, R., Frank, R., Blundell, P. A., and Macdonald-Bravo, H.,** Cyclin/PCNA is the auxiliary protein of DNA polymerase-δ, *Nature,* 326, 515, 1987.

24. **Bravo, R. and Macdonald-Bravo, H.,** Induction of the nuclear protein "cyclin" in quiescent mouse 3T3 cells stimulated by serum and growth factors. Correlation with DNA synthesis, *EMBO J.,* 3, 3177, 1984.

25. **Bravo, R.,** Epidermal growth factor inhibits the synthesis of the nuclear protein cyclin in A431 human carcinoma cells, *Proc. Natl. Acad. Sci. U.S.A.,* 81, 4848, 1984.

26. **Bravo, R. and Celis, J. E.,** Human proteins sensitive to neoplastic transformation in cultured epithelial and fibroblast cells, *Clin. Chem.,* 28, 949, 1982.

27. **Bravo, R. and Graf, T.,** Synthesis of the nuclear protein cyclin does not correlate directly with transformation in quail embryo fibroblasts, *Exp. Cell Res.,* 156, 450, 1985.

28. **Takasaki, Y., Fishwild, D., and Tan, E. M.,** Characterization of proliferating cell nuclear antigen PCNA recognized by autoantibodies in lupus sera, *J. Exp. Med.,* 159, 981, 1984.

29. **Ogata, K., Kurki, P., Celis, J. E., Nakamura, R. M., and Tan, E. M.,** Monoclonal antibodies to a nuclear protein (PCNA/cyclin) associated with DNA replication, *Exp. Cell Res.,* 168, 475, 1987.

30. **Celis, J. E. and Bravo, R.,** Synthesis of the nuclear protein cyclin in growing, senescent and morphologically transformed human skin fibroblasts, *FEBS Lett.,* 165, 21, 1984.

31. **Ogata, K., Ogata, Y., Nakamura, R. M., and Tan, E. M.,** Purification and N-terminal amino acid sequence of proliferating cell nuclear antigen (PCNA)/cyclin and development of ELISA for anti-PCNA antibodies, *J. Immunol.,* 135, 2623, 1985.

32. **Prelich, G., Kostura, M., Marshak, D. R., Mathews, M. B., and Stillman, B.,** The cell-cycle regulated proliferating cell nuclear antigen is required for SV40 DNA replication, *in vitro, Nature,* 326, 471, 1987.

33. **Matsumoto, K., Moriuchi, T., Koji, T., and Nakane, P. K.,** Molecular cloning of cDNA coding for rat proliferating cell nuclear antigen (PCNA)/cyclin, *EMBO J.,* 6, 637, 1987.

34. **Almendral, J. M., Huebsch, D., Blundell, P. A., Macdonald-Bravo, H., and Bravo, R.,** Cloning and sequence of the human nuclear protein cyclin: homology with DNA-binding proteins, *Proc. Natl. Acad. Sci. U.S.A.,* 84, 1575, 1987.

35. **Sadaie, M. R. and Mathews, M. B.,** Immunochemical and biochemical analysis of the proliferating cell nuclear antigen (PCNA) in HeLa cells, *Exp. Cell Res.,* 163, 423, 1986.

36. **Bravo, R. and Celis, J. E.,** Changes in the nuclear distribution of cyclin (PCNA) during S-phase are not triggered by post-translational modifications that are expected to moderately affect its charge, *FEBS Lett.,* 182, 435, 1985.

37. **Tan, C.-K., Castillo, C., So, A. G., and Downey, K. M.,** An auxiliary protein for DNA polymerase-δ from fetal calf thymus, *J. Biol. Chem.,* 261, 12310, 1986.

38. **Quinlan, M. P., Chen, L. B., and Knipe, D. M.,** The intranuclear location of a herpes simplex virus DNA-binding protein is determined by the status of viral DNA replication, *Cell,* 36, 857, 1984.

39. **Kornberg, A.,** DNA replication, *J. Biol. Chem.,* 263, 1, 1988.

40. **Fry, M. and Loeb, L. A.,** *Animal Cell DNA Polymerases,* CRC Press, Boca Raton, FL, 1986.

41. **Kelly, T. and Stillman, B., Eds.,** *Cancer Cells: Eukaryotic DNA Replication,* No. 6, Cold Spring Harbor Laboratory, Cold Spring Harbor, NY, 1988.

42. **Murakami, Y., Eki, T., Yamada, M.-A., Prives, C., and Hurwitz, J.,** Species-specific *in vitro* synthesis of DNA containing the polyoma virus origin of replication, *Proc. Natl. Acad. Sci. U.S.A.,* 83, 6347, 1986.

43. **Miller, M. R., Ulrich, R. G., Wang, T. S.-F., and Korn, D.,** Monoclonal antibodies against human DNA polymerase-α inhibit DNA replication in permeabilized human cells, *J. Biol. Chem.,* 260, 134, 1985.

44. **Kaczmarek, L., Miller, M. R., Hammond, R. A., and Mercer, W. E.,** A microinjected monoclonal antibody against human DNA polymerase-α inhibits DNA replication in human, hamster, and mouse cell lines, *J. Biol. Chem.,* 261, 10802, 1986.

45. **Murakami, Y., Yasuda, H., Miyazawa, H., Hanaska, F., and Yamada, M.,** Characterization of a temperature-sensitive mutant of mouse FM3A cells defective in DNA replication, *Proc. Natl. Acad. Sci. U.S.A.,* 82, 1761, 1985.

46. **Downey, K. M., Tan, C.-K., Andrews, D. M., Li, X., and So, A. G.,** Proposed roles for DNA polymerases alpha and delta at the replication fork, in *Cancer Cells: Eukaryotic DNA Replication,* No. 6, Cold Spring Harbor Laboratory, Cold Spring Harbor, NY, 1988, 403.

47. **Byrnes, J. J., Downey, K. M., Block, V. L., and So, A. G.,** A new mammalian DNA polymerase with 3′ to 5′ exonuclease activity: DNA polymerase δ, *Biochemistry,* 15, 2817, 1976.

48. **Goscin, L. P. and Byrnes, J. J.,** DNA polymerase δ: one polypeptide, two activities, *Biochemistry,* 21, 2513, 1982.

49. **Lee, M. Y. W. T., Tan, C.-K., Downey, K. M., and So., A. G.,** Further studies on calf thymus DNA polymerase δ purified to homogeneity by a new procedure, *Biochemistry,* 23, 1906, 1984.

50. **Byrnes, J. J.,** Differential inhibitors of DNA polymerases alpha and delta, *Biochem. Biophys. Res. Commun.,* 132, 628, 1985.

51. **Lee, M. Y. W. T., Toomey, N. L., and Wright, G. E.,** Differential inhibition of human placental DNA polymerases δ and α by BuPdGTP and BuAdATP, *Nucleic Acids Res.,* 13, 8623, 1985.

52. **Crute, J. J., Wahl, A. F., and Bambara, R. A.,** Purification and characterization of two new high molecular weight forms of DNA polymerase δ, *Biochemistry,* 25, 26, 1986.

53. **Marraccino, R. L., Wahl, A. F., Keng, P. C., Lord, E. M., and Bambara, R. A.,** Cell cycle dependent activities of DNA polymerases α and δ in Chinese hamster ovary cells, *Biochemistry,* 26, 7864, 1987.

54. **Lee, M. Y. W. T. and Toomey, N. L.,** Human placental DNA polymerase δ: identification of a 170-kilodalton polypeptide by activity staining and immunoblotting, *Biochemistry,* 26, 1076, 1987.

55. **Kunkel, T. A., Sabatino, R. D., and Bambara, R. A.,** Exonucleolytic proofreading by calf thymus DNA polymerase δ, *Proc. Natl. Acad. Sci. U.S.A.,* 84, 4865, 1987.

56. **Dresler, S. L. and Frattini, M. G.,** DNA replication and UV-induced DNA repair synthesis in human fibroblasts are much less sensitive than DNA polymerase α to inhibition by butylphenyl-deoxyguanosine triphosphate, *Nucleic Acids Res.,* 14, 7093, 1986.

57. **Nishida, C., Reinhard, P., and Linn, S.,** DNA repair synthesis in human fibroblasts requires DNA polymerase δ, *J. Biol. Chem.,* 263, 501, 1988.

58. **Fairman, M. P., Prelich, G., Tsurimoto, T., and Stillman, B.,** Cellular proteins required for SV40 DNA replication *in vitro,* in *Cancer Cells: Eukaryotic DNA Replication,* No. 6, Cold Spring Harbor Laboratory, Cold Spring Harbor, NY, 1988.

59. **Dean, F. B., Dodson, M., Borowiec, J. A., Ishimi, Y., Goetz, G. S., Bullock, P., Matson, S. W., Echols, H., and Hurwitz, J.,** Simian virus 40 (SV40) origin-dependent DNA unwinding and nucleoprotein complex formation, in *Cancer Cells: Eukaryotic DNA Replication,* No. 6, Cold Spring Harbor Laboratory, Cold Spring Harbor, NY, 1988.

60. **Wold, M. S., Li, J. J., Weinberg, D. H., Virship, D. M., Sherley, J. M., Verheyen, E., and Kelly, T.,** Cellular proteins required for SV40 DNA replication *in vitro,* in *Cancer Cells: Eukaryotic DNA Replication,* No. 6, Cold Spring Harbor Laboratory, Cold Spring Harbor, NY, 1988, 133.

61. **Prelich, G., Tan, C.-K., Kostura, M., Mathews, M. B., So, A. G., Downey, K. M., and Stillman, B.,** Functional identity of proliferating cell nuclear antigen and a DNA polymerase-δ auxiliary protein, *Nature,* 326, 517, 1987.

62. **Hammond, R. A., Byrnes, J. J., and Miller, M. R.,** Identification of DNA polymerase δ in CV-1 cells: studies implicating both DNA polymerase δ and DNA polymerase α in DNA replication, *Biochemistry,* 26, 6817, 1987.

63. **Decker, R. S., Yamaguchi, M., Possenti, R., Bradley, M. K., and DePamphilis, M. L.,** *In vitro* initiation of DNA replication in simian virus 40 chromosomes, *J. Biol. Chem.,* 262, 10863, 1987.

64. **Prelich, G. and Stillman, B.,** Coordinated leading and lagging strand synthesis during SV40 DNA replication *in vitro* requires PCNA, *Cell,* 53, 117, 1988.

65. **Bravo, R. and Macdonald-Bravo, H.,** Changes in the nuclear distribution of cyclin (PCNA) but not its synthesis depend on DNA replication, *EMBO J.,* 4, 655, 1985.
66. **Celis, J. E. and Celis, A.,** Cell cycle-dependent variations in the distribution of the nuclear protein proliferating cell nuclear antigen in cultured cells: subdivision of S phase, *Proc. Natl. Acad. Sci. U.S.A.,* 82, 3262, 1985.
67. **Kurki, P., Vanderlaan, M., Dolbeare, F., Gray, J., and Tan, E. M.,** Expression of the proliferating cell nuclear antigen during the cell cycle, *Exp. Cell Res.,* 166, 209, 1986.
68. **Kurki, P., Lotz, M., Ogata, K., and Tan, E. M.,** Proliferating cell nuclear antigen (PCNA)/cyclin in activated human T lymphocytes, *J. Immunol.,* 138, 4114, 1987.
69. **Celis, J. E. and Celis, A.,** Individual nuclei in polykaryons can control cyclin distribution and DNA synthesis, *EMBO J.,* 4, 1187, 1985.
70. **Madsen, P. and Celis, J. E.,** S-phase patterns of cyclin (PCNA) antigen staining resemble topographical patterns of DNA synthesis, *FEBS Lett.,* 193, 5, 1985.
71. **Bravo, R. and Macdonald-Bravo, H.,** Existence of two populations of cyclin/PCNA during the cell cycle association with DNA replication sites, *J. Cell Biol.,* 105, 1549, 1987.
72. **Celis, J. E., Madsen, P., and Lauridsen, J. B.,** Mid to late S-phase replication of the nucleolus in lymphoid human Molt-4 cells, *Leukemia,* 1, 568, 1987.
73. **Madsen, P., Ogata, K., and Celis, J. E.,** PCNA (Cyclin), autoantibodies and monoclonal antibodies reveal similar patterns of cyclin (PCNA) antigen staining in human cultured cells, *Leukemia,* 1, 220, 1987.
74. **Spector, D.,** unpublished data, 1987.
75. **Celis, J. E. and Madsen, P.,** Increased nuclear cyclin/PCNA antigen staining of non S-phase transformed human amnion cells engaged in nucleotide excision DNA repair, *FEBS Lett.,* 209, 277, 1986.
76. **Celis, J. E., Madsen, P., Nielsen, S. U., Gesser, B., Nielsen, H. V., Ratz, G. P., Lauridsen, J. B., and Celis, A.,** Cyclin (PCNA, auxiliary protein of DNA polymerase δ), dividin and progressin are likely components of the common pathway leading to DNA replication and cell division in human cells, in *Cancer Cells: Eukaryotic DNA Replication,* No. 6, Cold Spring Harbor Laboratory, Cold Spring Harbor, NY, 1988, 289.
77. **Zerler, B., Roberts, R. J., Mathews, M. B., and Moran, E.,** Different functional domains of the adenovirus E1A gene are involved in regulation of host cell cycle products, *Mol. Cell. Biol.,* 7, 821, 1986.
78. **Adams, R. L. P. and Lindsay, J. G.,** Hydroxyurea: reversal of inhibition in use as a cell-synchronizing agent, *J. Biol. Chem.,* 242, 1314, 1967.
79. **Macdonald-Bravo, H. and Bravo, R.,** Induction of the nuclear protein cyclin in serum-stimulated quiescent 3T3 cells is independent of DNA synthesis, *Exp. Cell Res.,* 156, 455, 1985.
80. **Stein, G. S., Plumb, M. A., Stein, J. L., Marashi, F. F., Sierra, L. F., and Baumbach, L. L.,** Expression of histone genes during the cell cycle in human cells, in *Recombinant DNA and Cell Proliferation,* Stein, G. S. and Stein, J. L., Eds., Academic Press, New York, 1984, 107.
81. **Johnson, L. F.,** Expression of dihydrofolate reductase and thymidylate synthase genes in mammalian cells, in *Recombinant DNA and Cell Proliferation,* Stein, G. S. and Stein, J. L., Eds., Academic Press, New York, 1984, 25.
82. **Bravo, R., Fey, S. J., and Celis, J. E.,** Gene expression in murine hybrids exhibiting different morphologies and tumorigenic properties, *Carcinogenesis,* 2, 769, 1981.
83. **Smetana, K., Gyorkey, F., Chan, P.-K., Tan, E., and Busch, H.,** Proliferating cell nuclear antigen (PCNA) and human malignant tumor nucleolar antigens (HMTNA) in nucleoli of human hematological malignancies, *Blut,* 46, 133, 1983.
84. **Takasaki, Y., Robinson, W. A., and Tan, E. M.,** Proliferating cell nuclear antigen in blast crisis cells of patients with chronic myeloid leukemia, *J. Natl. Cancer Inst.,* 73, 655, 1984.
85. **Moran, E.,** unpublished results, 1987.
86. **Moran, E. and Mathews, M. B.,** Multiple functional domains of the adenovirus E1A gene, *Cell,* 48, 177, 1987.
87. **Lillie, J. W., Loewenstein, P. M., Green, M. R., and Green, M.,** Functional domains of adenovirus type 5 E1a proteins, *Cell,* 50, 1091, 1987.
88. **Schneider, J. F., Fisher, F., Goding, C. R., and Jones, N. C.,** Mutational analysis of the adenovirus E1a gene: the role of transcriptional regulation in transformation, *EMBO J.,* 6, 2053, 1987.
89. **Moran, E. and Zerler, B.,** Interactions between cell growth regulating domains in the products of the adenovirus E1A oncogene, *Mol. Cell. Biol.,* 8, 1756, 1988.
90. **Ogata, K., Ogata, Y., Takasaki, Y., and Tan, E. M.,** Epitopes on proliferating cell nuclear antigen recognized by human lupus autoantibody and murine monoclonal antibody, *J. Immunol.,* 139, 2942, 1987.
91. **Tan, C.-K., Sullivan, K., Li, X., Tan, E. M., Downey, K. M., and So, A. G.,** Autoantibody to the proliferating cell nuclear antigen neutralizes the activity of the auxiliary protein for DNA polymerase delta, *Nucleic Acids Res.,* 15, 9299, 1987.
92. **Fritzler, M. J., McCarty, G. A., Ryan, J. P., and Kinsella, T. D.,** Clinical features of patients with antibodies directed against proliferating cell nuclear antigen, *Arthritis Rheum.,* 26, 140, 1983.

93. **Wang, E.,** Rapid disappearance of Statin, a nonproliferating and senescent cell-specific protein, upon reentering the process of cell cycling, *J. Cell Biol.,* 101, 1695, 1985.

94. **Celis, J. E., Madsen, P., Nielsen, S., and Gesser, B.,** Malignant transformation and expression of human cellular proteins, *Cancer Rev.,* 4, 1, 1986.

95. **Hirshhorn, R. R., Aller, P., Yuan, Z.-A., Gibson, C. W., and Baserga, R.,** Cell-cycle-specific cDNAs from mammalian cells temperature sensitive for growth, *Proc. Natl. Acad. Sci. U.S.A.,* 81, 6004, 1984.

96. **Lau, L. F. and Nathans, D.,** Identification of a set of genes expressed during the G0/G1 transition of cultured mouse cells, *EMBO J.,* 4, 3145, 1985.

97. **Bunn, C. C., Gharavi, A. E., and Hughes, G. R.,** Antibodies to extractable nuclear antigens in 173 patients with DNA-binding positive SLE: an association between antibodies to ribonucleoprotein and Sm antigens observed by counterimmunoelectrophoresis, *J. Clin. Lab. Immunol.,* 8, 13, 1982.

98. **Cotterill, S. M., Reyland, M. E., Loeb, L. A., and Lehman, I. R.,** A cryptic proofreading 3′→5′ exonuclease associated with the polymerase subunit of the DNA polymerase-primase from *Drosophila melanogaster, Proc. Natl. Acad. Sci. U.S.A.,* 84, 5635, 1987.

99. **Morris, G. F. and Mathews, M. B.,** Regulation of proliferaing cell nuclear antigen (PCNA/cyclin) during the cell cycle, submitted.

100. **Jaskulski, D., deRiel, K. J., Mercer, E. W., Calabretta, B., and Baserga, R.,** Inhibition of cellular proliferation by antisense oligodeoxynucleotides to PCNA cyclin, *Science,* 240, 1544, 1988.

101. **Zuber, M., Tan, E. M., and Ryoji, M.,** Involvement of proliferating cell nuclear antigen (cyclin) in DNA replication in living cells, *Mol. Cell. Biol.,* 9, 57, 1989.

102. **Wong, R. L., Katz, M. E., Ogata, K., Tan, E. M., and Cohen, S.,** Inhibition of nuclear DNA synthesis by an autoantibody to proliferating cell nuclear antigen/cyclin, *Cell. Immunol.,* 110, 443, 1987.

103. **Lee, S.-H., Ishimi, Y., Kenny, M. K., Bullock, P., Dean, F. B., and Hurwitz, J.,** An inhibitor of the *in vitro* elongation reaction of simian virus 40 DNA replication is overcome by proliferating-cell nuclear antigen, *Proc. Natl. Acad. Sci. U.S.A.,* 85, 9469, 1988.

104. **Burgers, P. M. J.,** Mammalian cyclin/PCNA (DNA polymerase δ auxiliary protein) stimulates processive DNA synthesis by yeast DNA polymerase III, *Nucleic Acids Res.,* 16, 6297, 1988.

105. **Bauer, G. A. and Burgers, P. M. J.,** The yeast analog of mammalian cyclin/proliferating-cell nuclear antigen interacts with mammalian DNA polymerases δ, *Proc. Natl. Acad. Sci. U.S.A.,* 85, 7506, 1988.

106. **Sitney, K. C., Budd, M. E., and Campbell, J. L.,** DNA polymerase III, a second essential DNA polymerase, is encoded by the *S. cerevisiae CDC2* gene, *Cell,* 56, 599, 1989.

107. **O'Reilly, D.R., Crawford, A. M., and Miller, L. K.,** Viral proliferating cell nuclear antigen, *Nature,* 337, 606, 1989.

108. **Moore, K. S., Sullivan, K., Tan, E. M., and Prystowsky, M. B.,** Proliferating cell nuclear antigen/cyclin is an interleukin 2-responsive gene, *J. Biol. Chem.,* 262, 8447, 1987.

109. **Jaskulski, D., Gatti, C., Travali, S., Calabretta, B., and Baserga, R.,** Regulation of the proliferating cell nuclear antigen cyclin and thymidine kinase mRNA levels by growth factors, *J. Biol. Chem.,* 263, 10175, 1988.

Section III. Negative Control of Cell Proliferation

Chapter 8

CELLULAR SENESCENCE: THE RESULT OF A GENETIC PROGRAM

Olivia M. Pereira-Smith, Andrea L. Spiering, and James R. Smith

TABLE OF CONTENTS

I. INTRODUCTION

Normal human and animal cells in culture exhibit limited division potential, in striking contrast to virus- or carcinogen-transformed cells or tumor-derived cells which divide indefinitely *in vitro*. The limited division potential of normal cells has been proposed as a model for aging at the cellular level.[1] The cells initially divide logarithmically and accumulate a variable number of population doublings, depending on the age and species of donor.[2-6] The rate of cell division then slows and the cells eventually cease dividing, entering the stage referred to as senescence. At this time, the cells are still metabolically active[6-10] and can be maintained in a viable state for 2 years or more,[11,33] though these senescent cells are unable to synthesize DNA and divide. In rodent cultures, spontaneous immortalization can occur either in the logarithmic growth phase or the senescent phase, and these immortal variants can proliferate indefinitely. In human cells, however, such spontaneous immortalization has never been observed. In a few rare instances following treatment with viruses, carcinogens, or cobalt irradiation, immortal cell lines have been obtained from cultures of human cells.[12-16] However, it must be emphasized that these human immortal cell lines are obtained at extremely low frequencies.

The mechanisms that limit the division potential of normal human cells in culture and the changes that occur in normal human cells to allow the generation of immortal variants are unknown. There are two broad categories into which all the various theories that have been proposed to account for cellular aging can be separated. One group of hypotheses presumes that random accumulation of cellular damage is responsible for cellular senescence. This could involve accumulation of errors, mutations, and free radical and DNA damage. The other group of theories postulates that a genetic program similar to differentiation is responsible for the aging phenotype. In this paper, we will describe experiments that have allowed us to distinguish between these two categories of mechanisms responsible for cellular aging and which support the idea that a genetic program is involved.

II. CELL HYBRIDIZATION EXPERIMENTS

We decided to study the proliferative potential of hybrids obtained following fusion of normal cells with immortal cells, postulating that if random accumulation of damage is responsible for the *in vitro* aging process, the assumption must be that immortal cells have somehow acquired the ability to deal better with the damage or to avoid the damage, thereby permitting them to divide indefinitely. One would expect that if this is the case, fusion of immortal cells with normal cells will generate hybrids that are also immortal.

The selection system used for isolating the hybrids was one that had been used by many other investigators, one which involved the generation of double-mutant parent cell lines that had both a recessive and a dominant mutation. The recessive mutation used was a deficiency in one of two enzymes, hypoxanthine phosphoribosyl transferase (HPRT$^-$) or thymidine kinase (TK$^-$), and the dominant mutation was either ouabain or neomycin resistance. Such a cell line could then be fused with any wild-type cell that was HPRT$^+$ or TK$^+$ and sensitive to ouabain or neomycin, and hybrids subsequently could be selected in medium containing hypoxanthine-aminopterin-thymidine (HAT) plus the drug of interest.

We began studying hybrids obtained from fusion of normal human cells with immortal SV40-transformed human cells,[17] since the viral large T antigen has been implicated in transformation and is thought to be involved in the immortalization process. Because one can readily detect the large T antigen by immunofluorescence staining, hybrids may be analyzed for the presence and expression of the viral genome. Following cell fusion, we inoculated approximately one to four cells per square centimeter into selective medium. After 2 weeks, large clones of 200 cells or more were isolated and carried in culture until

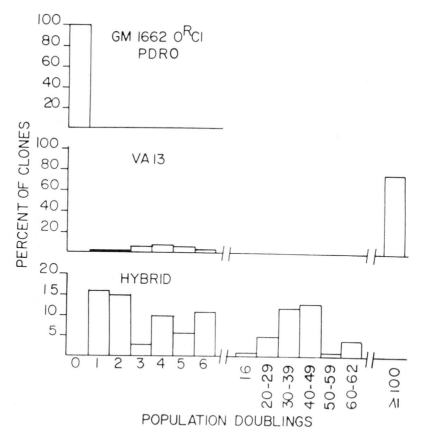

FIGURE 1.   Life span distributions of parental cell clones and hybrid clones. The GM1662 ouabain-resistant clone, PDR 0, is a senescent human cell line; VA13 is an immortal SV40-transformed human cell line. The distribution of hybrid life span is derived from small hybrid clone sizes scored microscopically and from the life span achieved by extensively dividing isolated clones before they entered the nonproliferative state. (From Pereira-Smith, O. M. and Smith, J. R., *Som. Cell Genet.*, 7, 411, 1981. With permission.)

they either ceased division or achieved at least 100 population doublings (our criterion for cellular immortality). The culture dishes were fixed, stained, and microscopically scored for the number of cells in the small clones. We considered a single cell to be a clone that was unable to divide at all, two cell clones to have achieved one population doubling (PD), four cell clones to have achieved two PD, and so on. The results of one such experiment are shown in Figure 1. In this case, the normal parent was senescent, and essentially all cells in the population were unable to achieve any population doublings. In the immortal SV40-transformed parent, the majority (about 70% of the population) were able to achieve greater than 100 PD. The majority of the hybrids (ca. 60%) had very limited proliferation potential (less than 6 PD), and about 40% of the hybrids went on to proliferate more extensively, achieving between 16 and 62 PD. However, even after achieving as many as 60 PD, these hybrids ceased dividing. The cells in the hybrid populations that had ceased doubling expressed SV40 large T antigen that was detectable by immunofluorescence and capable of inducing DNA synthesis in quiescent young human cells.[18] These results led us to conclude that the phenotype of cellular senescence is dominant in the presence of an integrated, expressed viral genome.

To confirm the generality of this phenomenon, we proceeded to fuse normal human cells with a variety of immortal human cell lines, including other immortal SV40-transformed

TABLE 1
Cell Lines Fused to Normal Human Cells

Cell line	Description
GM639	SV40 transformed skin fibroblasts
VA13	SV40 transformed lung fibroblasts
GM847	SV40 transformed skin fibroblasts (HPRT⁻ Lesch Nyhan)
HT1080	Fibrosarcoma (N-*ras*⁺)[a]
108021A	APRT⁻ clone of HT1080 (N-*ras*⁺)[a]
HeLa	Cervical carcinoma
T98G	Glioblastoma
143BTK⁻	Osteosarcoma secondarily transformed by Kirsten mouse sarcoma virus (K-*ras*⁺)[a]

[a] + = known to contain activated oncogene.

cell lines, and with tumor-derived cell lines (Table 1). All of the hybrids obtained from such fusions exhibited a limited division potential, supporting the earlier result indicating that the phenotype of immortality is recessive.[19] Our current hypothesis proposes the involvement of a series of programmed processes involving genes or sets of genes in normal cell growth control, and it also proposes that immortal cells arise through recessive changes in any one of these genes or sets of genes. Therefore, when normal cells are fused with immortal cells, the recessive change is complemented by the existing gene or sets of genes in the normal cell, and the hybrid behaves like a normal cell exhibiting limited division.

We exploited the fact that immortality is recessive to determine the complexity and the number of genes or sets of genes that are involved in normal growth control. We accomplished this by fusing different immortal cell lines with each other. If the two parent cells had become immortal by the same processes, then the genotype of the hybrid would involve two recessive sets of genes which would result in the phenotype of an immortal hybrid (Figure 2A). On the other hand, if one were to fuse immortal cells that had become immortal because of different sets of changes, there would be complementation of the recessive genes to produce a hybrid that would behave like a normal cell and would have limited division potential (Figure 2B). We have analyzed 20 immortal cell lines to date using the strategy shown in Figure 3. We began with an immortal SV40-transformed parent cell line, arbitrarily assigned to Group A. It was then fused with all other available immortal cell lines. If the hybrids obtained had unlimited life spans, they were assigned to Group A, along with the SV40-transformed line. If hybrids had a limited life span they were designated "not Group A". In these experiments, we restricted analysis to the large hybrid clones that could proliferate for more than 15 PD. Because we were dealing with aneuploid cell lines, we could not be sure that smaller hybrid clones obtained were truly representative of complementation rather than some other nonspecific genetic effect. A cell line from "not Group A" was chosen as the representative for Group B and was fused with the other cell lines. The parent cells involved in fusions that resulted in hybrids of unlimited life span were assigned to Group B, along with the parent cell line. The remaining cell lines were designated "not Group A, not Group B" and so on.

We chose a large variety of cell lines in order to assess the parameters affecting complementation group assignment. We used a variety of immortal SV40-transformed cell lines to determine if cell type or embryonal layer of origin affected complementation group assignment. We included other DNA tumor virus-containing cell lines to determine if they would be assigned to the same group as the SV40-transformed cell lines. We included cell lines that had been obtained following carcinogen treatment or cobalt irradiation to determine their group assignment and a large battery of tumor-derived cell lines (bladder carcinomas and lung carcinomas) to determine if the tumor types would be assigned to the same group.

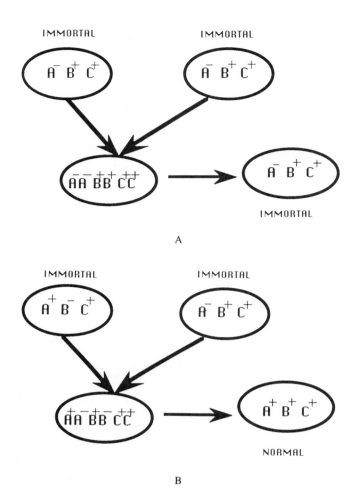

FIGURE 2. Schematic diagrams of cell fusion between two immortal cells. (A) Each immortal cell has become immortal by the same processes; (B) each immortal cell has obtained immortality by different changes.

We also included lines containing activated oncogenes to determine whether their presence would affect complementation group assignment.

We have identified a limited number of complementation groups (four) to which we have assigned the cell lines to date. Other fusion results are still in progress. The most striking correlation that we have found is that all tested immortal SV40-transformed cell lines assign to Group A, indicating that this virus immortalizes different human cells by the same processes.[18] We are currently pursuing the question of whether any SV40 viral genes or gene products are involved in the maintenance of immortalization of human cells. T antigen-expressing cells that exhibit limited division potential have been observed in our hybrid studies, as well as in other studies.[12,15-17,19] We know that T antigen alone is not sufficient to maintain immortalization of these cells. Therefore, two possibilities remain: either the SV40 virus acts in cooperation with cellular genes to maintain immortalization or the virus initiates some events, such as chromosome rearrangement and karyotypic abnormalities, that are actually reponsible for maintenance of immortalization and are independent of viral genes. The cell lines EJ and HT1080, which contain activated *ras* oncogenes, assign to the same group as the SV40 cell lines. If we find T antigen is involved in maintenance of immortalization, candidate cellular genes that could cooperate with T antigen would include

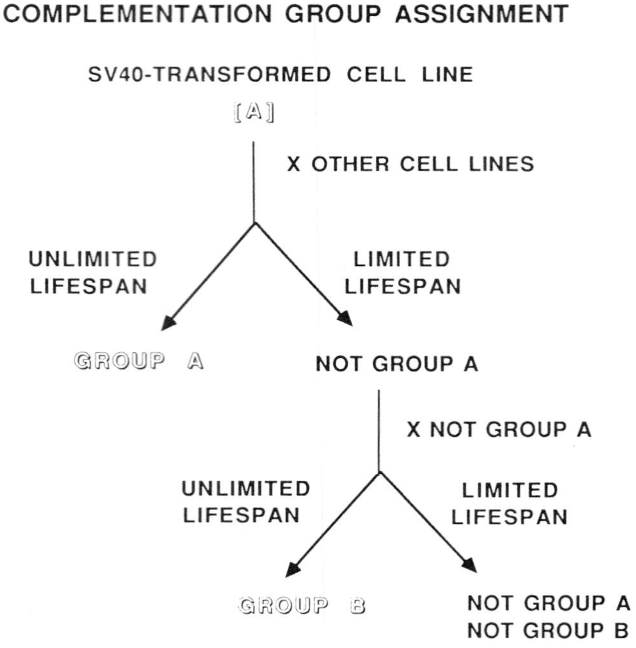

COMPLEMENTATION GROUP ASSIGNMENT

SV40-TRANSFORMED CELL LINE
[A]

X OTHER CELL LINES

UNLIMITED LIFESPAN

LIMITED LIFESPAN

GROUP A   NOT GROUP A

X NOT GROUP A

UNLIMITED LIFESPAN

LIMITED LIFESPAN

GROUP B   NOT GROUP A
NOT GROUP B

FIGURE 3.   Strategy for complementation group assignment.

the *ras* family of oncogenes. By assigning different immortal cell lines to specific complementation groups, we can begin a focused approach to determine the changes which have
occurred in different cell lines within a single complementation group to result in the immortal
phenotype.

III. DNA SYNTHESIS INHIBITORS

A theory on the mechanisms that may be involved in normal cell growth control has
evolved from heterokaryon and cell fusion studies done by the groups of Norwood and Stein
and by ourselves.[20-25] These results indicate that senescent cells express a surface membrane-
associated protein inhibitor of initiation of DNA synthesis not produced in young proliferation-competent cells.[26,27] Our current hypothesis is that cellular senescence is the result of
a genetic program, the end point of which is the production of this inhibitor protein.

Heterokaryon and cell fusion studies also indicate that a protein inhibitor of DNA synthesis
having properties similar to the senescent cell inhibitor is present in the surface membranes
of cells made nondividing (quiescent) by growth factor deprivation.[26-28] However, the quiescent cell inhibitor protein does differ in some properties from that expressed in senescent
cells. We do not know if the inhibitory proteins are the same protein constitutively expressed
in senescent cells and reversible in quiescent cells or if they are totally different proteins;
only purification of the proteins will allow this to be determined.

Treatment of quiescent and senescent fibroblast monolayers (human foreskin fibroblasts,
CSC303) with the detergent octyl-β-D-glucopyranoside results in solubilization of these
inhibitor proteins. Young proliferation-competent cells are inhibited from entering S phase
by these extracted proteins in a dose-dependent manner (Figure 4). We have concentrated
on purification of the quiescent inhibitor protein due to the difficulties involved in obtaining
sufficient numbers of senescent human fibroblasts. Figure 5 shows the effect of duration of
quiescence on inhibitory activity. Maintenance of quiescence for more than 5 weeks does

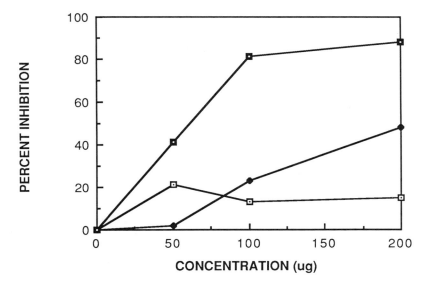

FIGURE 4. Dose-dependent inhibition of DNA synthesis by quiescent and senescent monolayer extracts. Young proliferation-competent cells (2×10^4) are seeded on 12-mm glass coverslips in 24-well plates. After attachment, the cells are exposed to ^{3}H-thymidine-containing media $+/-$ cell extracts. After 24 h, the cells are fixed and processed for autoradiography. Percent inhibition is defined as (percent label in control nuclei minus percent label in treated nuclei) divided by percent label in control nuclei. ⊡ , young cell extract; ◆ , quiescent cell extract; ◼ , senescent cell extract.

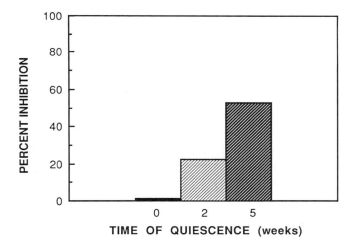

FIGURE 5. Effect of duration of quiescence on inhibitory activity. Percent inhibition of DNA synthesis was determined as described in Figure 4, using 100 μg/ml extract. ◼, nonquiescent; ▨ , 2-week quiescent; ▨ , 5-week quiescent.

not greatly increase inhibitory activity. Using 5-week quiescent human fibroblasts, we have purified the inhibitor protein fourfold utilizing coupled ammonium sulfate extractions (Figure 6).

If one accepts the hypothesis that the inhibitor of DNA synthesis that is expressed in senescent cells is the end point of a genetic program that is responsible for the aging process, a number of mechanisms can be postulated for escape from senescence. A cell may become

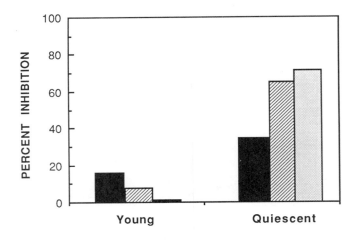

FIGURE 6.   Recovery of quiescent cell inhibitory activity during ammonium sulfate extractions. ■, crude extract; ▨ , supernatant from 10%
ammonium sulfate; □ , pellet from 30% ammonium sulfate from young
and quiescent cell monolayers. Percent inhibition was determined as described for Figure 4, using 100 µg/ml extract.

immortal because it no longer produces the inhibitor (a gene is lost or transcription is blocked),
because it produces an inactive inhibitor (defects in post-transcriptional or post-translational
modifications), or because it fails to respond to an active inhibitor (perhaps through loss of
a receptor).

To investigate whether any one of these postulated mechanisms occurs, we analyzed the
immortal cell line SUSM-1 for production of an active inhibitor of DNA synthesis. This
cell line was derived from the human liver fibroblast line AD387 following treatment with
the chemical carcinogen 4-nitroquinoline-1-oxide.[34] We found that SUSM-1 cells produce
an extremely potent inhibitor that is able to inhibit young cells at concentrations 20-fold
lower than the quiescent or senescent cell inhibitors (Figure 7). The SUSM-1 cell line is
itself incapable of responding to the inhibitor protein produced by either senescent or quiescent human cells, as well as to the inhibitor protein produced by itself. This result suggests
that loss of ability to respond to exogenous or endogenous inhibitors may indeed be one of
the mechanisms by which cells are able to proliferate indefinitely.

To test whether mRNA isolated from senescent cells could inhibit DNA synthesis, proliferation-competent cells made quiescent by short-term growth factor deprivation were
microinjected with mRNA that was isolated by guanidine isothiocyanate extraction and
passage over an oligo(dT) column.[29] Fetal bovine serum (10%) was then returned to the cell
culture medium to stimulate cell division. The ability of the injected mRNA to inhibit the
stimulation of DNA synthesis was scored by ^{3}H-thymidine autoradiography. The result of
an experiment in which poly-A[+] RNA isolated from senescent human fibroblasts was injected
into proliferation-competent cells is shown in Figure 8. Inhibition ranged from 50 to 80%.
This inhibitory activity could be eliminated by treating the mRNA with RNAse prior to
injection. The non-poly-A[+] RNA fraction obtained from the oligo(dT) column had no
inhibitory effect. Young cell poly-A[+] RNA also had no inhibitory activity. We injected
poly-A[+] RNA isolated from cells made quiescent by removal of serum growth factors and
observed low but significant inhibition.[29] Next, we did a series of dilution experiments to
determine the concentration at which activities of these poly-A[+] RNAs would be undetectable. We found that poly-A[+] RNA isolated from senescent cells retained high inhibitory
activity, even at concentrations of 0.03 mg/ml or less. Thus, senescent cell poly-A[+] RNA
was 100-fold more active than the quiescent cell poly-A[+] RNA, and poly A[+] RNA extracted

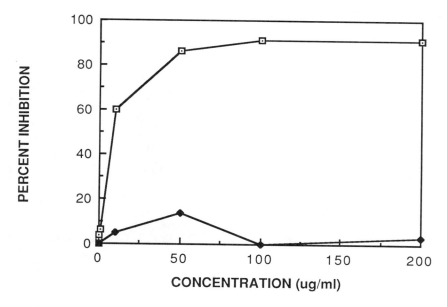

FIGURE 7.   Dose-dependent inhibition of DNA synthesis by SUSM-1 monolayer extracts.
⊡ , SUSM-1 crude extract; ◆ , young fibroblast crude extract.

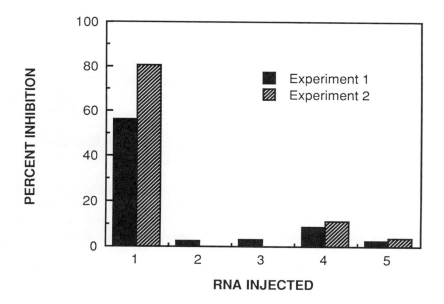

FIGURE 8.   RNA microinjection. RNAs isolated from senescent, quiescent, and young
human diploid fibroblasts were microinjected into quiescent cells, as described in the text.
1, senescent cell poly-A$^+$ RNA, 1 mg/ml; *2*, senescent cell poly-A$^+$ RNA (1 mg/ml) +
RNAse; *3*, senescent cell non-poly-A$^+$ RNA, 2 mg/ml; *4*, quiescent cell poly-A$^+$ RNA, 1.7
mg/ml; *5*, young cell poly-A$^+$ RNA, 1 mg/ml.

from young cells had essentially no activity, even at a concentration of 5 mg/ml. From the
amount of RNA required for inhibition, we calculated that the abundance of inhibitory
mRNAs in senescent cells was on the order of 0.1 to 1% of all of the poly-A$^+$ RNA in the
cells.[29] This result suggests that we should easily be able to identify senescent-specific cDNA
clones coding for this inhibitor of DNA synthesis by plus/minus screening of a cDNA library

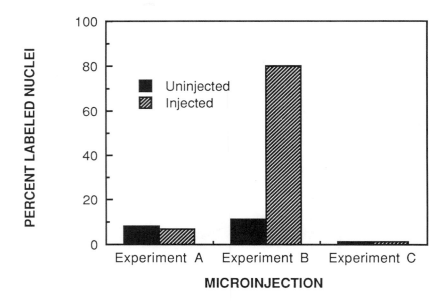

FIGURE 9.   The effect of microinjected pBR322 ad c-H-*ras*(EJ) DNAs on DNA synthesis in human diploid fibroblasts. DNA (1 mg/ml) was microinjected into quiescent or senescent cells, as described in the text. Experiment A, pBR322 DNA microinjected into quiescent cells; Experiment B, c-H-*ras*(EJ) DNA microinjected into quiescent cells; Experiment C, c-H-*ras*(EJ) microinjected into senescent cells.

generated from senescent cells. Once these cDNA probes become available to us, we will be able to analyze different tissues and make correlations between *in vitro* and *in vivo* aging that have previously not been possible. We can also probe the various immortal cell lines within different complementation groups to determine whether immortalization involves the inhibitor gene or other genes that may control the expression of the inhibitor.

IV. EFFECT OF EXOGENOUS *ras* DNA ON SENESCENT AND QUIESCENT CELLS

The isolation of DNA synthesis inhibitors from senescent cells, combined with reports discussing suppressor genes or anti-oncogenes,[30] led us to determine whether oncogenes could override the action of the inhibitor of DNA synthesis produced in senescent cells.[31] We studied the C-H-*ras* oncogene because it is extremely well characterized. Cells made quiescent either by removal of serum growth factors or by growth to high density were stimulated to synthesize DNA following microinjection of plasmid pBR322 carrying the c-H-*ras* DNA from the EJ carcinoma cell line (Figure 9). Senescent cells, although expressing the injected DNA (as determined by immunofluorescence), were not stimulated to synthesize DNA (Figure 9). These results further suggest differences between the quiescent and senescent cell inhibitors. Additionally, we microinjected senescent human cells with a combination of oncogenes known to not only transform rodent cells but to convert them to the tumorigenic phenotype. We used the c-H-*ras* oncogene, along with the DNA coding for the EIA region of adenovirus. As shown in Table 2, this combination of oncogenes was unable to initiate DNA synthesis in the senescent cells, despite the fact that they were expressing the introduced DNAs. To eliminate the possibility that senescent cells were more sensitive to the microinjection procedure and therefore did not respond to introduced DNA, we microinjected a plasmid containing SV40 virus DNA (pBSV1). SV40 DNA is the only agent known to initiate DNA synthesis in senescent cells, though the cells that enter S phase, as

TABLE 2
Microinjection of Oncogenes into Senescent Cells

	Concentration (mg/ml)	% Label	
		Injected	Uninjected
c-H-*ras*	1.5	<1	<1
c-H-*ras* + pAd12E1A	1.5 + 1.5	<1	<1
pBSV-1	0.07	12	<1

measured by ^{3}H-thymidine uptake, do not enter mitosis.[32] Microinjected SV40 DNA was capable of stimulating DNA synthesis in the senescent cells, showing that senescent cells are able to respond to microinjected DNA (Table 2). These results further demonstrate the dominant action of the senescent cell inhibitor.

V. DISCUSSION

The mechanisms that limit the growth potential of normal cells have been the subject of intensive study for many years. The majority of the studies have been predicated on the general idea that cessation of cell division is the result of defects in the cell machinery. A widely held idea is that these defects result from accumulation of various sorts of damage (e.g., various kinds of DNA damage have been proposed).

The observation that hybrids formed between normal human fibroblasts and immortal human cells have a finite life span in culture argues for a dominant process that results in cellular senescence. This suggests that the accumulation of various kinds of errors is not a significant contributing factor in cellular aging. The ability to assign immortal cells to different complementation groups further strengthens this interpretation. If a lower error rate or the ability to proliferate carrying high levels of errors was a mechanism of cellular immortality, then one would not expect hybrids formed from the fusion of two immortal cell lines to have a finite proliferative potential. We have found many different combinations of fusions of immortal cells that result in the cessation of proliferation many population doublings after fusion (i.e., the limited life span phenotype). Further evidence that loss of cell proliferation is an active process comes from experiments in which cytoplasts prepared from senescent cells have been fused with proliferation competent young cells. In such fusion products the ability to synthesize DNA is inhibited. If the cytoplasts derived from senescent cells are treated with trypsin or protein synthesis inhibitors prior to fusion, the inhibitory activity is abolished. These results argue strongly that senescent cells are producing an inhibitor of DNA synthesis. We feel that it is likely that the production of this inhibitor is the event that triggers cellular senescence; however, this remains to be proven. An interesting finding in this regard is the production of an inhibitor by young human diploid fibroblasts made quiescent by long-term deprivation of serum growth factors. This inhibitor is rapidly lost upon stimulation of quiescent cells with 10% fetal bovine serum. However, the inhibitor detected in senescent cells is constitutively produced, even in the presence of high levels of serum growth factors. It is not clear at this time whether these are the same or different inhibitors. If they are the same, then senescent cells have lost the ability to down regulate its expression in response to growth factor signals. In either case, a thorough study of the mechanism of expression of these inhibitor genes is warranted.

We feel that the experimental results discussed in this chapter provide strong evidence that cellular senescence is an active process resulting from the genetic program of normal cells. The nature of the program and the mechanism of action of the senescent cell inhibitor are not known at this time and await the results of further studies.

ACKNOWLEDGMENT

This work was supported by NIH grants AGO4749 and AG5333 and by USPHS Training Grant T32-CA09197.

REFERENCES

1. **Hayflick, L.,** The limited *in vitro* lifetime of human diploid cell strains, *Exp. Cell Res.,* 37, 614, 1965.
2. **Martin, G. M., Sprague, C. A., and Epstein, C. J.,** Replicative lifespan of cultivated human cells. Effects of donor age, tissue and genotype, *Lab. Invest.,* 23, 86, 1979.
3. **Goldstein, S., Littlefield, J. W., and Soeldner, J. S.,** Diabetes mellitus and aging: diminished plating efficiency of cultured human fibroblasts, *Proc. Natl. Acad. Sci. U.S.A.,* 64, 155, 1969.
4. **LeGuilly, Y., Simon, M., Lenoir, P., and Bourel, M.,** Long term culture of human adult liver cells. Morphological changes related to *in vitro* senescence and effect of donors age on growth potential, *Gerontologia,* 19, 303, 1973.
5. **Schneider, E. L. and Mitsui, Y.,** The relationship between *in vitro* cellular aging and *in vivo* human age, *Proc. Natl. Acad. Sci. U.S.A.,* 73, 3584, 1976.
6. **Hayflick, L.,** The cellular basis for biological aging, in *Handbook of the Biology of Aging,* Finch, C. E. and Hayflick, L., Eds., D Van Nostrand, New York, 1977.
7. **Cristofalo, V. J., Parris, N., and Kritchevsky, D.,** Enzyme activity during the growth and aging of human cells *in vitro, J. Cell. Physiol.,* 69, 263, 1967.
8. **Pitha, J., Adams, R., and Pitha, P. M.,** Viral probe into the events of cellular (*in vitro*) aging, *J. Cell. Physiol.,* 83, 211, 1974.
9. **Razin, S., Pfendt, E. A., Matsumura, T., and Hayflick, L.,** Comparison by autoradiography of macromolecular biosynthesis in young and old human diploid fibroblasts, *Mech. Ageing Dev.,* 6, 379, 1977.
10. **Paz, M. A. and Gallop, P. M.,** Collagen synthesized and modified by aging fibroblasts in culture, *In Vitro,* 11, 302, 1975.
11. **Matsumura, T., Zerrudo, Z., and Hayflick, L.,** Senescent human diploid cells in culture: survival, DNA synthesis and morphology, *J. Gerontol.,* 34, 328, 1979.
12. **Gaffney, E. V., Fogh, J., Ramos, L., Loveless, J. D., Fogh, H., and Dowling, A.,** Established lines of SV40 transformed human amnion cells, *Cancer Res.,* 1668, 1970.
13. **Kakunaga, T.,** Neoplastic tranformation of human diploid cells by chemical carcinogens, *Proc. Natl. Acad. Sci. U.S.A.,* 75, 1334, 1978.
14. **Namba, M., Nishitani, K., and Kimoto, T.,** Neoplastic transformation of a normal human diploid cell strain WI-38 with CO-60 gamma rays, *Jpn J. Exp. Med.,* 48, 303, 1978.
15. **Shein, H. M., Enders, J. F., Palmer, L., and Grogan, E.,** Further studies on SV40 induced transformation in human renal cell cultures. I. Eventual failure of subcultivation despite a continuing high rate of cell division, *Proc. Soc. Exp. Biol. Med.,* 115, 618, 1964.
16. **Gotoh, S., Gelb, L., and Schlessinger, D.,** SV40-transformed human diploid cells that remain transformed throughout their limited lifespan, *J. Gen. Virol.,* 43, 409, 1979.
17. **Pereira-Smith, O. M. and Smith, J. R.,** Expression of SV40 T antigen in finite lifespan hybrids of normal SV40 transformed fibroblasts, *Som. Cell Genet.,* 7, 411, 1981.
18. **Pereira-Smith, O. M. and Smith, J. R.,** Functional SV40 T antigen is expressed in hybrid cells having finite proliferative potential, *Mol. Cell. Biol.,* 7, 1541, 1987.
19. **Pereira-Smith, O. M. and Smith, J. R.,** Evidence for the recessive nature of cellular immortality, *Science,* 221, 964, 1983.
20. **Norwood, T. H., Pendergrass, W. R., Sprague, C. A., and Martin, G. M.,** Dominance of the senescent phenotype in heterokaryons between replicative and post-replicative human fibroblast-like cells, *Proc. Natl. Acad. Sci. U.S.A.,* 71, 2231, 1974.
21. **Stein, G. H. and Yanishevsky, R. M.,** Entry into S phase is inhibited in two immortal cell lines fused to senescent human diploid cells, *Exp. Cell Res.,* 120, 155, 1979.
22. **Yanishevsky, R. M. and Stein, G. H.,** Ongoing DNA synthesis continues in young human diploid cells (HDC) fused to senescent HDC, but entry into S phase is inhibited, *Exp. Cell Res.,* 126, 469, 1980.
23. **Burmer, G. C., Motulsky, H., Ziegler, C. J., and Norwood, T. H.,** Inhibition of DNA synthesis in young cycling human diploid fibroblast-like cells upon fusion to enucleate cytoplasts from senescent cells, *Exp. Cell Res.,* 145, 79, 1983.

24. **Drescher-Lincoln, C. K. L. and Smith, J. R.,** Inhibition of DNA synthesis in proliferating human diploid fibroblasts by fusion with senescent cytoplasts, *Exp. Cell Res.,* 144, 455, 1983.
25. **Drescher-Lincoln, C. K. L. and Smith, J. R.,** Inhibition of DNA synthesis in senescent-proliferating human cybrids is mediated by endogenous proteins, *Exp. Cell Res.,* 153, 208, 1984.
26. **Pereira-Smith, O. M., Fisher, S. F., and Smith, J. R.,** Senescent and quiescent cell inhibitors of DNA synthesis: membrane associated protein(s), *Exp. Cell Res.,* 160, 297, 1985.
27. **Stein, G. H. and Atkins, L.,** Membrane associated inhibitor of DNA synthesis in senescent human diploid fibroblasts. Characterization and comparison to quiescent cell inhibitor, *Proc. Natl. Acad. Sci. U.S.A.,* 83, 9030, 1986.
28. **Stein, G. H., Atkins, L., Beeson, M., and Gordon, L.,** Quiescent human diploid fibroblasts: common mechanism for inhibition of DNA replication in density-inhibited and serum-deprived cells, *Exp. Cell Res.,* 162, 255, 1986.
29. **Lumpkin, C. K. L., McClung, K., Pereira-Smith, O. M., and Smith, J. R.,** Existence of high abundance antiproliferative mRNAs in senescent human diploid fibroblasts, *Science,* 232, 393, 1986.
30. **Sager, R., Tanaka, K., Lau, C. C., Ebina, Y., and Anisowicz, A.,** Resistance of human cells to tumorigenesis induced by cloned transforming genes, *Proc. Natl. Acad. Sci. U.S.A.,* 80, 7601, 1983.
31. **Lumpkin, C. K. L., Knepper, J. E., Butel, J. S., Smith, J. R., and Pereira-Smith, O. M.,** Mitogenic effects of the proto-oncogene and oncogene forms of c-H-*ras* DNA in human diploid fibroblasts, *Mol. Cell. Biol.,* 6, 2990, 1986.
32. **Gorman, S., Hoffman, E., Nichols, W. W., and Cristofalo, V. J.,** Spontaneous transformation of a cloned cell line of normal diploid bovine vascular endothelial cells, *In Vitro,* 20, 339, 1984.
33. **Pereira-Smith, O. M., Spiering, A. L., and Smith, J. R.,** unpublished data.
34. **Namba, M.,** personal communication.

Chapter 9

INHIBITORS OF DNA SYNTHESIS IN SENESCENT AND QUIESCENT HUMAN DIPLOID FIBROBLASTS

Gretchen H. Stein

TABLE OF CONTENTS

I. INTRODUCTION

Human diploid fibroblasts (HDF) have a finite proliferative life span in culture.[1] This means that at first, young HDF can proliferate vigorously and accomplish about four population doublings (PD) per week in medium containing 10% fetal bovine serum. As the cells undergo successive PD, their proliferative capacity declines until they cannot accomplish even one PD in a reasonable amount of time, which is usually 1 to 4 weeks. The cessation of net PD defines the finite proliferative life span phenotype. We call this the FPL$^+$ phenotype, using the convention that the phenotypes of normal HDF are always designated " + "[2] (see Table 1 for a summary of the phenotypes of human fibroblasts). Cells that are transformed to "immortality", i.e., that have an unlimited proliferative capacity, express the FPL$^-$ phenotype. When senescent HDF reach the end of their proliferative life span, they enter a G_1-arrested senescent state in which they can be kept alive for many months.[3,4] We call this the S$^+$ phenotype, where S stands for senescence and not for S phase.[2] Cells that exhibit the FPL$^+$ phenotype can also exhibit a different behavior (the S$^-$ phenotype) at the end of their life span.[2,5] The S$^-$ phenotype means that individual cells in the population continue to synthesize DNA even though the population as a whole has ceased to increase in cell number (FPL$^+$ phenotype). As expected, there is an increase in cell death in FPL$^+$, S$^-$ populations at the end of their life span.[5] Thus, FPL$^+$ populations of cells may cease to proliferate at the end of their life span, owing to either a lack of cell birth with no appreciable cell death (S$^+$ phenotype) or a balance of cell death and cell birth (S$^-$ phenotype). Later in this paper, we will discuss evidence for the existence of the S$^-$ phenotype.

The ability to enter a viable, reversible, G_1-arrested quiescent state (Q$^+$ phenotype) is another aspect of control of cell proliferation in normal HDF.[5,6] These cells express the Q$^+$ phenotype whenever they experience poor growth conditions, such as cell crowding or mitogen deprivation. In contrast, transformed cells frequently express the Q$^-$ phenotype, which means that they continue to traverse the cell cycle even when they are crowded or mitogen deprived. Nevertheless, they reach a terminal cell density in any given culture. This is achieved through a balance of cell death and cell birth rather than through a lack of cell birth.[8] There are obvious parallels between the S$^+$ and Q$^+$ phenotypes because both involve arrest with G_1-phase DNA content[3,6] and maintenance of viability for many months.[4,7] Likewise, the S$^-$ and Q$^-$ phenotypes have in common continued DNA synthesis and decreased viability. One of the issues that we will deal with in this paper is the possibility that there is a common mechanism for inhibition of DNA synthesis in HDF in both the S$^+$ senescent state and the Q$^+$ quiescent state.

Our working hypothesis concerning the mechanisms for the FPL$^+$, S$^+$, and Q$^+$ phenotypes has two tenets.[2,3,9,10] The first tenet is that the aging process involves a progressively decreasing ability of cells to recognize or respond to mitogens. Cells that undergo this aging process express the FPL$^+$ phenotype. The second tenet is that HDF produce or activate an inhibitor of entry into S phase whenever they experience poor growth conditions, such as cell crowding or mitogen deprivation. Cells that produce this inhibitor express the Q$^+$ phenotype. Taken together, these two tenets imply that (1) as FPL$^+$ cells age, they gradually become functionally mitogen deprived, even though they are cultured in mitogen-containing medium; and (2) functional mitogen deprivation in old HDF triggers production of the same inhibitor of entry into S phase as does environmental mitogen deprivation of young HDF. Thus, this hypothesis suggests that senescent and quiescent HDF contain the same inhibitor of entry into S phase. This is compatible with the reversibility of the inhibitor in quiescent HDF and its irreversibility in senescent HDF because in the former case, production of the inhibitor is triggered by poor growth conditions, which can be reversed, and in the latter case, production of the inhibitor is triggered by a change in the cells, which cannot be reversed by manipulating the culture conditions. In the subsequent sections of this chapter,

TABLE 1
Phenotypes of Human Fibroblasts

FPL^+:   finite proliferative life span
FPL^-:   unlimited proliferative life span
   S^+:   arrest in G_1 phase; high viability at end of life span
   S^-:   continued DNA synthesis; increased cell death at end of life span
   Q^+:   arrest in G_1 phase; high viability when crowded or mitogen deprived
   Q^-:   continued DNA synthesis; increased cell death when crowded or mitogen deprived

we will summarize the data that led to this hypothesis, describe our more recent studies of the membrane-associated inhibitors of DNA synthesis in senescent and quiescent HDF, and describe experiments that test the genetic predictions of this hypothesis.

II. HETEROKARYON STUDIES

Studies of the behavior of senescent HDF in heterokaryons formed with replicative cells of various types have suggested that senescent HDF contain an inhibitor of entry into S phase.[11,12] When senescent HDF are fused to replicating young HDF, DNA synthesis is not reinitiated in the senescent HDF nuclei in heterodikaryons. Rather, the young HDF nuclei in the heterodikaryons are prevented from synthesizing DNA. In our experiments, the cessation of DNA synthesis in the young replicating HDF nuclei did not take place immediately after fusion.[12] Instead, the young HDF nuclei in S phase at the time of fusion were able to continue DNA synthesis in the heterokaryons in a manner that suggested that they could complete the ongoing round of DNA replication. These results suggest that senescent HDF contain an inhibitor of entry into S phase. This conclusion is consistent with the fact that the senescent HDF themselves are arrested with G_1-phase DNA content.

When senescent HDF are fused to transformed cells, two alternative behaviors are observed in heterodikaryons. Human cells transformed by SV40 or adenovirus serotype 5 (Ad5) are able to induce DNA synthesis in senescent HDF nuclei in heterodikaryons.[10,13] In contrast, carcinogen-transformed HDF are not able to induce DNA synthesis in senescent HDF nuclei; rather, the transformed nuclei in these heterodikaryons are inhibited from entering S phase.[10] Human cells transformed by Rous sarcoma virus and most human tumor cells are similarly inhibited by fusion to senescent HDF. HeLa is the only human tumor cell line (out of six tested) that was capable of inducing DNA synthesis in senescent HDF.[13] These data indicate that transformation to the FPL^- phenotype does not necessarily confer the ability to override or inactivate the senescent HDF inhibitor of entry into S phase. However, transformation by a DNA virus may involve the ability to override the senescent HDF inhibitor because this is a common attribute of SV40-transformed cells, Ad5-transformed cells, and HeLa cells, which have been shown to contain human papilloma virus 18 DNA sequences.

Quiescent HDF also appear to have an inhibitor of entry into S phase, as measured by the heterodikaryon assay.[9,14] Furthermore, the quiescent HDF inhibitor has many of the same attributes as the senescent HDF inhibitor. Briefly, it inhibits replicating HDF, carcinogen-transformed HDF, and several human tumor cell lines, but it is overridden by SV40-transformed HDF, Ad5-transformed cells, and HeLa cells in heterodikaryons.[9] In addition, the kinetics of DNA synthesis inhibition or induction are the same in heterodikaryons formed between a given type of replicative cell (e.g., HDF or HeLa) and either senescent or quiescent HDF.[14] However, there is an exception to the parallel behavior of senescent and quiescent HDF in heterodikaryons; cycloheximide treatment of serum-deprived quiescent HDF does not abolish their inhibitory activity in heterokaryons and cybrids as it does the inhibitory activity of senescent HDF.[15,16] It is not known whether this difference in response to cycloheximide reflects an inherent difference between the senescent HDF inhibitor and the

TABLE 2
Serum Stimulation of Young and Old Quiescent WI-38

	% labeled nuclei	
Stage when nuclei were labeled	PDL[a] = 16—24	PDL[a] = 48—57
At quiescence	4	3
After refeeding with medium + 10% serum	36	16
After refeeding with medium + 30% serum	57	39
During exponential growth	95	73

Note: Young and old WI-38 cells were labeled with 1 μCi/ml ³H-thymidine for 24 h during various phases of growth. The percentage labeled nuclei was determined by autoradiography.

[a] PDL = population doubling level.

quiescent HDF inhibitor or a difference in rates of protein degradation in the high-serum-containing medium in which the senescent HDF were grown vs. the low-serum-containing medium in which the quiescent HDF were grown.

The heterokaryon data described above first suggested that senescent and quiescent HDF might contain the same inhibitor of entry into S phase. One way that this could happen would be if the HDF inhibitor was produced or activated in response to two different signals, one from senescent HDF and one from quiescent HDF. However, another possibility is that it is actually the same signal in both cases.[9] We have incorporated this latter possibility into our hypothesis by postulating that an effect of the aging process is to make old HDF functionally mitogen deprived. This idea is consistent with the following data. Table 2 shows a comparison of the percentage of ³H-thymidine-labeled nuclei in young and old HDF under various growth conditions. During logarithmic growth phase, the labeling index of the young HDF was higher than that of the old HDF, as had been shown previously by Cristofalo and Sharf.[17] Similarly, when both young and old HDF were made quiescent and then stimulated with fresh medium containing 10 or 30% serum, the percentage of cells synthesizing DNA was always greater in the young HDF populations than in the old ones (36 vs. 16% at 10% serum and 57 vs. 39% at 30% serum). However, the interesting aspect of this experiment was that when the old HDF were given 30% serum, their proliferative response was as good as that of the young HDF given 10% serum (39 vs. 36%). Furthermore, the difference between the responsiveness of the young and old cells was narrowed from a ratio of 2.15 at 10% serum to a ratio of 1.46 at 30% serum. These results suggested that an important change in old HDF might be in their ability to recognize or respond to mitogens, such that they require a progressively stronger mitogenic signal to elicit the same proliferative response as they age. More recently, Ohno has shown in a very thorough way that as HDF age from 28 PD to the end of their life span at 70 PD, they have a progressively increasing serum requirement to maintain the same rate of proliferation.[18] Eventually, even 100% serum is insufficient for proliferation. These data support the hypothesis that HDF cease to proliferate in a given amount of serum or other mitogen when their ability to respond to that mitogen decreases to the point where the aged cells "feel" as mitogen deprived as do young HDF in a mitogen-deficient medium.

III. MEMBRANE-ASSOCIATED INHIBITORS OF DNA SYNTHESIS

Recently, we and others have been looking for the inhibitor (or inhibitors) of DNA synthesis that have been predicted from the heterokaryon experiments.[15,19-23] We have found

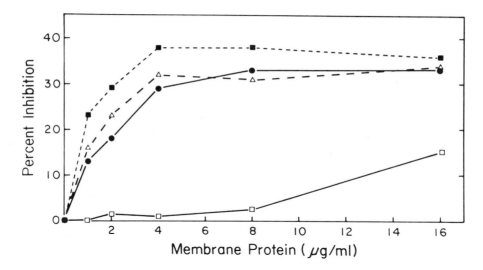

FIGURE 1. Inhibition of entry into S phase by membrane fractions prepared from senescent HDF (●), serum-deprived quiescent young HDF (△), density-inhibited quiescent young HDF (■), and replicating HDF (□). Percentage inhibition was determined by comparing the percentage of ^{3}H-thymidine-labeled nuclei in membrane-treated cultures of replicating young HDF vs. untreated cultures of replicating young HDF. The percentage labeled nuclei in the untreated cultures of replicating young HDF varied from 38 to 46% in these experiments. (From Stein, G. H. and Atkins, L., *Proc. Natl. Acad. Sci. U.S.A.*, 83, 9030, 1986. With permission.)

that both senescent and quiescent HDF contain plasma membrane-associated inhibitors of DNA synthesis.[21,22] When plasma membranes were prepared from senescent HDF and added to the culture medium of young replicating HDF, they caused a reduction in the percentage of young cells synthesizing DNA (Figure 1). The inhibitory activity of the senescent HDF plasma membranes reached a plateau of about 35% inhibition at 4 to 8 μg of membrane protein per milliliter. When plasma membranes were prepared from quiescent HDF, they had virtually the same inhibitory effect on DNA synthesis in the recipient cells. This was true regardless of whether the quiescent cells were serum deprived or density inhibited. Plasma membranes prepared from replicating HDF had a negligible inhibitory effect at the doses where senescent and quiescent HDF membranes had their maximal effect (Figure 1). The modest inhibitory effect of replicating HDF at a dose of 16 μg/ml may mean that replicating HDF cells contain a low level of the inhibitor, or it may mean that the replicating cell population contains a small fraction of nonreplicative cells that possess the inhibitor.

Characterization of the senescent HDF membrane-associated inhibitory activity has shown that it is sensitive to inactivation by trypsin and periodate, which cleave proteins and carbohydrates, respectively.[22] These data suggest that the senescent HDF inhibitory activity depends on a glycoprotein. The inhibitory activity is also inactivated by heating at 80°C for 10 min. This result implies that the senescent HDF inhibitor is not the same as the transforming growth factor beta/growth inhibitor, which is heat stable and typically inhibitory to epithelial cells rather than to mesenchymal cells. On the other hand, the quiescent HDF membrane-associated inhibitor shows the same sensitivity to heat, trypsin, and periodate as the senescent HDF inhibitor. Therefore, these data are in keeping with the possibility that senescent and quiescent HDF have the same inhibitor.

Because quiescence is reversible and senescence is not, we examined the effect of serum stimulation on the membrane-associated inhibitory activity in both types of cells.[22] When young density-inhibited quiescent HDF were refed with fresh serum-containing medium, most of the membrane-associated inhibitory activity was still present 12 h later. By 20 h

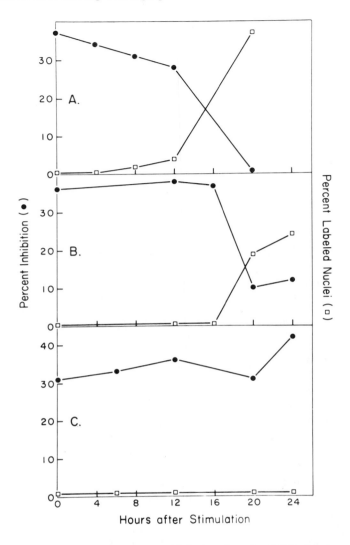

FIGURE 2.   Effect of refeeding with fresh medium plus 10% fetal bovine serum on density-inhibited quiescent young HDF (A), density-inhibited quiescent old HDF (B), and senescent HDF (C). At various times after stimulation, cells were harvested for analysis of their membrane-associated inhibitory activity. The dose-response curve for inhibition of DNA synthesis in replicating recipient cells was determined for each membrane preparation, as described in Figure 1. The average percentage inhibition at 4, 8, and 16 μg of membrane protein per milliliter is given here (●). The average percentage inhibition was used because it more accurately reflected the overall differences between the dose-response curves at different time points than did any single value. The percentage of cells in S phase in the membrane donor cell cultures was also measured by a pulse of ³H-thymidine at each time point (□). (From Stein, G. H. and Atkins, L., *Proc. Natl. Acad. Sci. U.S.A.*, 83, 9030, 1986. With permission.)

poststimulation, all of the inhibitory activity was gone (Figure 2A). Measurements of DNA synthesis in parallel populations of serum-stimulated quiescent HDF showed that there was little induction of DNA synthesis during the first 12 h after stimulation, but by 20 h, 37% of the cells were in S phase. Previous studies of our own and others have shown that the peak of DNA synthesis is usually reached by 20 h poststimulation in quiescent young HDF.[24] Furthermore, entry into S phase by 35 to 40% of the cells is a typical result in our system. Thus, these results suggest that loss of the membrane-associated inhibitor of DNA synthesis may be a necessary, but not a sufficient, prerequisite for DNA synthesis in these cells because serum stimulation abolished all of the inhibitory activity while only 37% of the cells entered S phase. In addition, the timing of the loss of inhibitory activity in these cells suggests that abolition of the inhibitor is a late G_1 event.

When comparable serum stimulation experiments were carried out with senescent HDF that had 0 PD remaining, the results were quite different (Figure 2C). In this case, serum stimulation caused no loss of the membrane-associated inhibitory activity, and none of the senescent cells entered S phase.[22] These results suggest that at least one of the reasons that senescent HDF fail to enter S phase is that they do not inactivate or eliminate their membrane-associated inhibitor of DNA synthesis.

The effect of serum stimulation on quiescent old HDF (2 to 3 PD remaining) was examined to determine whether there was a transition between the all (100% removal) or none (0% removal) responses of quiescent young and senescent HDF, respectively (Figure 2B). The data showed that about two thirds of the inhibitory activity was abolished in serum-stimulated quiescent old HDF and that the fraction of cells that entered S phase was about two thirds as great as that of serum-stimulated quiescent young HDF. These results further support the hypothesis that as HDF approach the end of their life span, they become progressively less able to remove the membrane-associated inhibitor under the same conditions that lead to complete removal of the inhibitor in younger cells. The timing of the loss of the membrane-associated inhibitor in quiescent old HDF further supports the idea that this is a late-G_1 event whose timing is related to the timing of entry into S phase, because both events occur a few hours later in quiescent old HDF than in quiescent young HDF.

The failure of senescent HDF to eliminate their membrane-associated inhibitor following serum stimulation supports our hypothesis that a key aspect of the aging process is a decreased ability to respond to serum, such that the senescent cells are functionally serum deprived. However, several laboratories have shown that senescing HDF (defined as cells with greater than 90% of their life span completed) can carry out a number of events in response to serum stimulation, even though only a small fraction of the cells enter S phase in response to the stimulus.[25-27] For example, Rittling et al.[27] studied 11 genes expressed in the G_0 to S phase transition in serum-stimulated quiescent HDF and found that each one was expressed at a comparable level in serum-stimulated senescing HDF. Likewise, others have shown that the levels of DNA polymerase alpha,[26] thymidine kinase,[25] and total thymidine triphosphate[25] also increase as much in serum-stimulated senescing HDF as in serum-stimulated quiescent HDF. These results indicate that senescing HDF are still responsive to at least some of the signals in serum.

Do the data cited above contradict the hypothesis that there is a progressive loss of responsiveness to serum with age? We suggest that they refine the hypothesis by indicating that not all aspects of response to serum are lost. Indeed, only one or a few key processes may be affected. The data concerning the membrane-associated inhibitor of DNA synthesis show that its abolition in response to serum stimulation is an event that is increasingly defective in old HDF and absent in fully senescent HDF. Thus, it may be one of the key processes that is affected by cellular aging. It is possible that abolition of the inhibitor is on a different pathway of response to serum stimulation than are the prereplicative events studied by Rittling et al.,[27] Olashaw et al.,[25] and Pendergrass et al.[26] Alternatively, abolition of the inhibitor may occur at a later time than any of the events measured in the above studies. This would require that the abolition of the inhibitor take place very close to the boundary between G_1 and S phase. Such a possibility is consistent with the observed kinetics of abolition of the inhibitor in quiescent cells. In either case, failure to abolish the membrane-associated inhibitor of DNA synthesis in senescent HDF has the potential for being responsible for the failure to enter S phase in senescent HDF. Therefore, an important goal for the future is to learn more about the identity and regulation of this inhibitor.

IV. GENETIC STUDIES OF THE RELATIONSHIP BETWEEN SENESCENCE AND QUIESCENCE

Two simple predictions of our hypothesis are (1) that the FPL^+ and S^+ phenotypes of normal HDF are separable and (2) that the S and Q phenotypes are linked, i.e., FPL^+ cells express either S^+ and Q^+ or S^- and Q^-. We have tested this hypothesis by forming somatic cell hybrids between normal HDF (FPL^+, S^+, Q^+) and carcinogen-transformed HDF (FPL^-, Q^-). The resulting cell hybrid clones expressed the FPL^+ phenotype of their normal parent, the Q^- phenotype of their transformed parent, and the new S^- phenotype at the end of their life span. As discussed earlier, the S^- phenotype means that the cells continued to synthesize DNA at the end of their life span. For example, in the case of cell hybrids formed between 75-16 human foreskin fibroblasts (FPL^+, S^+, Q^+) and SUSM-1 carcinogen-transformed HDF (FPL^-, Q^-), the hybrid clones had an average of 33% ^3H-thymidine-labeled nuclei per day when analyzed 4 weeks after there was cessation of net growth in the cultures. In contrast, clones of the HDF parent had 0% labeled nuclei at the end of their life span. These results show that the FPL^+ phenotype is dominant in cell hybrids formed with carcinogen-transformed HDF and that cells that are FPL^+ do not necessarily express the S^+ phenotype at the end of their life span. They also support, but do not prove, our contention that FPL^+ cells are either S^+ and Q^+ or S^- and Q^-.

As a further test of the genetic predictions of our hypothesis, we examined the FPL, S, and Q phenotypes of precrisis SV40-transformed HDF.[5] Previous studies had shown that precrisis SV40-transformed HDF have a limited proliferative capacity that is greater than that of untreated HDF and that ends in a crisis in which the cells die.[28] (On rare occasions, a variant cell type survives crisis and gives rise to an FPL^-, postcrisis SV40-transformed cell line.) Although investigators had long speculated that crisis in SV40-transformed HDF might be related to senescence in normal HDF, the nature of this relationship was obscure.[29,30] Our discovery of the S^- phenotype in cell hybrids suggested the obvious possibility that precrisis SV40-transformed cells express the FPL^+, S^-, Q^- combination of phenotypes. We examined the aging process in precrisis SV40-transformed HDF and found that it was like that of HDF in overall growth kinetics, clonal heterogeneity, and progressively decreasing ability to respond to serum, as measured by the "Ohno assay".[18] We confirmed that the precrisis cells expressed the Q^- phenotype, and we found that the cells continued to synthesize DNA at the end of their life span, with an average of 72% labeled nuclei 4 weeks after cessation of net growth in the population (i.e., they expressed the S^- phenotype). Our data also showed that there was a dramatic increase in the relative rate of cell death in the populations as they approached the end of their life span and entered crisis after a few more weeks. These data confirmed the prediction that the S^- phenotype would involve an increase in cell death to balance (and ultimately exceed) the continued cell birth in an S^- population. These data are all consistent with the interpretation that SV40 transformation of HDF creates cells that are FPL^+ and Q^-, as a consequence of which they express the S^- phenotype at the end of their life span.

V. SUMMARY AND THOUGHTS ABOUT THE FUTURE

In this chapter we have presented a two-part hypothesis on the mechanisms of senescence, quiescence, and finite proliferative life span, and we have presented evidence that is supportive of that hypothesis. Chances are that such a hypothesis is too simplistic. However, at present it has the virtue of organizing a body of knowledge into a conceptual framework that can help guide our future studies. An obvious need for the future is to identify the inhibitor or inhibitors of DNA synthesis that are present in senescent and quiescent HDF. Although our present studies have emphasized plasma membrane-associated inhibitors, it is unlikely that they are the whole story. Logically, one would expect a second factor (or factors) to be involved in communication with the nucleus. In addition, we have preliminary evidence that "quiescent" CT-1 carcinogen-transformed HDF have a membrane-associated inhibitor, even though they do not have inhibitory activity in the heterokaryon assay. This result also suggests the presence of at least one other DNA synthesis inhibitory factor in cells such as senescent and quiescent HDF, which do have inhibitory activity in the heterokaryon assay.

Another equally important need for the future is to try to learn about the aging process itself. Our hypothesis predicts that production of an inhibitor of DNA synthesis in senescent HDF is an end-stage phenomenon. Even when we know what that inhibitor is, we will not know why it is produced. Many hypotheses have been advanced about the aging process. These fall into the categories of stochastic mechanisms, genetically programmed mechanisms, or combinations of the two. It has been difficult to test the best of these hypotheses because (1) it is difficult to measure low levels of errors and (2) the idea of a genetic program has no detailed form at present.[31] Nevertheless, knowledge about the aging process will be helpful in interpreting the ever more detailed molecular information that should be forthcoming about senescent HDF.

One of the most impressive changes that HDF and other FPL$^+$ mammalian cells undergo is an age-related progressive demethylation of their DNA.[32] The magnitude of the change is large, such that the cells have lost approximately half of their methylated cytosines by the end of their life span; it occurs throughout the life span, and it is absent in FPL$^-$ cells. Furthermore, Holliday[33] and Fairweather et al.[34] have recently shown that acceleration of the DNA demethylation process in HDF shortens their life span. What if DNA demethylation is a key element in the aging process in HDF through its effect on patterns of gene expression? What implications might this have for studies of the differences between senescent and young HDF?

Previous studies on DNA methylation patterns of individual genes suggest that hypomethylation typically allows gene expression,[35] even though there are exceptions to this generalization.[36,37] If we assume that the progressive DNA demethylation during aging of HDF is a stochastic process,[38] then we might expect that (1) senescent HDF will express a variety of genes that are not expressed in young HDF and (2) there will be some heterogeneity among individual senescent HDF in the genes that they express at the end of their life span. Is this sort of stochastic mechanism consistent with the possibility that senescent HDF cease to proliferate because they contain an inhibitor of entry into S phase? Some years ago, Norwood et al.[39] showed that HDF treated with either an amino acid analog or mitomycin C were transiently nonreplicative. When these nonreplicative cells were analyzed in the heterokaryon assay, they were able to inhibit DNA synthesis in replicating HDF nuclei. One interpretation of these data is that even though a variety of steps in the mitogenic response pathway were probably damaged in the treated cells, they all (or almost all) triggered production of an inhibitor of DNA synthesis. We suggest that progressive DNA demethylation could likewise create a variety of defects or inefficiencies in the mitogenic response pathway of aging HDF, which would eventually lead to production of the same inhibitor of DNA synthesis.

In conclusion, we feel that it will be important to continue to investigate both the aging process and the end state of senescent cells in order to fully understand the molecular mechanism for the phenomenon of cellular aging.

ACKNOWLEDGMENTS

The work described in this chapter is the result of a team effort. In particular, I would like to acknowledge the contributions of Dr. Rosalind Yanishevsky, Ms. Laura Atkins, Ms. Mary Beeson, and Ms. Lena Gordon. I am also grateful to Dr. Laurel Donahue for her critical reading of the manuscript. This work was supported by research grants AG00947 and AG04811 from the National Institute on Aging.

REFERENCES

1. **Hayflick, L. and Moorhead, P. S.,** The serial cultivation of human diploid cell strains, *Exp. Cell Res.,* 25, 585, 1961.
2. **Stein, G. H., Namba, M., and Corsaro, C. M.,** Relationship of finite proliferative lifespan, senescence, and quiescence in human cells, *J. Cell. Physiol.,* 122, 343, 1985.
3. **Yanishevsky, R. M., Mendelsohn, M. L., Mayall, B. H., and Cristofalo, V. J.,** Proliferative capacity and DNA content of aging human diploid cells in culture: a cytophotometric and autoradiographic analysis, *J. Cell. Physiol.,* 84, 165, 1974.
4. **Bell, E., Marek, L. F., Levinstone, D. S., Merrill, C., Sher, S., Young, I. T., and Eden, M.,** Loss of division potential in vitro: aging or differentiation, *Science,* 202, 1158, 1978.
5. **Stein, G. H.,** SV40-transformed human fibroblasts: evidence for cellular aging in precrisis cells, *J. Cell. Physiol.,* 125, 36, 1985.
6. **Wiebel, F. and Baserga, R.,** Early alterations in amino acid pools and protein synthesis of diploid fibroblasts stimulated to synthesize DNA by the addition of serum, *J. Cell. Physiol.,* 74, 191, 1969.
7. **Dell'Orco, R. T., Crissman, H. A., Steinkamp, J. A., and Kraemer, P. M.,** Population analysis of arrested human diploid fibroblasts by flow microfluorimetry, *Exp. Cell Res.,* 92, 271, 1975.
8. **Schiaffonati, L. and Baserga, R.,** Different survival of normal and transformed cells exposed to nutritional conditions nonpermissive for growth, *Cancer Res.,* 37, 541, 1977.
9. **Stein, G. H. and Yanishevsky, R. M.,** Quiescent human diploid cells can inhibit entry into S phase in replicative nuclei in heterodikaryons, *Proc. Natl. Acad. Sci. U.S.A.,* 78, 3025, 1981.
10. **Stein, G. H., Yanishevsky, R. M., Gordon, L., and Beeson, M.,** Carcinogen-transformed human cells are inhibited from entry into S phase by fusion to senescent cells but cells transformed by DNA tumor viruses overcome the inhibition, *Proc. Natl. Acad. Sci. U.S.A.,* 79, 5287, 1982.
11. **Norwood, T. H., Pendergrass, W. R., Sprague, C. A., and Martin, G. M.,** Dominance of the senescent phenotype in heterokaryons between replicative and post-replicative human fibroblast-like cells, *Proc. Natl. Acad. Sci. U.S.A.,* 71, 2231, 1974.
12. **Yanishevsky, R. M. and Stein, G. H.,** Ongoing DNA synthesis continues in young human diploid cells (HDC) fused to senescent HDC but entry into S phase is inhibited, *Exp. Cell Res.,* 126, 469, 1980.
13. **Norwood, T. H., Pendergrass, W. R., and Martin, G. M.,** Reinitiation of DNA synthesis in senescent human fibroblasts upon fusion with cells of unlimited growth potential, *J. Cell Biol.,* 64, 551, 1975.
14. **Rabinovitch, P. S. and Norwood, T. H.,** Comparative heterokaryon study of cellular senescence and the serum-deprived state, *Exp. Cell Res.,* 130, 101, 1980.
15. **Pereira-Smith, O. M., Fisher, S. F., and Smith, J. R.,** Senescent and quiescent cell inhibitors of DNA synthesis: membrane-associated proteins, *Exp. Cell Res.,* 160, 297, 1985.
16. **Burmer, G. C., Rabinovitch, P. S., and Norwood, T. H.,** Evidence for differences in the mechanism of cell cycle arrest between senescent and serum-deprived human fibroblasts: heterokaryon and metabolic inhibitor studies, *J. Cell. Physiol.,* 118, 97, 1984.
17. **Cristofalo, V. J. and Sharf, B. B.,** Cellular senescence and DNA synthesis: thymidine incorporation as a measure of population age in human diploid cells, *Exp. Cell Res.,* 76, 419, 1973.
18. **Ohno, T.,** Strict relationship between dialyzed serum concentration and cellular lifespan *in vitro, Mech. Ageing Dev.,* 11, 179, 1979.
19. **Lieberman, M. A., Raben, D., and Glaser, L.,** Cell surface-associated growth inhibitory proteins. Evidence for conservation between mouse and human cell lines, *Exp. Cell Res.,* 133, 413, 1981.
20. **Wieser, R. J., Heck, R., and Oesch, F.,** Involvement of plasma membrane glycoproteins in the contact-dependent inhibition of growth of human fibroblasts, *Exp. Cell Res.,* 158, 493, 1985.
21. **Stein, G. H., Atkins, L., Beeson, M., and Gordon, L.,** Quiescent human diploid fibroblasts: common mechanism for inhibition of DNA replication in density-inhibited and serum-deprived cells, *Exp. Cell Res.,* 162, 255, 1986.

22. **Stein, G. H. and Atkins, L.,** Membrane-associated inhibitor of DNA synthesis in senescent human diploid fibroblasts: characterization and comparison to quiescent cell inhibitor, *Proc. Natl. Acad. Sci. U.S.A.,* 83, 9030, 1986.
23. **Wieser, R. J. and Oesch, F.,** Contact inhibition of growth of human diploid fibroblasts by immobilized plasma membrane glycoproteins, *J. Cell Biol.,* 103, 361, 1986.
24. **Augenlicht, L. H. and Baserga, R.,** Changes in the G_0 state of WI-38 fibroblasts at different times after confluence, *Exp. Cell Res.,* 89, 255, 1974.
25. **Olashaw, N. E., Kress, E. D., and Cristofalo, V. J.,** Thymidine triphosphate synthesis in senescent WI38 cells: relationship to loss of replicative capacity, *Exp. Cell Res.,* 149, 547, 1983.
26. **Pendergrass, W. R., Saulewicz, A. C., Salk, D., and Norwood, T. H.,** Induction of DNA polymerase alpha in senescent cultures of normal and Werner's syndrome cultured skin fibroblasts, *J. Cell. Physiol.,* 124, 331, 1985.
27. **Rittling, S. R., Brooks, K. M., Cristofalo, V. J., and Baserga, R.,** Expression of cell cycle-dependent genes in young and senescent WI-38 fibroblasts, *Proc. Natl. Acad. Sci. U.S.A.,* 83, 3316, 1986.
28. **Girardi, A. J., Jensen, F. C., and Koprowski, H.,** SV40-induced transformation of human diploid cells: crisis and recovery, *J. Cell. Comp. Physiol.,* 65, 69, 1965.
29. **Koprowski, H., Jensen, F., Girardi, A., and Koprowska, I.,** Neoplastic transformation, *Cancer Res.,* 26, 1980, 1966.
30. **Huschtscha, L. I. and Holliday, R.,** Limited and unlimited growth of SV40-transformed cells from human diploid MRC-5 fibroblasts, *J. Cell Sci.,* 63, 77, 1983.
31. **Holliday, R.,** The unsolved problem of cellular ageing, *Monogr. Dev. Biol.,* 17, 60, 1984.
32. **Wilson, V. L. and Jones, P. A.,** DNA methylation decreases in aging but not in immortal cells, *Science,* 220, 1055, 1983.
33. **Holliday, R.,** Strong effects of 5-azacytidine on the *in vitro* lifespan of human diploid fibroblasts, *Exp. Cell Res.,* 166, 543, 1986.
34. **Fairweather, D. S., Fox, M., and Margison, G. P.,** The *in vitro* lifespan of MRC-5 cells is shortened by 5-azacytidine-induced demethylation, *Exp. Cell Res.,* 168, 153, 1987.
35. **Razin, A. and Riggs, A. D.,** DNA methylation and gene function, *Science,* 210, 604, 1980.
36. **McKeon, C., Ohkubo, H., Pastan, I., and deCrombrugghe, B.,** Unusual methylation pattern of the alpha 2(I) collagen gene, *Cell,* 29, 203, 1982.
37. **Kunnath, L. and Locker, J.,** Developmental changes in the methylation of the rat albumin and alpha-fetoprotein genes, *EMBO J.,* 2, 317, 1983.
38. **Goldstein, S. and Shmookler Reis, R. J.,** Methylation patterns in the gene for the alpha subunit of chorionic gonadotropin are inherited with variable fidelity in clonal lineages of human fibroblasts, *Nucleic Acids Res.,* 13, 7055, 1985.
39. **Norwood, T. H., Pendergrass, W., Bornstein, P., and Martin, G. M.,** DNA synthesis of sublethally injured cells in heterokaryons and its relevance to clonal senescence, *Exp. Cell Res.,* 119, 15, 1979.

Chapter 10

GROWTH CONTROL IN CULTURED 3T3 FIBROBLASTS: MOLECULAR PROPERTIES OF A GROWTH REGULATORY FACTOR ISOLATED FROM CONDITIONED MEDIUM

John L. Wang, Quan Sun, and Patricia G. Voss

TABLE OF CONTENTS

I. INTRODUCTION

There is well-documented evidence that in populations of normal cells cultured *in vitro* the growth rate diminishes as a critical cell density is approached. Cell growth and division are renewed when the cell density is decreased, either by removal of a portion of the cells or by provision of an increased surface area for growth. This phenomenon, appropriately termed density-dependent inhibition of growth, was first described and most extensively studied in the 3T3 cell line.[1,2]

Although the mechanisms responsible for this phenomenon are not understood, a number of lines of recent evidence suggest that inhibitory factors released into the medium by the 3T3 cells themselves may be responsible for at least part of the observed suppression of cell division. Studies of Canagaratna and Riley[3] on the patterns of nuclear incorporation of radioactive thymidine in cultures with local cell densities between 0.2×10^4 and 6.2×10^4 cells per square centimeter indicated that DNA synthesis in these cells was critically dependent on the local cell density. More detailed analyses of the data showed that there was an inverse relationship between the local cell density and the proportion of labeled cells and that this density-dependent regulation of DNA synthesis was exhibited in relatively sparse cultures, well before the onset of cell-to-cell contact.[4,5] Finally, Harel et al. showed that phosphate metabolism and cell growth in sparse cultures of 3T3 cells were inhibited when they shared the same medium with dense cultures of the same cells.[6]

In this chapter we report that media conditioned by exposure to cultures of density-inhibited 3T3 cells contain a growth inhibitory activity that acts on sparse, proliferating cultures of the same cell line. We have purified this activity and have shown that it is an endogenous 3T3 cell product consisting of a single polypeptide of M_r ca. 13,000. The results suggest a system for studying the mechanism of density-dependent growth control at the level of direct interaction of an endogenous growth inhibitor with its target cells.

II. COCULTURE OF SPARSE AND DENSE 3T3 CELLS

The proliferative properties of sparse 3T3 cells cocultivated with density-inhibited 3T3 cells were studied using the experimental scheme of Harel et al.,[6] in which two coverslips inoculated at different cell densities were placed in the same petri dish containing fresh medium (Figure 1). DNA synthesis on a per cell basis was measured 24 h after coculture by determining the amount of ³H-thymidine incorporated into acid-precipitable DNA and then normalized to the total amount of protein present on the respective coverslips using ¹⁴C-labeled amino acids incorporated into the cells prior to the coculture period. The results showed that DNA synthesis in cells on a coverslip containing a sparse culture was suppressed when the coverslip was cocultivated with another coverslip containing a dense culture (Table 1). In contrast, this inhibition was not observed when two coverslips, both containing sparse cultures, were incubated together. The levels of DNA synthesis in the sparse and dense cultures incubated alone served as the controls in this experiment.

To ascertain that the diminution of ³H-thymidine incorporation in cells on the sparse coverslip was due to a true suppression of DNA synthesis rather than to dilution, pool size changes, or altered transport of the ³H-thymidine label, we analyzed DNA synthesis at the level of individual cells by autoradiography. Considering the fact that the generation time and length of S phase for the 3T3 cells used in the present study were estimated to be about 22 h and 9 h,[7,8] respectively, exposure of these cells to a 3-h ³H-thymidine pulse was expected to result in the labeling of about 50% of the cells. In agreement with this calculation, we found (Table 1) that approximately 52% of the cells grown on the sparse coverslip (S alone) were labeled with ³H-thymidine. In contrast, about 5% of the cells on the dense coverslip (D alone) were labeled.

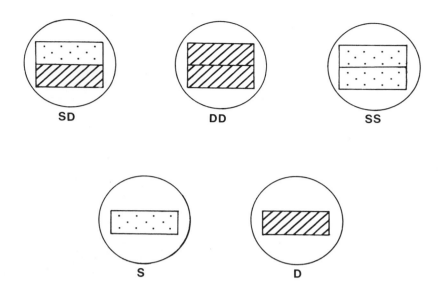

FIGURE 1.   Schematic diagram illustrating the experimental design for culturing two cover-slips next to each other in a tissue culture dish. The letter S represents a coverslip containing a sparse culture (ca. 10^4 cells/cm^2) of 3T3 fibroblasts, and the letter D represents a coverslip containing a dense culture (ca. 5×10^4 cells/cm^2) of the same cells. The dimensions of the coverslip were 2.1×1 cm. The coverslips were cultured in a dish (10 cm^2 growth surface) containing 1.5 ml of Dulbecco Modified Eagle's Medium with 10% calf serum.

TABLE 1
DNA Synthesis in Sparse and Dense Cultures of 3T3 Cells

Conditions[a]	Number of coverslips	DNA synthesis/ protein content[b]	Number of cells counted	% labeled cells[c]
S alone	4	21.0 ± 0.9	240	52.0 ± 1.0
D alone	4	3.5 ± 0.6	211	5.2 ± 0.3
S of SS	8	20.5 ± 1.5	434	48.3 ± 2.5
D of DD	10	3.7 ± 0.4	435	8.0 ± 3.0
S of SD	5	13.4 ± 1.9	515	36.8 ± 2.1
D of SD	5	3.0 ± 0.2	410	2.5 ± 0.8

[a] The abbreviations used are S, sparse coverslip (ca. 10^4 cells/cm^2); D, dense coverslip (ca. 5×10^4 cells/cm^2).

[b] DNA synthesis was measured by the incorporation of ^{3}H-thymidine (1 μCi/culture, 3 h); protein content was measured by the amount of ^{14}C radioactivity due to ^{14}C-labeled amino acids incorporated prior to the coculture period (0.75 μCi/culture, 24 h). The data are expressed as the average of results obtained from the number of coverslips ± SEM.

[c] For the autoradiographic experiments, no ^{14}C-labeled amino acids were added to the sparse and dense cultures prior to the coculture period, and the amount of ^{3}H-thymidine used in assays of DNA synthesis was 0.1 μCi/culture. After the incorporation period, the cells were washed five times with cold phosphate-buffered saline and once with 5% trichloroacetic acid and then fixed in absolute ethanol. The coverslips were then dipped in NTB-2 Nuclear Track emulsion and exposed for 72 h. After final fixation, slides were stained in giemsa. The percent labeled cells was determined by visually counting the number of labeled nuclei among the total number of cells counted. The data are expressed as the average of results obtained from the number of coverslips ± SEM.

More importantly, only 37% of the cells on the sparse coverslip cocultivated with a dense coverslip (S of the SD pair) were labeled, well below the values obtained for all the other sparse coverslips (Table 1). Therefore, the percentage of labeled cells decreased in a similar fashion with the total incorporation of acid-precipitable radioactivity. Above all, however, we found that the distributions of grain counts for the labeled cells, as well as the average number of grains per labeled cell, were similar for all the sparse cultures, irrespective of whether they were from the S, SD, or SS combinations (Figure 2). Chi-square analyses were performed on the data shown in Figure 2; the results indicated that the grain count distributions were the same at the 90% confidence level. All of these results strongly suggested that the inhibition of ^{3}H-thymidine incorporation in the sparse cultures by the presence of a dense culture reflected a true suppression of cellular DNA synthesis rather than any alterations of the specific activity or transport of the radioactive label. The results raise the possibility that medium exposed to dense cultures may acquire a growth inhibitory activity that acts on sparse cells.

III. THE EFFECT OF CONDITIONED MEDIUM ON THE PROLIFERATIVE PROPERTIES OF TARGET CULTURES

Treatment of sparse, proliferating cultures of 3T3 cells with medium conditioned by exposure to density-inhibited 3T3 cultures resulted in an inhibition of growth and division in the target cells when compared to similar treatment with unconditioned medium. The conditioned medium (CM) was prepared according to the protocol described previously.[8] Source cells were grown to a monolayer, and fresh growth medium was then incubated with the monolayer for 24 h at 37°C. The collected medium was centrifuged and used as CM. Unconditioned medium (UCM) was prepared in parallel and was incubated under the same conditions, but in the absence of any cells. Target cells, on which the effect of CM was to be tested, were seeded at a density of 2.5×10^3 cells per square centimeter and were used well before the onset of confluence in the cultures.

The differential effect of CM and UCM on target cells was demonstrated using three assay systems: (1) assessment of total cell number, (2) measurement of ^{3}H-thymidine incorporated into acid-precipitable DNA, and (3) determination of the percentage of radioactively labeled nuclei in individual cells after incorporation of ^{3}H-thymidine. DNA synthesis in these CM-treated cells was markedly inhibited when compared to parallel cultures treated with UCM (Figure 3a). The ^{3}H-thymidine incorporation was also assayed by autoradiography at the level of individual cells. In UCM-treated cultures, approximately 56% of the nuclei were labeled, as compared to only 35% in CM-treated cultures. The fraction of labeled nuclei decreased to 29% when cultures were treated with CM for 48 h.

More importantly, the average number of grains per labeled cell was invariant within the error of estimation for CM- and UCM-treated cultures. The distributions of grain counts for labeled cells of both cultures appeared to be similar. Therefore, the inhibition of ^{3}H-thymidine incorporation in target cells treated with CM reflected a reduction in the percentage of cells undergoing DNA synthesis rather than alterations of the transport or pool sizes of the label.

These results were confirmed by comparing the number of cells in target cultures treated with UCM and CM (Figure 3b). Target cells exposed to UCM continued to proliferate up to the characteristic saturation density (5×10^4 cells per square centimeter). In contrast, cultures exposed to CM showed much smaller increases in their cell numbers.

All of these results suggest that medium conditioned by exposure to density-inhibited 3T3 cells may contain a growth inhibitory activity. We have carried out other experiments to show that this growth inhibitory activity in CM, prepared and tested in the 3T3 system, has the following key properties:[8]

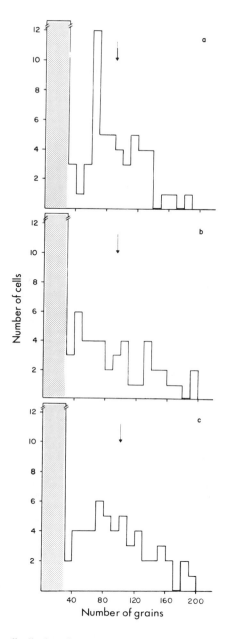

FIGURE 2. Grain count distributions for sparse proliferating 3T3 cells cultured under various conditions: (a) two sparse cultures together (S of SS), (b) sparse culture in the presence of dense culture (S of SD), and (c) sparse culture alone (S). Slides were scanned systematically, and at least 100 labeled cells were counted in every case. Multiple counts of the grains in the same cell indicated that the standard error of the grain counting method was 5 to 10%. The average number of grains per labeled cell for each of the cultures is indicated by the arrow in the grain count distribution graph. The hatched portion of the bar graph indicates the number of unlabeled cells. The grain distributions were subjected to chi-square analysis.

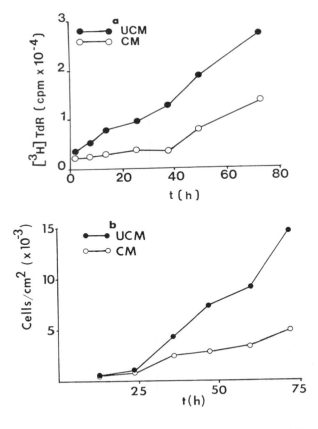

FIGURE 3.   (a) The kinetics of ^{3}H-thymidine incorporation in 3T3 cells cultured in the presence of CM (○) and UCM (●). At various times, parallel cultures were pulsed with 1 μCi of ^{3}H-thymidine for 3 h. (b) The kinetics of the increase in cell density of 3T3 cells cultured in the presence of CM (○) and UCM (●). From Wang, J. L., Steck, P. A., and Kurtz, J. W., in *Growth of Cells in Hormonally Defined Media*, Vol. 9, Cold Spring Harbor Conf. Cell Proliferation, Sato, G., Sirbasku, D., and Pardee, A., Eds., Cold Spring Harbor Laboratory, Cold Spring Harbor, NY, 1982, 305. With permission.)

1. It is not cytotoxic, and its effects on cell growth are reversible.
2. The inhibitory activity can be accumulated in the medium before the onset of extensive cell-to-cell contact.
3. The inhibitor has a more pronounced effect on target cells at high density than it does on cells at lower density.
4. The activity can be collected in the absence of serum and can be demonstrated despite the presence of freshly added serum.
5. The activity is decreased upon prolonged exposure to target cell cultures.
6. The inhibitory factor can be concentrated by precipitation and fractionated by gel filtration.

IV. PURIFICATION OF THE GROWTH INHIBITORY ACTIVITY IN CONDITIONED MEDIUM

Gel filtration of the ammonium sulfate precipitate of CM on a column of Sephadex®

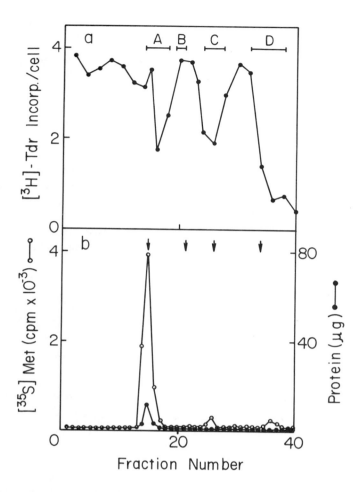

FIGURE 4. Chromatography of the growth inhibitory activity derived from CM on a column (90 × 1 cm) of Sephadex® G-50 equilibrated with Dulbecco modified Eagle's medium. The CM was collected from source cells that had been cultured in the presence of ^{35}S-methionine for 24 h. Fractions of 1.6 ml were collected. (a) Profile of the growth inhibitory activity assayed by the inhibition of ^{3}H-thymidine (^{3}H-Tdr) incorporation in target cells. The horizontal bars containing the letters A, B, C, and D denote fractions which were pooled for analysis by PAGE. The fractions in pool D lysed the target cells, resulting in no incorporation of ^{3}H-Tdr. (b) Profile of the protein content assayed by the Lowry method (●—●) or assayed by counting the trichloroacetic acid-precipitable radioactivity due to ^{35}S-methionine (^{35}S-Met, ○—○). The vertical arrows indicate the positions of elution of molecular weight markers bovine serum albumin (68,000), myoglobin (17,000), ribonuclease T (11,000), and bacitracin (1,400). (From Wang, J. L., Steck, P. A., and Kurtz, J. W., in *Growth of Cells in Hormonally Defined Media*, Vol. 9, Cold Spring Harbor Conf. Cell Proliferation, Sato, G., Sirbasku, D., and Pardee, A., Eds., Cold Spring Harbor Laboratory, Cold Spring Harbor, NY, 1982, 305. With permission.)

G-50 partitioned the inhibitory activity into two major components (components A and C, Figure 4a). Component A, which was eluted at the void volume of the column, contained the major portion of the sample's protein (Figure 4b). In contrast, the second component of growth inhibitory activity, component C, was associated with a minute amount of protein material (Figure 4b). The position of elution of component C on Sephadex® G-50 suggested that the material contained polypeptide chains with molecular weights of approximately 12,000.[9]

The materials containing growth inhibitory activity at various stages of fractionation were subjected to analysis by SDS-polyacrylamide gel electrophoresis (SDS-PAGE). After electrophoresis, the gel was stained with Coomassie blue and then subjected to fluorography to reveal ^{35}S-labeled protein components. The gel revealed a large number of proteins (as detected by fluorography) in CM, in the ammonium sulfate precipitate, and in component A of the Sephadex® G-50 column. In contrast, component C yielded two major bands on the fluorograph; the molecular weights estimated for the two bands in component C were 13,000 and 10,000. We have designated this fraction (component C, Figure 4) FGR-s, for a fibroblast growth regulator that is secreted or released into the medium in a soluble form. Hereafter, the M_r = 13,000 polypeptide (pI $\approx$ 10)[10] in this fraction will be referred to as FGR-s (13 kDa). Similarly, the M_r = 10,000 polypeptide (pI $\approx$ 7)[10] will be designated FGR-s (10 kDa).

DEAE-cellulose chromatography of a ^{35}S-methionine-labeled preparation of FGR-s resulted in the separation of several components (Figure 5). When the fractions eluting from the ion-exchange column were assayed for growth inhibitory activity, only component A (Figure 5b) exhibited activity; the remainder of the components failed to show any appreciable activity (Figure 5a). The sum of the growth inhibitory activity in component A (Figure 5b) accounted for ca. 80% of the total activity applied to the column. There was a sixfold enrichment in terms of specific activity in this fractionation step.

After SDS-PAGE and fluorography, component A (Figure 5b) yielded a single polypeptide, migrating at a position corresponding to a molecular weight of 13,000 (Figure 5c, lane 2). Identical results were obtained irrespective of whether the polyacrylamide gel was electrophoresed under reducing (with β-mercaptoethanol) or nonreducing conditions. In addition to fluorography, we also subjected the gel to staining with the silver technique. Again, component A predominantly yielded a single polypeptide of M_r = 13,000. This material (component A, Figure 5) is designated FGR-s (13 kDa).[11]

V. EVIDENCE THAT FGR-s (13 kDa) IS RESPONSIBLE FOR GROWTH INHIBITORY ACTIVITY

Using the partially purified fraction FGR-s as the immunogen, we have carried out *in vitro* immunization of rat splenocytes and have generated hybridoma lines secreting monoclonal antibodies directed against components of the FGR-s preparation. One such antibody, designated 2A4, specifically bound the M_r = 13,000 polypeptide.[12] To test the possibility that antibody 2A4 could neutralize the growth inhibitory activity of FGR-s, the effect of the inhibitor preparation on ^{3}H-thymidine incorporation in target 3T3 cells was assayed in the presence and absence of the purified antibody. In the present assay, 60% inhibition was obtained with approximately 5 μl of the FGR-s preparation (Figure 6). When the activity of FGR-s was assayed in the presence of antibody 2A4, however, there was a higher level of ^{3}H-thymidine incorporation (i.e., a reduced level of growth inhibition). For example, the addition of 25 ng/ml of antibody 2A4 reduced the inhibitory effect of 5 μl of FGR-s from 60 to 30%. When the amount of FGR-s added was increased tenfold (to 50 μl), the same 25 ng/ml of antibody 2A4 was ineffective in reversing the inhibitory effect.

Particularly striking was the observation that antibody 2A4 (25 ng/ml) also increased the level of DNA synthesis in 3T3 cultures in the absence of any exogenously added FGR-s (Figure 6).[12,13] Similarly, when a small amount of FGR-s (0.5 μl) was used, resulting in 25% inhibition, the addition of antibody 2A4 raised the level of DNA synthesis even beyond that of control cultures, which lacked FGR-s and antibody 2A4. Therefore, over the entire range of FGR-s concentration tested, the addition of antibody 2A4 resulted in a higher level of DNA synthesis.

These effects of antibody 2A4 on DNA synthesis of 3T3 cells and on the activity of FGR-s were specific. Antibody 104, which was not reactive with FGR-s polypeptides, failed to yield the same effects.[12] These observations suggest that the results obtained with antibody 2A4 were most probably not due to a growth factor contaminating the immunoglobulin fraction. This conclusion is further supported by experiments that showed the same effects

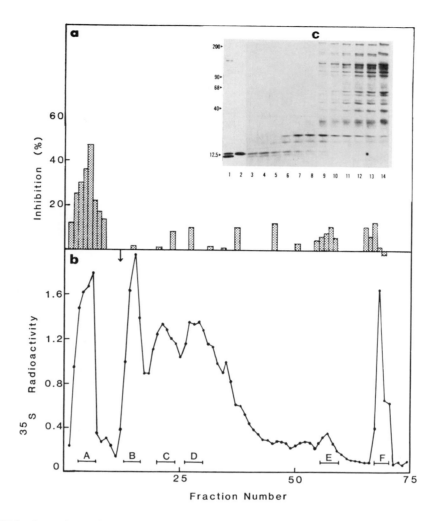

FIGURE 5. Ion-exchange chromatography of a ³⁵S-methionine-labeled FGR-s preparation (1.5×10^6 cpm) on a column (0.8×2 cm) of DEAE-cellulose equilibrated with 5 mM Tris, pH 8.0. At the point indicated by the arrow, a linear gradient (0 to 0.5 M NaCl, 100 ml total volume) was used to elute the material bound on the column. 1.7-ml fractions were collected. (a) Profile of the growth inhibitory activity assayed by the inhibition of ³H-thymidine incorporation into target cells. The data on the ordinate axis are expressed as the percentage of inhibition relative to control cultures and represent the averages of triplicate determinations. (b) Profile of the protein content assayed by counting the radioactivity due to ³⁵S-methionine. The horizontal bars marked A to F denote the fractions that were pooled for further analysis. (c, inset) SDS-PAGE of various fractions from the column. The acrylamide concentration of the running gel was 16%. The samples applied to the gel contained β-mercaptoethanol (4% v/v). The radioactive polypeptides were revealed by fluorography; lanes 1 and 2 were developed on a different fluorogram than the remainder of the lanes. The arrows indicate the positions of migration of molecular weight (in thousands) markers. Lane 1, ³⁵S-methionine-labeled FGR-s; lane 2, fraction 6; lane 3, fraction 9; lane 4, fraction 15; lane 5, fraction 21; lane 6, fraction 27; lane 7, fraction 29; lane 8, fraction 33; lane 9, fraction 35; lane 10, fraction 38; lane 11, fraction 46; lane 12, fraction 52; lane 13, fraction 57; lane 14, fraction 68. (Reproduced from *The Journal of Cell Biology*, 1986, vol. 102, pp. 362-369, by copyright permission of The Rockefeller University Press.)

of antibody 2A4 when the assays were carried out in the presence of freshly added calf serum (5%). Finally, the growth inhibitory activity of a highly purified preparation of FGR-s (13 kDa) was depleted when the material was passed through an affinity column containing antibody 2A4.[11] All of these results strongly suggest that FGR-s (13 kDa) was directly responsible for the observed growth inhibitory activity.

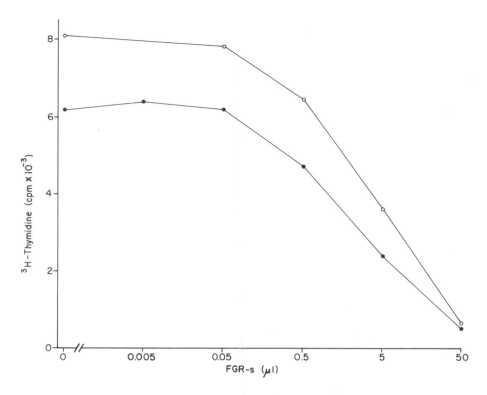

FIGURE 6.   Dose-response curve for the effect of FGR-s on the incorporation of ^{3}H thymidine by 3T3 cells in the absence (●–●) and presence (○–○) of antibody 2A4 (25 ng/ml). (From Wang, J. L. and Hsu, Y.-M., in *Growth, Cancer, and the Cell Cycle*, Skehan, P. and Friedman, S. J., Eds., Humana Press, Clifton, NJ, 1984, 177. With permission.)

The growth inhibitory effect of FGR-s (13 kDa) on target cells was dependent on the concentration of ligand added (Figure 7). The concentration of inhibitor required for 50% inhibition was ca. 3 ng/ml (corresponding to ca. 0.23 n*M*). The inhibition was reversible within 20 h after the removal of FGR-s (13 kDa).[11]

The inhibitory effect of FGR-s (13 kDa) on cell proliferation was also reflected in assays of cell number after treatment with the inhibitory fraction. Target cells treated with a control fraction continued to proliferate; in contrast, cultures treated with FGR-s (13 kDa) failed to increase in cell number at the same rate.[11] These data provided a confirmation of the growth inhibitory activity of FGR-s (13 kDa) using an assay independent of ^{3}H-thymidine incorporation.

The main features of the inhibition of 3T3 cell proliferation by FGR-s (13 kDa) include[11] (1) the reversible effect of FGR-s (13 kDa) on target cells, which cannot be ascribed to cytotoxicity; (2) the effect of FGR-s (13 kDa) on DNA synthesis being most prominent at high ($>5 \times 10^3$ cells/cm^2) target cell density, consistent with previous observations, made with conditioned medium,[8] that a minimum target cell density may be required to observe the inhibitory effect; and (3) the inhibition by FGR-s (13 kDa) on 3T3 cells being most potent at low ($<5\%$) serum concentrations, in agreement with previous results that the binding of radioactive FGR-s is decreased by increasing concentrations of serum.[14]

VI. RELATIONSHIP BETWEEN FGR-s (13 kDa) AND OTHER NEGATIVE REGULATORS OF CELL GROWTH

Our present results on the molecular properties of FGR-s (13 kDa) should be compared to several other negative regulators of cell growth that have been purified. Transforming

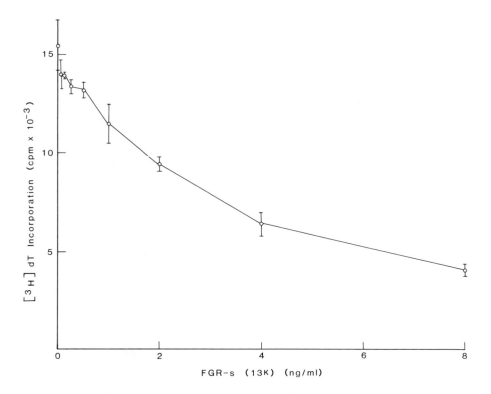

FIGURE 7. Dose-response curve of the growth inhibitory activity of FGR-s (13 kDa) on target 3T3 cells (2 × 10⁴ cells/cm²). The protein concentration of FGR-s (13 kDa) was determined by the silver-staining technique on SDS gels. The target cells were treated with FGR-s (13 kDa) for 20 h and then assayed for the incorporation of ³H-thymidine (³H-dT). (Reproduced from *The Journal of Cell Biology*, 1986, vol. 102, pp. 362-369, by copyright permission of The Rockefeller University Press.)

growth factor β (TGF-β) is a protein originally isolated on the basis of its capacity to reversibly induce non-neoplastic cells to express the transformed phenotype, as measured by loss of density-dependent inhibition of growth and acquisition of anchorage-independent growth.[15,16] Although TGF-β was originally purified on the basis of its transforming property, a number of lines of evidence[17] indicate that it is related, if not identical, to a growth inhibitor isolated from the conditioned medium of BSC-1 African green monkey kidney epithelial cells.[18] Indeed, TGF-β can either stimulate or inhibit growth; the expression of these two activities is modulated by other growth factors and is not solely dependent on cell type or conditions of anchorage-dependent vs. anchorage-independent growth.[19,20]

TGF-β now stands as a paradigm for several important features of growth regulation: (1) it is one of the first peptide growth inhibitors purified to homogeneity,[21,22] its amino acid sequence has been determined by molecular cloning techniques,[23] and its cell surface receptor has been identified;[24] (2) it is a negative regulator of cell growth that may function in the autocrine pathway;[20] and (3) it is bifunctional (i.e., it exhibits both stimulatory and inhibitory growth regulatory activities).[19,20] In comparing the properties of FGR-s (13 kDa) and TGF-β, it should be noted that the latter is active as a dimer of two identical polypeptides (M_r = 12,500) under nonreducing conditions. Upon reduction of the disulfide linkage, the biological activity is lost.[15] In contrast, FGR-s (13 kDa) is active as a single polypeptide (M_r = 13,000).

A growth inhibitor for Ehrlich ascites mammary carcinoma cells *in vitro* has been purified from bovine mammary gland.[25,26] The activity consists of a single polypeptide (M_r = 13,000) whose sequence has been determined.[27] Antisera raised in mice and in rabbits specifically

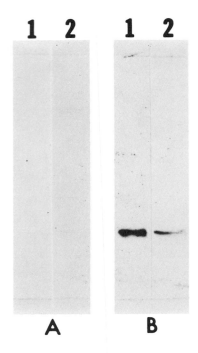

FIGURE 8.   Immunoblotting analysis of bovine MDGI and mouse FGR-s (13 kDa). Panel A: preimmune control serum; panel B: rabbit antiserum against MDGI. Lane 1: MDGI (ca. 700 ng); lane 2: FGR-s (13 kDa) (ca. 350 ng). The protein samples were electrophoresed on a 15% polyacrylamide gel and then transferred to nitrocellulose paper. After incubation with a 1:250 dilution of the appropriate serum, the nitrocellulose papers were incubated with horseradish peroxidase-conjugated goat anti-rabbit immunoglobulin and developed using 4-chloronaphthol as substrate.

bind the mammary-derived growth inhibitor (MDGI) and neutralize its growth inhibitory activity. Analyses of the tissue distribution of the inhibitor showed a high concentration of MDGI in lactating, but not in nonlactating, bovine mammary glands. Milk fat globule membranes and lung tissue also showed reactivity with the antisera directed against MDGI.

When samples of FGR-s (13 kDa) were subjected to immunoblotting analysis with rabbit anti-MDGI, a single polypeptide (M_r = 13,000), co-migrating with MDGI, was observed (Figure 8B, lanes 1 and 2). Preimmune control experiments on a separate nitrocellulose blot were negative (Figure 8A, lanes 1 and 2). These results suggest that MDGI and FGR-s (13 kDa) share common structural features. This conclusion is consistent with the previous observation that MDGI[25,26] and FGR-s (13 kDa)[9,11] behave very similarly during their respective purification steps. Perhaps these growth inhibitory polypeptides derived from bovine mammary gland and from a cultured mouse fibroblast line would provide examples of another family of negative regulators of cell growth,[16,28] in analogy to the family of polypeptides related to TGF-β,[15,16] as well as the well-known growth factor families. It remains to be determined whether these similarities (and differences) can be demonstrated at the level of biological activity: cross-inhibition by MDGI and FGR-s (13 kDa) on their respective target cells vs. strict cell line specificity of growth inhibition.

Using procedures similar to our previous studies,[8,9] Wells and Mallucci[29] have shown that secondary cultures of mouse embryo fibroblasts release into the medium a growth inhibitory activity whose physicochemical behavior and polypeptide composition closely parallel those of FGR-s. The molecular weights of the polypeptides in their active fractions are 11,000 and 14,000. In this connection, it should be noted that both soluble and plasma membrane-associated growth inhibitory fractions, derived from 3T3 cells and with properties similar to FGR-s, have been reported by several laboratories.[30-33] The molecular identity and properties of these active fractions and their relationship to FGR-s (13 kDa) remain to be elucidated.

ACKNOWLEDGMENT

This work was supported by grants GM 27203 from the National Institutes of Health and FRA-221 from the American Cancer Society.

REFERENCES

1. **Todaro, C. J. and Green, H.,** Quantitative studies of the growth of mouse embryo cells in culture and their development into established lines, *J. Cell Biol.*, 17, 299, 1963.
2. **Stoker, M. G. P. and Rubin, H.,** Density-dependent inhibition of cell growth in culture, *Nature*, 215, 171, 1967.
3. **Canagaratna, M. C. P. and Riley, P. A.,** The pattern of density-dependent growth inhibition in murine fibroblasts, *J. Cell. Physiol.*, 85, 271, 1975.
4. **Harel, L. and Jullien, M.,** Evaluation of proximity inhibition of DNA synthesis in 3T3 cells, *J. Cell. Physiol.*, 88, 253, 1976.
5. **Canagaratna, M. C. P., Chapman, R., Ehrlich, E., Sutton, P. M., and Riley, P. A.,** Evidence for long-range effects in density-dependent inhibition of proliferation (DDIP) in 3T3 cells, *Differentiation*, 9, 157, 1977.
6. **Harel, L., Jullien, M., and DeMonti, M.,** Diffusible factor(s) controlling density inhibition of 3T3 cell growth: a new approach, *J. Cell. Physiol.*, 96, 327, 1978.
7. **Paul, D., Brown, K., Rupniak, H. T., and Ristow, H. J.,** Cell cycle regulation by growth factors and nutrients in normal and transformed cells, *In Vitro*, 14, 76, 1978.
8. **Steck, P. A., Voss, P. G., and Wang, J. L.,** Growth control in cultured 3T3 fibroblasts. Assays of cell proliferation and demonstration of a growth inhibitory activity, *J. Cell Biol.*, 83, 562, 1979.
9. **Steck, P. A., Blenis, J., Voss, P. G., and Wang, J. L.,** Growth control in cultured 3T3 fibroblasts. II. Molecular properties of a fraction enriched in growth inhibitory activity, *J. Cell Biol.*, 92, 523, 1982.
10. **Wang, J. L., Steck, P. A., and Kurtz, J. W.,** Growth control in cultured 3T3 fibroblasts: molecular properties of a growth regulatory factor isolated from conditioned medium, in *Growth of Cells in Hormonally Defined Media*, Cold Spring Harbor Conf. Cell Proliferation, Vol. 9, Sato, G., Sirbasku, D., and Pardee, A. B., Eds., Cold Spring Harbor Laboratory, Cold Spring Harbor, N.Y, 1982, 305.
11. **Hsu, Y.-M. and Wang, J. L.,** Growth control in cultured 3T3 fibroblasts. V. Purification of an M_r 13,000 polypeptide responsible for growth inhibitory activity, *J. Cell Biol.*, 102, 362, 1986.
12. **Hsu, Y.-M., Barry, J. M., and Wang, J. L.,** Growth control in cultured 3T3 fibroblasts: neutralization and identification of a growth inhibitory factor by a monoclonal antibody, *Proc. Natl. Acad. Sci. U.S.A.*, 81, 2107, 1984.
13. **Wang, J. L. and Hsu, Y.-M.,** Isolation and characterization of a growth regulatory factor from 3T3 cells, in *Growth, Cancer, and the Cell Cycle*, Skehan, P. and Friedman, S. J., Eds., Humana Press, Clifton, NJ, 1984, 177.
14. **Voss, P. G., Steck, P. A., Calamia, J. C., and Wang, J. L.,** Growth control in cultured 3T3 fibroblasts. III. Binding interactions of a growth inhibitory activity with target cells, *Exp. Cell Res.*, 138, 397, 1982.
15. **Sporn, M. B., Roberts, A. B., Wakefield, L. M., and Assoian, R. K.,** Transforming growth factor-β: biological function and chemical structure, *Science*, 233, 532, 1986.
16. **Keski-Oja, J. and Moses, H. L.,** Growth inhibitory polypeptides in the regulation of cell proliferation, *Med. Biol.*, 65, 13, 1987.
17. **Tucker, R. F., Shipley, G. D., Moses, H. L., and Holley, R. W.,** Growth inhibitor from BSC-1 cells closely related to platelet type β transforming growth factor, *Science*, 226, 705, 1984.
18. **Holley, R. W., Armour, R., Baldwin, J. H., and Greenfield, S.,** Activity of a kidney epithelial cell growth inhibitor on lung and mammary cells, *Cell Biol. Int. Rep.*, 7, 141, 1983.
19. **Roberts, A. B., Anzano, M. A., Wakefield, L. M., Roche, N. S., Stern, D. F., and Sporn, M. B.,** Type β transforming growth factor: a bifunctional regulator of cellular growth, *Proc. Natl. Acad. Sci. U.S.A.*, 82, 119, 1985.

20. **Sporn, M. B. and Roberts, A. B.,** Autocrine growth factors and cancer, *Nature,* 313, 745, 1985.
21. **Assoian, R. K., Komoriya, A., Myers, C. A., Miller, D. M., and Sporn, M. B.,** Transforming growth factor-β in human platelets. Identification of a major storage site, purification and characterization, *J. Biol. Chem.,* 258, 7155, 1983.
22. **Roberts, A. B., Anzano, M. A., Meyers, C. A., Wideman, J., Blacker, R., Pan, Y.-C., Stein, S., Lehrman, S. R., Smith, J. M., Lamb, L. C., and Sporn, M. B.,** Purification and properties of a type β transforming growth factor from bovine kidney, *Biochemistry,* 22, 5692, 1983.
23. **Derynck, R., Jarrett, J. A., Chen, E. Y., Eaton, D. H., Bell, J. R., Assoian, R. K., Roberts, A. B., Sporn, M. B., and Goeddel, D. V.,** Human transforming growth factor-β complementary DNA sequence and expression in normal and transformed cells, *Nature,* 316, 701, 1985.
24. **Massagué, J. and Like, B.,** Cellular receptors for type β transforming growth factor. Ligand binding and affinity labeling in human and rodent cell lines, *J. Biol. Chem.,* 260, 2636, 1985.
25. **Böhmer, F. D., Lehman, W., Schmidt, H. E., Langen, P., and Grosse, R.,** Purification of a growth inhibitor for Ehrlich ascites mammary carcinoma cells from bovine mammary gland, *Exp. Cell Res.,* 150, 466, 1984.
26. **Böhmer, F. D., Lehman, W., Noll, F., Samtleben, R., Langen, P., and Grosse, R.,** Specific neutralizing antiserum against a polypeptide growth inhibitor for mammary cells purified from bovine mammary gland, *Biochim. Biophys. Acta,* 846, 145, 1985.
27. **Böhmer, F. D., Kraft, R., Otto, A., Wernsdedt, C., Hellman, U., Kurtz, A., Müller, T., Rohde, K., Etzold, G., Lehman, W., Langen, P., Heldin, C. H., and Grosse, R.,** Identification of a polypeptide growth inhibitor from bovine mammary gland (MDGI) — sequence homology to fatty acid and retinoid binding proteins, *J. Biol. Chem.,* 262, 15137, 1987.
28. **Wang, J. L. and Hsu, Y.-M.,** Negative regulators of cell growth, *Trends Biochem. Sci.,* 11, 24, 1986.
29. **Wells, V. and Mallucci, L.,** Properties of a cell growth inhibitor produced by mouse embryo fibroblasts, *J. Cell. Physiol.,* 117, 148, 1983.
30. **Harel, L., Chatelain, G., and Golde, A.,** Density-dependent inhibition of growth: inhibitory diffusible factors from 3T3 and Rous sarcoma virus (RSV) transformed 3T3 cells, *J. Cell. Physiol.,* 119, 101, 1984.
31. **Natraj, C. V. and Datta, P.,** Control of DNA synthesis in growing Balb/c 3T3 mouse cells by a fibroblast growth regulatory factor, *Proc. Natl. Acad. Sci. U.S.A.,* 75, 6115, 1978.
32. **Peterson, S. W., Lerch, V., Moynahan, M. E., Carson, M. P., and Vale, R.,** Partial characterization of a growth-inhibitory protein in 3T3 cell plasma membranes, *Exp. Cell Res.,* 142, 447, 1982.
33. **Whittenberger, B., Raben, D., Lieberman, M. A., and Glaser, L.,** Inhibition of growth of 3T3 cells by extract of surface membranes, *Proc. Natl. Acad. Sci. U.S.A.,* 75, 5457, 1978.

Chapter 11

PROGRAMMED GENE EXPRESSIONS SUGGEST MULTIPLE BLOCKS TO REPLICATION DURING CELL AGING

Eugenia Wang

TABLE OF CONTENTS

I. INTRODUCTION

In 1957, Swim and Parker[1] reported that normal embryonic and adult human fibroblasts in culture have a limited life span, indicating that cells, like organisms, are mortal and susceptible to aging. A large-scale study describing the same phenomenon was reported later by Hayflick.[2] Subsequently, Hayflick[3] proposed that the limited replicative potential of human cells *in vitro* is the expression of senescence at the cellular level; these observations were later confirmed by many workers for cells from a variety of tissues and organisms. Cultured human fibroblasts have since become the most widely studied cellular model of aging.

However, the most crucial aspect of cellular aging is still a mystery today: what causes cells to stop traversing their replicative cycle at the end of their *in vitro* life span? In other words, what is the regulatory mechanism permanently holding the senescent fibroblasts at the resting stage and rendering them incapable of advancing through the transition between resting (G_0) stage and the S phase of the cell cycle?

We hypothesize that the cessation of proliferation in senescent fibroblasts is regulated by a family of genes whose expressions are nonproliferation dependent. Thus, the onset of cellular senescence manifested by termination of replication may be related to the activation of a specific set of nonproliferation-dependent gene expressions; i.e., the incapacity to replicate in aging fibroblasts may be negatively controlled by a group of growth-suppressing genes.

Our hypothesis is in contrast to the familiar speculation on the mechanism controlling the transition between G_0 and S phases: growth-promoting genes coding for growth factors, growth factor receptors, or enzymes needed for the DNA synthesizing machinery and necessary for the success of cell cycle traverse. However, our hypothesis proposing the existence of nonproliferating- or quiescent-specific genes does not contradict that of growth-promoting genes, if on the larger scale the transition between G_0 and S phases is controlled by two counteracting genetic mechanisms, namely the growth-inhibiting yin and growth-promoting yang families of genes.[4] The final termination of cell cycle traverse for fibroblasts in culture may be simply explained as the activation of the yin family or the accumulation of the yin family overriding the function of the yang family.

The first step in supporting our hypothesis of negative controls for cell cycle progression is to obtain evidence indicating the existence of nonproliferating-specific genes or their protein products, followed by a demonstration that these genes function to counteract the growth-promoting mechanism. Recent research on anti-oncogenes will certainly provide the strongest support for the hypothesis of negative growth control. However, in discussing normal growing cells vs. nongrowing cells, we will focus on lines of evidence from (1) cell fusion studies and (2) unique nonproliferating-specific gene expression.

A series of studies on heterokaryons, produced by fusion between senescent fibroblasts and their young counterparts or transformed derivatives, suggests that the cessation of proliferation in the former cells is controlled by "dominant factors".[5-8] This impression is based on the observation that in such hybrids, DNA synthesis in the replicating cells is turned off by factors associated with the nuclei of senescent fibroblasts. Smith and his coworkers report that the dominant factors may be endogenous proteins associated with the plasma membrane.[9,10] The existence of these proteinaceous factors is further evidenced by the inhibitory activity upon DNA synthesis exerted by mRNA derived from senescent fibroblasts.[11] Recently, O'Brien et al.[12] showed that *in vitro* senescence in virally transformed human fibroblasts restricts tumor-forming ability.

II. CELL BIOLOGY OF STATIN, A NONPROLIFERATION-DEPENDENT GENE EXPRESSION

Because we do not yet know what genes do indeed control cell proliferation, it is expedient first to identify all those genes whose expression is cell-cycle dependent, and in particular, quiescent (G_0) phase dependent, in the hope that some of these genes may turn out to be those we are searching for. For this reason we have attempted to identify, by the production of monoclonal antibodies, senescence- and nonproliferating-specific proteins.

1. Unique expression of statin in nonproliferating senescent cells. Mouse monoclonal antibody S-30 was produced from hybridomas prepared from mice injected with the cytoskeleton extract of an *in vitro* aged culture of human fibroblasts (3529) derived from a 66-year-old donor. The antibody stains positively the nuclei of the nonproliferating cells present predominantly in the senescent cultures of five selected fibroblast strains derived from donors of different age groups[13,14] (Figure 1, Table 1), whereas a negative reaction is observed in cultures of their young counterparts. In the intermediate stage of the *in vitro* life span of these cell strains, a heterogeneous positive reaction for staining with S-30 antibody is observed in different subfractions of cell cultures (Figure 2).

2. Presence of statin in growth-arrested young fibroblasts, and its rapid disappearance upon reentering the process of cell cycling. The presence of S-30 can be induced in young fibroblasts at an early stage of their life span; this induced expression of statin will decline rapidly once the block to cell cycle traverse is removed.[15,16] The rapid deexpression is observed in fibroblasts involved in the *in vitro* wound healing process, by addition of serum to serum-starved cultures,[15,16] or by subculturing after trypsinization and replating from confluent cultures. Kinetic analysis shows that 50% of the cell population loses its statin expression by 12 h after these culturing manipulations. Also, fluorescence studies show that the pattern of statin disappearance is not homogeneous along the entire nuclear periphery; while it is being lost, there seem to be residual sites of statin distribution, seen as patches or caps along the nuclear periphery (Figure 3). These results suggest that heterogeneity exists in the area of the nuclear envelope, as judged from statin distribution.

3. The loss of statin expression precedes the activation of DNA synthesis. In serum-starved cultures, more than 85% of the cells stain positively for statin. After stimulation with serum, the expression of statin declines rapidly within the first 12 to 14 h.[16] On the other hand, an increase in the level of DNA synthesis, signifying entry into S phase, is observed initially at 18 h after serum stimulation and reaches maximal levels 6 h later. Immunoprecipitation of statin derived from cells harvested at different intervals after serum stimulation reveals that the level of statin synthesis is reduced as soon as 4 h after serum addition.[16] These results suggest that a rapid decline in statin synthesis occurs immediately after cells are induced to proliferate and well before the transition from G_1 to S phase. The relationship between statin expression and activity of DNA synthesis can be best described in the confluent vs. sparse culture system shown in Diagram A.

4. Difference of statin expression in nonproliferating young and senescent cells. Although both senescent and growth-arrested young cells have statin present in their nuclei, there is a significant difference in protein quantity. As seen in Figure 4, results of the immunoprecipitation of senescent extracts with S-30 antibody show a pronounced band at 57,000 Da, whereas the same amount of extract derived from young confluent cultures shows only a faint band at the same molecular weight. Both immunofluorescence and immunoelectron microscopic observations show that statin in growth-arrested young cells is primarily located at the nuclear envelope region, whereas statin in senescent cells is located not only at the envelope, but also throughout the nucleoplasm (see Figure 4 of Wang[13,15]).

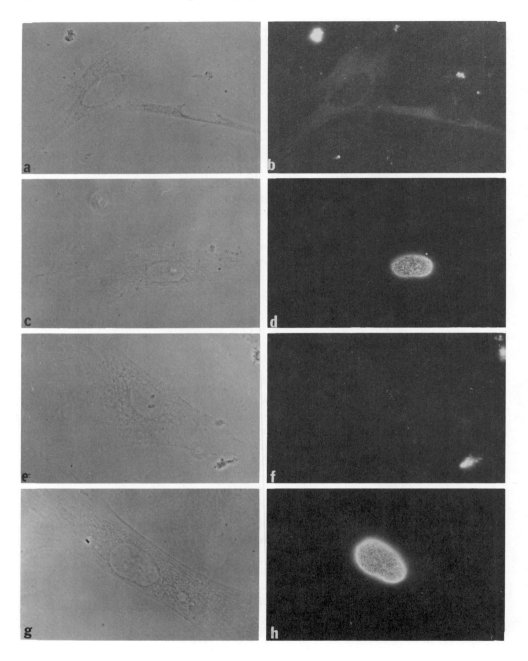

FIGURE 1.   The expression of statin as identified by monoclonal antibody S-30, in young and old fibroblast cultures of two different cell strains, by phase-contrast (a, c, e, g) and immunofluorescence (b, d, f, h) microscopy. (a, b) Lack of S-30 staining activity in the nuclei of 0011 fibroblasts at an early stage of their life span (cumulative population doubling level [CPDL] < 10). (c, d) The positive staining reaction of S-30 antigen in senescent fibroblasts of the 0011 cell strain (CPDL > 65). Approximately 95% of the cell population at this stage shows the positive nuclear staining pattern, as seen here. (e, f) Lack of S-30 staining in young cells of 3529 strain (CPDL < 10). There are a few fluorescent granules located in the nuclei indicating a minute presence of S-30. (g, h) Positive staining reaction of S-30 in senescent fibroblasts of the 3529 cell strain (CPDL > 28). Again, the staining reaction was found in approximately 95% of the cell cultures. (Magnification × 510.)

TABLE 1
Strains of Human Fibroblasts Used

Repository number	Donor age	Donor sex	CPDL[a]	
			Mean	Range
GM0011	8 fetal weeks	M	65	57—72
GM2936B	20 d	M	57	54—60
GM0038A	9 years	F	56	52—61
GM2912A	26 years	M	31	27—38
GM3529	66 years	M	28	24—32

[a] CPDL = cumulative population doubling levels.

III. STATIN EXPRESSION IN TERMINALLY DIFFERENTIATED SYSTEMS

1. Histochemical studies of statin in different tissues. Our next question is whether statin is specific to nonproliferating fibroblasts, or are all nonproliferating cells, whether fibroblasts or not, statin positive? To answer this question, we performed a survey of statin presence in various tissues derived from humans, dogs, rats, mice, and chickens.[17]

Figure 5 shows an example of positive tissue staining for statin. This panel shows a frozen section cut perpendicularly through the epidermis of human biopsy material. Statin staining is absent in the basal layer, which is composed of rapidly dividing keratinocytes. In contrast, statin is expressed as a nuclear ringlink structure in the nondividing keratinocytes of the suprabasal cells. The results of our tissue survey, as shown in Table 2, demonstrate that, in general, statin is present in those tissues composed of cells that are terminally differentiated and no longer proliferating; in tissues composed of mostly replicating cells, statin is not present. The exception to this rule is the existence of fibrocytes in the dermis layer of the skin region, where statin-positive and statin-negative cells are found next to each other in the same area (Figure 6). Quantitative analysis of statin presence in different tissues reveals that liver, heart, and brain are the tissues containing the greatest amount of statin.

2. Activation of statin expression during the process of terminal differentiation. In the well-developed *in vitro* muscle system, primary myoblasts from chick embryonic pectoral muscle are positive for S-132 (cyclin-like) antigen, and they are statin negative.[18] In early myotubes (3 d in culture), nuclei remain cyclin positive and statin negative, whereas older cells (16 d in culture) become statin positive (Figure 7). When intermediate stages are examined, it is found that cyclin gradually disappears while statin appears over this 2-week period. Rat myoblasts are cyclin positive and statin negative; in the rat myotube, however, the shift from cyclin positive to statin positive occurs at a much more rapid rate than in the chicken system.

Neuroblastoma (strain Neuro 2A) cells are cyclin positive and statin negative. However, after 30 h of serum starvation-induced differentiation, nuclei are shifted to become statin positive and they lose their cyclin expression. Similarly, in erythroleukemia (MEL) cells, nuclei are cyclin positive and statin negative; upon induction to differentiation by dimethyl sulfoxide (DMSO) treatment, statin staining appears in the nuclei and cyclin staining begins to disappear. Taken together, we believe that these results suggest that statin appearance is a marker for cells upon commitment to terminal differentiation *in vitro*.

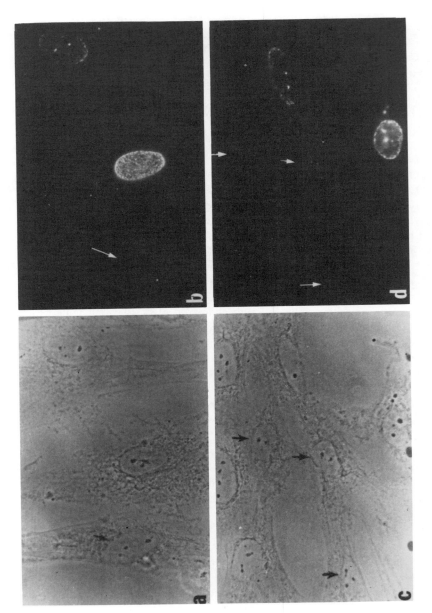

FIGURE 2. The heterogeneous expression of statin in cultures at the intermediate stage of their life span, as recognized by S-30 antibody by phase-contrast (a, c) and immunofluorescence (b, d) microscopy. (a, b) Staining pattern of S-30 in the 0011 cell strain at CPDL = 40. Approximately 30 to 40% of the cell population shows S-30 staining activity, while the remaining subfraction of the cell culture is still negative (arrows). (c, d) Staining pattern of S-30 in the 3529 cell strain at CPDL = 15. Approximately 20 to 30% of the cell population show the positive reaction with S-30 antibody, while the remaining subfraction is negative (arrows). (Magnification × 450.) (Reproduced from *The Journal of Cell Biology*, 1985, Vol. 100, pp. 545-551, by copyright permission of the Rockefeller University Press.)

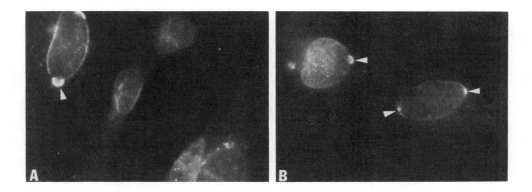

FIGURE 3. Immunofluorescence of the heterogeneous nuclear appearance of statin in quiescent young cells (0011; CPDL = 12) that have been stimulated to reenter the cell cycle traverse by adding 10% serum back into a serum-starved quiescent culture. (A) 6 h after addition of serum, as evidenced by weaker staining intensity, some cells have already lost statin, with some nuclei showing unique patchy localized staining for statin (arrows). (B) 12 h after addition of serum, most cells have lost statin from their nuclei and exhibit only localized polar residual presence of statin, as shown by the large arrows. (Magnification × 570.)

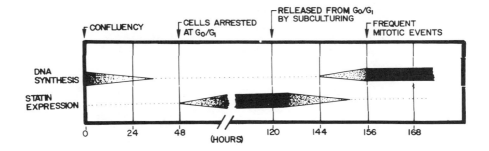

DIAGRAM A. Confluency arrest of cell cycling.

IV. CHARACTERIZATION OF THE DISTRIBUTION OF STATIN AT THE NUCLEAR ENVELOPE REGION

Results of both immunofluorescence and immunogold labeling show that statin is localized at the region of the nuclear envelope in nonproliferating cells. However, these histochemical experiments were performed with detergent-extracted samples. For exact localization of statin we chose another statin-positive system, rat liver nuclei, for the sake of quick access to large experimental quantities and to avoid detergent extraction during sample preparation so that the intact morphology at the nuclear envelope region is preserved. Figure 8 shows that statin remains associated with the isolated rat liver nuclei; in particular, the protein is localized at the envelope region, as evidenced by the bright fluorescence at the periphery. Ultrastructural localization by gold labeling shows that statin is specifically localized at the nuclear lamina region, facing the nucleoplasm (Figure 9). Little labeling occurs in either the area of the cisternae of the nuclear envelope or in the outskirts of the envelope facing the cytoplasm; some labeling is seen in the nuclear matrix region. We further isolated the nuclear lamina with the pore complex from purified nuclei to demonstrate by immunoblotting assays that statin is indeed associated with the lamina fraction.

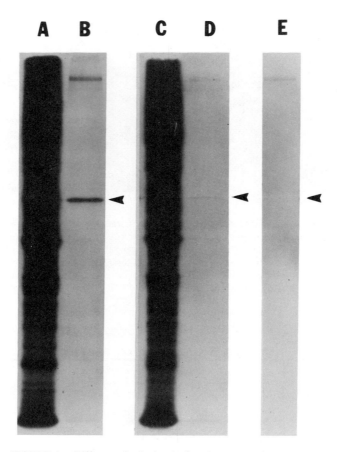

FIGURE 4. Difference in the level of statin presence in senescent fibroblasts (3529; CPDL > 30) and young confluency-arrested cells (3529; CPDL < 10), measured by immunoprecipitation assays. Insoluble cell extracts with the same radioactive counts were obtained from senescent cultures (A) and confluent young cells (C) and were used for incubation with the same amount of S-44 antibody, another monoclonal antibody to statin. The resulting precipitates were run in parallel to the starting materials, as demonstrated in (B) for the precipitating band for (A) and in (D) for the precipitating band for (C). Note the presence of a prominent band at 57,000 Da in lane B in contrast to the faint band seen in lane D. Lane E represents the result when S-44 antibody was replaced with a control mouse ascites serum in assays for lane B.

V. CHARACTERIZATION OF SEVERAL *IN VITRO* SENESCENCE-DEPENDENT SECRETORY PROTEINS

The above discussions show that similarity in statin expression exists in both nonproliferating young cells and senescent fibroblasts. Both cell types can be promoted to leave G_0 and enter the G_1 phase; however, obviously senescent cells cannot enter the S phase and resume replication. This then led us to ask, what is the difference between young growth-arrested cells and senescent cells which renders the latter permanently arrested in G_1? (Our observation of the late G_1 block for senescent cells has also been suggested by recent work of Rittling et al.[19])

We hypothesize that the G_1 block for permanent growth arrest in senescent cells is in

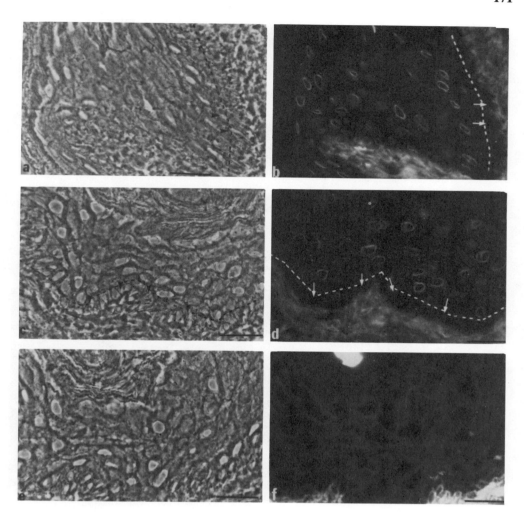

FIGURE 5. *In situ* localization of statin expression in frozen rat skin sections (a — d). Statin is seen as a ringlike nuclear pattern in the suprabasal level, with very weak activity of antibody reaction in the basal layer (arrows). The fluorescence seen in the area of dermis beneath the basal layer is nonspecific binding activity; similar intensity is seen in control specimens, where the initial monoclonal antibody was replaced with the growing medium of hybridoma cells (e, f). (a, c, e) Phase-contrast and (b, d, f) fluorescence microscopy of the same specimens. (Magnification × 240.) (From Wang, E. and Krueger, J. G., *J. Histochem. Cytochem.*, 33, 587, 1985. With permission.)

part related to the absence and/or presence of secreted proteins acting as autocrine growth factors.[20] To identify these proteins, we analyzed the cellular components secreted into the medium by cells under many different growth conditions; we found two groups of secretory proteins whose expressions are considered to be specifically dependent on the *in vitro* aging state. One group consists of three proteins (molecular weight range, 80 to 87 kDa) found in the media of sparse, confluent, and serum-starved cultures of young fibroblasts; these proteins disappear from the medium as soon as the cell culture reaches the nonproliferative senescent state. The other group consists of two proteins (51 and 57 kDa) detected in the medium of senescent cultures. Immunoprecipitation assays show that the statin antibody cannot react with the 57-kDa secreted protein, suggesting that the 57-kDa polypeptide identified in the medium is not statin even though the two proteins share the same molecular weight. Alternative expression between the two groups of secretory proteins is found in both

TABLE 2
Immunolocalization of S-30 Antibody in Different Tissues

Tissue sites	Activity for S-30 antibody	Relative index of proliferation activity
Suprabasal layers of skin epidermis	+ + + +	−
Esophagus epithelium (suprabasal layers)	+ + + +	−
Tongue epithelium	+ + + +	−
Hair epithelium	+ + + +	−
Kidney tubules (simple epithelium)	−	+ + +
Gastrointestinal epithelium	−	+ + +
Gastrointestinal smooth muscle	+	−
Heart smooth muscle	+ + +	−
Vascular smooth muscle	−	+
Chondrocytes	+ + +	−
Hepatocytes	+ + +	−
Brain	+ / −	+ / −

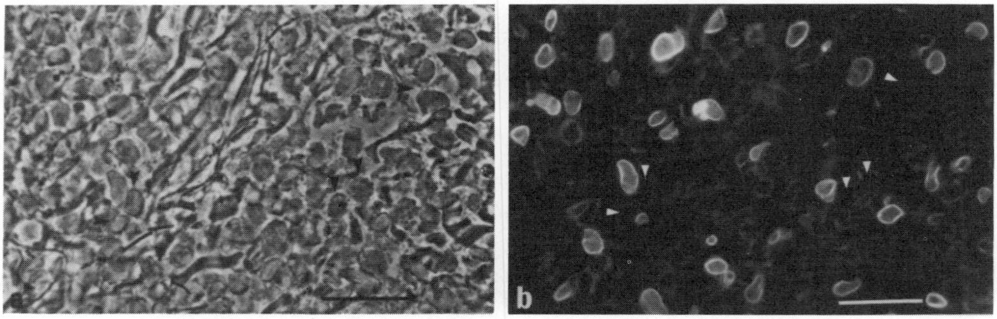

FIGURE 6.   The presence of statin as seen by staining with S-30 antibody in rat dermis. As in other tissues, the staining pattern is seen as a nuclear ringlike appearance in some cells, most probably fibroblasts. However, some cells (arrows) seen by phase-contrast microscopy (a) do not show the presence of statin, attested by the absence of intensity seen with fluorescence optics (b). (Magnification × 270.) (From Wang, E. and Krueger, J. G., *J. Histochem. Cytochem.*, 33, 587, 1985. With permission.)

the 0011 and 3529 cell strains under the same growth conditions (Figure 10). These proteins were detected by ^{35}S-methionine labeling to avoid misidentification with serum proteins; the half-life for both groups of proteins is as short as 4 h. Preliminary results show that both groups of the secreted proteins are able to bind heparin sulfate, suggesting that they are proteoglycans. Varying concentrations of serum in the medium, ranging from 0.5 to 10%, have no effect on the synthesis of these proteins. These results suggest that when human fibroblasts reach the state of senescence, they begin to secrete two new proteins, while they stop secreting three other proteins found only in cultures of young cells.

Therefore, alternative expression between the group of specific secreted proteins from young cells and another group of senescent cell-specific secreted proteins may play the key role in the G_1 block of cell cycle traverse in *in vitro* aged fibroblasts.

VI. SUMMARY AND DISCUSSION

1. Statin is present in the nucleus of most nonproliferating cells.
2. Statin is only expressed in the G_0 phase, and it is the first known cellular marker for this phase of the cell cycle.

PANEL A

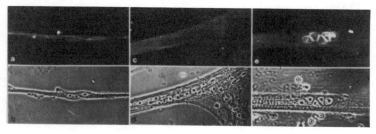

PANEL B

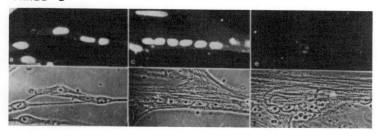

PANEL C

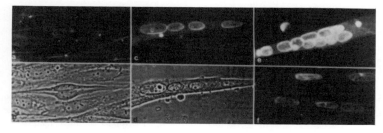

PANEL D

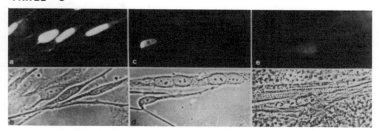

FIGURE 7.   Alternative expression between statin and the cyclin-like S-132 antigen in the *in vitro* system of muscle differentiation. In Panel A are fluorescence (a, c, e) and phase-contrast (b, d, f) micrographs showing the absence of statin in chick myogenic cells before fusion (a, b) and remaining absence in myotubes 3 d after fusion (c, d); statin does not appear until 7 d after fusion (e, f). In contrast, Panel B shows that the cyclin-like antigen is present in myoblasts and early fused myotubes at stages similar to a, b, c, and d of Panel B. The shift from cyclin-like nuclear antigen to statin antigen occurs at a much faster rate in the rat myoblast system. Panel C demonstrates the absence of statin in myoblasts; it appears immediately after fusion into myotubes. Panel D shows the rapid loss of cyclin-like S-132 antigen at the same time; in other words, the high level of this antigen in the unfused myoblasts is immediately lost after fusion into the myotube. (Magnification × 360.) (From Connolly, J. A., Sarabia, V. E., Kelvin, D. J., and Wang, E., *Exp. Cell Res.*, 174, 461, 1988. With permission.)

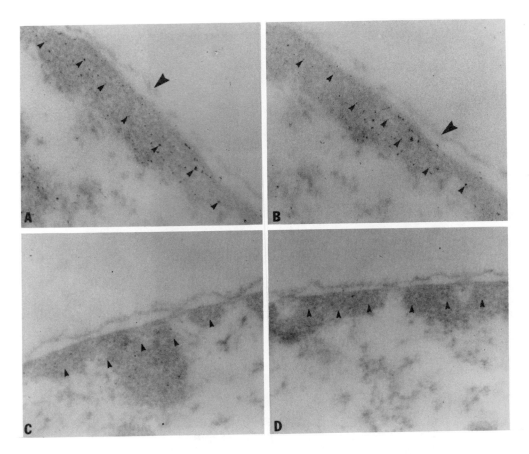

FIGURE 8.   Ultrastructural localization of statin in isolated rat liver nuclei. The nuclei were isolated according to the procedure described by Dwyer and Blobel.[21] (A) Localization of statin in isolated nuclei when clone S-30 monoclonal antibody was used as the primary antibody; as reflected by the positions of the gold particles (small arrows), statin is mostly localized at the region internal to the envelope and is not present in the area of the outer envelope (large arrow) or in the cisternae. (B) Similar statin localization was found when another statin antibody, clone S-44, was used as the primary antibody. In both A and B, some statin staining was observed in the internal area of the nuclei. The specificity of the statin localization, as shown by protein-A gold labeling, was verified by the absence of concentrated gold particles in the same region of nuclear lamina (arrows) when (C) the primary antibody was omitted from the immunogold labeling procedure or (D) the primary antibody was replaced with the same amount of a control ascites monoclonal antibody. A few gold particles, 5 nm in size, are found randomly scattered in the nucleoplasm because of nonspecific reaction between gold particles and the specimens. (Magnification × 75,000.)

3.   Upon stimulation with serum or growth factors, both young and senescent cells can be stimulated to leave G_0 and enter the G_1 phase of the cell cycle.

4.   Statin is present in most terminally differentiated cells, and it can be used as a marker for the final stage of differentiation.

5.   Statin is associated with the lamina component of the nuclear envelope.

6.   Alternative expression between two groups of secreted proteins is found to be related specifically to the senescence-associated G_1 block to progress through the cell cycle.

The above discussion and the current literature suggest that a two-stage block for cell cycle traverse may be in operation to produce permanent growth arrest in senescent cells. So far, these two stages are identified at the G_0 and late G_1 phases of the cell cycle; our results show that statin expression is related to the G_0 phase, while the alternative expression of two groups of secreted proteins is related to the late G_1 phase (Diagram B).

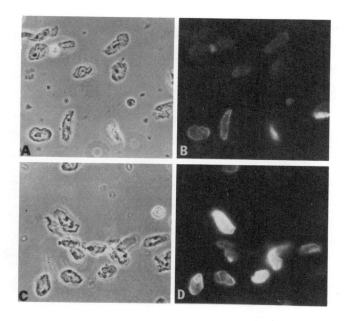

FIGURE 9. Immunofluorescence identification of statin presence in isolated nuclei purified from rat liver. Isolated nuclei were immobilized and fixed with formaldehyde before processing for incubation with antibody S-30. Comparison of phase-contrast (A, C) and fluorescence (B, D) images of the same specimens indicates that statin remains associated with the nuclei through the various steps of nuclear isolation. Bright fluorescence staining can be seen at the nuclear envelope. (Magnification × 720.)

Evidence suggests that the classical G_0 block, as typified by the removal of nutrients from the culture environment, is a reversible growth arrest. In other words, when nutrients and/or growth factors such as serum are added to the culture, cells are able to leave the block and reenter the G_1 phase of the cell cycle. Due to this fact, we may term the G_0 block as the *regulatory* block. By manipulating the concentration of serum, one can regulate cells to stay in or leave the G_0 growth block. Recently, Rittling et al.[19] reported that even senescent fibroblasts are able to leave the G_0 phase and resume the traverse of G_1 after addition of fresh serum to the culture.

The late G_1 block, on the other hand, was recently identified as the block responsible for the irreversible growth arrest; we term it the *obligatory* block for preventing cells from entering into the S phase. So far, knowledge of *in vitro* culturing has not provided us the ability to overcome the irreversible nature of this growth block. For example, *in vitro* aged fibroblasts, while retaining statin expression, are able to leave the G_0 phase and travel along most of the early G_1 phase, but they fail to complete the traverse and remain nonreplicative.

What then might cause senescent fibroblasts to irreversibly halt during the DNA synthesis phase of the cell cycle? Whatever the causes may be, they can be categorized into two areas: (1) growth modifier — the absence of growth promoters and/or the presence of growth inhibitors — and (2) growth responder — the presence or absence of cellular responders to receive the signal for cell cycle traverse. Our finding a difference in the expressions of secretory proteins between growth-arrested young cells and nonreplicative senescent fibroblasts may provide one of the first few examples that can be developed further as the marker for the late G_1 block. However, much more work must be performed to understand whether statin and the secretory proteins function either as the growth modifier or as the growth responder, or simply as the byproduct of the reactions toward them.

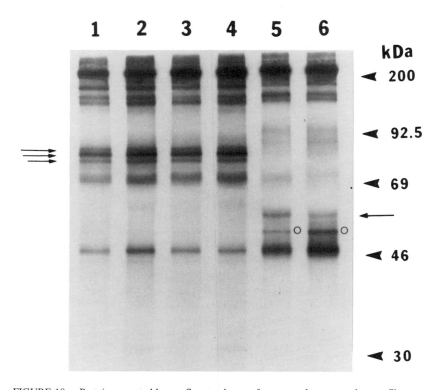

FIGURE 10.   Proteins secreted by confluent cultures of young and senescent human fibroblasts. Cultures were grown to confluence in Eagle's minimal essential medium supplemented with 10% fetal bovine serum and 1% nonessential amino acids at 37°C in a humidified atmosphere of 5% CO_2. A duplicate set of confluent cultures was further serum starved in the same medium supplemented with 0.5% fetal bovine serum for 2 d prior to the labeling experiment. Cells were labeled with 50 μCi/ml of ^{35}S-methionine (1148 Ci/mmol, New England Nuclear) in methionine-free Eagle's minimal essential medium supplemented with 0.5% fetal bovine serum, 0.5% nonessential amino acids, and 0.8 mg/1 of methionine at 37°C for 17 h. The cell-free conditioned media were collected and precipitated with trichloracetic acid (TCA). An equal number of TCA-precipitated counts from each sample were analyzed on 10% SDS-polyacrylamide gel under reducing conditions. Lane 1, young fibroblasts of 0011 (CPDL = 15); lane 3, young fibroblasts of 3529 (CPDL = 7); lane 5, senescent fibroblasts of 3529 (CPDL = 24). Lanes 2, 4, and 6 were duplicate cultures of lanes 1, 3, and 5, respectively, that were serum starved. Arrows indicate the presence of three proteins with molecular weights of 80, 84, and 87 kDa in lanes 1 through 4, and two proteins of 51 and 57 kDa in lanes 5 and 6. The open circle indicates the 51-kDa protein that may not be cell aging specific (see discussion in Ching and Wang[20]). (From Ching, G. and Wang, E., *Proc. Natl. Acad. Sci.*, 85, 151, 1988. With permission.)

Needless to say, the two growth blocks (G_0 and late G_1) discussed here are by no means the only possible gates to halt the cell cycle traverse. Subdivision of the G_0-G_1→S transition into more stages besides those described here will become possible in the future by recognizing differential gene expressions at different time intervals along the route of this transition. Initially these gene expressions, as statin and the two groups of secreted protein described here, may be used as markers to denote the different stages; later, they can be used as probes to investigate control of irreversible growth arrest during *in vitro* aging.

ACKNOWLEDGMENTS

The author wishes to thank Ms. Francine Mantha for typing this manuscript and Mr. Alan N. Bloch for critically reviewing the text. This work was performed using a grant

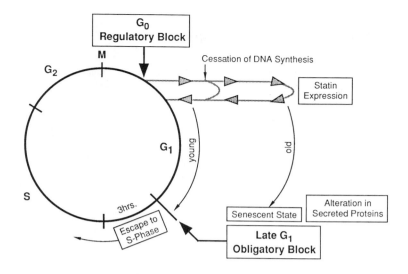

DIAGRAM B. Two possible stages of growth block in aging fibroblasts.

from the U.S. National Institute on Aging (RO1 AGO3020) and private funding from the Bloomfield Family Foundation.

REFERENCES

1. **Swim, H. E. and Parker, R. F.,** Culture characteristics of human fibroblasts propagated serially, *Am. J. Hyg.,* 66, 235, 1957.
2. **Hayflick, L.,** Recent advances in the cell biology of aging, *Mech. Ageing Dev.,* 14, 58, 1980.
3. **Hayflick, L.,** The limited *in vitro* life time of human diploid cell strains, *Exp. Cell Res.,* 37, 614, 1965.
4. **Sager, R.,** Genetic suppression of tumor formation: a new frontier in cancer research, *Cancer Res.,* 46, 1573, 1986.
5. **Stein, G. H. and Yanishevsky, R. M.,** Quiescent human diploid cells can inhibit entry into S-phase in replicative nuclei in heterodikaryons, *Proc. Natl. Acad. Sci. U.S.A.,* 78, 3025, 1981.
6. **Rabinovitch, P. S. and Norwood, T. H.,** Comparative heterokaryon study of cellular senescence and the serum-deprived state, *Exp. Cell Res.,* 130, 101, 1980.
7. **Pereira-Smith, O. M. and Smith, J. R.,** Expressin of SV4O T antigens in finite lifespan hybrids of normal and SV40-transformed fibroblasts, *Som. Cell Genet.,* 7, 411, 1981.
8. **Matsumura, T., Pfendt, A., Zerrudo, Z., and Hayflick, L.,** Senescent human diploid fibroblast (WI38) attempted induction of proliferation by infection with SV4O and by fusion with UV-irradiated continuous cell lines, *Exp. Cell Res.,* 125, 453, 1980.
9. **Dreschler-Lincoln, C. K. and Smith, J. R.,** Inhibition of DNA synthesis in senescent-proliferating human hybrids is mediated by endogenous proteins, *Exp. Cell Res.,* 153, 208, 1984.
10. **Pereira-Smith, O. M., Fisher, S. F., and Smith, J. R.,** Senescent and quiescent cell inhibitors of DNA synthesis, *Exp. Cell Res.,* 160, 297, 1985.
11. **Lumpkin, C. K., Jr., McGlung, J. K., Pereira-Smith, O. M., and Smith, J. R.,** Existence of high abundance antiproliferative mRNA's in senescent human diploid fibroblasts, *Science,* 232, 393, 1986.
12. **O'Brien, W., Stenman, G., and Sager, R.,** Suppression of tumor growth by senescence in virally transformed human fibroblasts, *Proc. Natl. Acad. Sci. U.S.A.,* 83, 8659, 1986.
13. **Wang, E.,** A 57,000-molecular-weight protein uniquely present in nonproliferating cells and senescent human fibroblasts, *J. Cell Biol.,* 100, 545, 1985.
14. **Wang, E.,** Are cross-bridging structures involved in the bundle formation of intermediate filaments and the decrease in locomotion that accompany cell aging?, *J. Cell Biol.,* 100, 1466, 1985.
15. **Wang, E.,** Rapid disappearance of statin, a nonproliferating and senescent-cell specific protein upon reentering the process of cell cycling, *J. Cell Biol.,* 101, 1695, 1985.

16. **Wang, E. and Lin, S. L.,** Disappearance of statin, a protein marker for nonproliferating and senescent cells, following serum-stimulated cell cycle entry, *Exp. Cell Res.,* 167, 135, 1986.
17. **Wang, E. and Krueger, J. G.,** Application of a unique monoclonal antibody as a marker for nonproliferating subpopulations of cells of some tissues, *J. Histochem. Cytochem.,* 33, 587, 1985.
18. **Connolly, J. A., Sarabia, V. E., Kelvin, D. J., and Wang, E.,** The disappearance of a cyclin-like protein and the appearance of statin are correlated with the onset of differentiation during myogenesis *in vitro, Exp. Cell Res.,* 174, 461, 1988.
19. **Rittling, S. R., Brooks, K. M., Cristofalo, V. J., and Baserga, R.,** Expression of cell cycle-dependent genes in young and senescent WI-38 fibroblasts, *Proc. Natl. Acad. Sci. U.S.A.,* 83, 3316, 1986.
20. **Ching, G. and Wang, E.,** Absence of three secreted proteins and presence of a 57-KDa protein related to irreversible arrest of cell growth, *Proc. Natl. Acad. Sci. U.S.A.,* 85, 151, 1988.
21. **Dwyer, N. and Blobel, G.,** *J. Cell Biol.,* 70, 581, 1976.

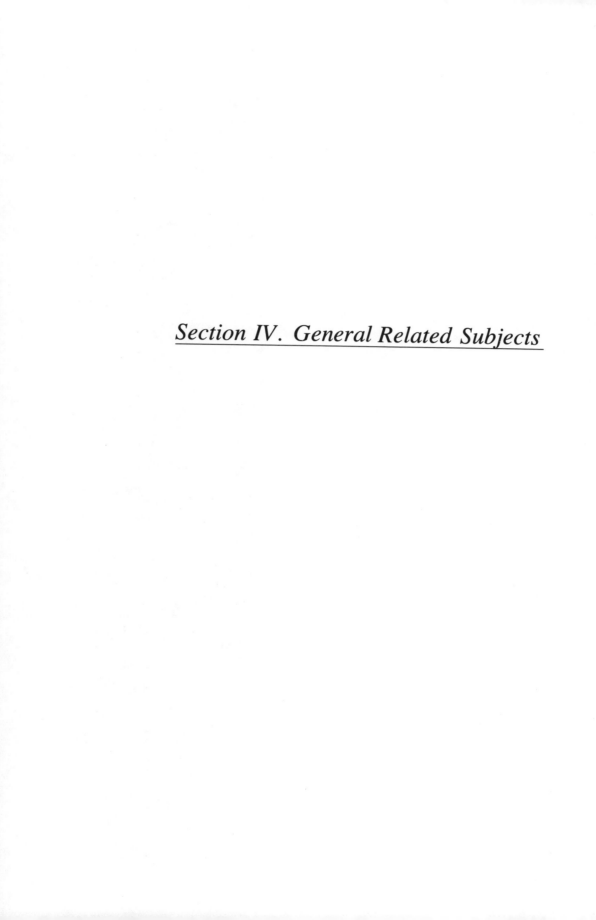

Section IV. General Related Subjects

Chapter 12

A POTENTIAL ROLE FOR INTERSPERSED REPETITIVE SEQUENCE ELEMENTS IN NEGATIVE GROWTH REGULATION

Bruce H. Howard

TABLE OF CONTENTS

I. INTRODUCTION

Gene transfer and molecular cloning technologies represent powerful tools with which to gain new insights into the molecular mechanisms that underlie the aging process. Over the past several years, our laboratory has applied these technologies in an effort to identify genes that negatively regulate mammalian cell growth. In the course of this work, it has become apparent that our results have interesting, albeit theoretical, implications for cellular senescence. The purpose of this chapter is to discuss several hypotheses concerning cellular senescence that derive from our work. Where possible, these hypotheses will be related to findings published by other workers in the aging field.

The genetic determinants of human aging are likely to be very complex. Indeed, it has been suggested that as much as 7% of the genome may have the potential to impact on some aspect of the aging process.[1] In an effort to work within an experimentally tractable system, many investigators have studied the senescence of human embryo fibroblast-like cells in tissue culture (see Norwood and Smith[2] for review). Although the precise relevance of *in vitro* cellular senescence to aging in the intact organism remains uncertain,[3,4] it can be hoped that the mechanism which limits fibroblast life span *in vitro* is related to that by which senescence occurs in specific critical tissues *in vivo*. It seems plausible, for example, that neuroendocrine and immunological aspects of aging reflect senescence, at the cellular level, of critical "pacemaker" cells in those organs.

With respect to *in vitro* senescence and immortalization of human cells, two very interesting observations have emerged over the past several years. First, immortalization of most human cell types by dominant oncogenes is relatively inefficient; in particular, there has been a notable lack of success in acutely immortalizing human fibroblasts or lymphocytes by individual or combinations of cellular oncogenes.[5-7] Even a relatively "complete" viral oncogene, such as SV40 T antigen, yields immortalized human fibroblast lines only after the culture has passed through a crisis phase.[8-10] These findings are consistent with somatic cell hybrid experiments which indicate that immortalization is determined primarily by recessive genetic events.[11-13] Second, the process of *in vitro* senescence, like differentiation, appears to possess stochastic characteristics. Although it is almost certainly an oversimplification to argue that senescence is purely stochastic in nature, several published studies suggest that there is a strong probabilistic component to this process.[14-16] From these observations, we have been led to pose two questions: (1) are there one or more classes of "senescence" genes whose growth suppression function is dominant over cellular oncogenes and (2) could the expression of such senescence genes be controlled by mechanisms that are regulated at least in part in a stochastic manner?

The gene transfer approach that we have pursued to clone growth suppression genes is the inverse of the strategy used to clone oncogenes. Rather than transfecting genomic DNA from malignantly transformed cell lines into a nontransformed recipient indicator line such as mouse NIH/3T3, we transfected genomic DNA from normal cells into a malignantly transformed cell line. WI-38 and MRC-5 human embryo fibroblasts were selected as normal cell strains; genomic DNA from these strains was cotransfected with a neo selectable marker plasmid. HeLa cells were chosen as the transformed indicator line. Our experimental protocol involved a double selection: an initial BrdUrd/Hoechst fluorescent light suicide selection against cells that failed to exhibit a transient inhibition of replication and a subsequent selection for expression of the cotransfected selectable marker.

The results of this gene transfer assay for growth suppression genes have been published elsewhere.[17] Briefly summarized, they include several rather unexpected findings. First, the growth inhibitory activity in genomic DNA from human embryo fibroblasts was observed under certain circumstances to be very strong; i.e, from 3 to >20% of stably transfected cells, as scored by the selectable marker, exhibited growth inhibition. Similar activity was not detected in control DNAs from *Escherichia coli* or HeLa cells. Second, this activity appeared to be influenced by epigenetic control mechanisms, since it could be enhanced by forcing the fibroblasts into a deeply quiescent state prior to harvesting DNA. Third, higher growth suppression activity was detected in a cosmid library prepared from WI-38 genomic DNA than from control *E. coli* or HeLa cosmid libraries. This result suggested that the observed activity resided in specific DNA sequences that could be characterized by molecular cloning. Finally, random fractionation of the WI-38 cosmid library (sib selection) suggested that sequences with the potential to exhibit inhibitory activity could be dispersed and that they have a copy number of from several thousand to >10,000.

In attempting to interpret these results, we recognized a potential relationship to studies on cellular aging. Specifically, we considered the possibility that growth suppression genes detected in the BrdUrd/Hoechst transfection assay might be candidates for senescence genes. This supposition was based primarily on a parallel between results obtained with the transfection assay and results derived from the heterokaryon growth regulatory assay developed by Norwood et al.[18] This heterokaryon assay has been reported to detect inhibition of DNA replication in normal (and some transformed) cells fused to human embryo fibroblasts in either of two states: deep quiescence or senescence.[18-25] We observed that our transfection assay, like the heterokaryon assay, detected induction of growth inhibitory activity subsequent to maintenance of fibroblasts in low serum conditions, i.e., in a deeply quiescent state. In addition, we found a possible correlation between cellular senescence and transfectable growth suppression activity; thus, the strongest activity was detected in DNA harvested from WI-38 embryo fibroblast cultures that exhibited two characteristics of early *in vitro* senescence, increased doubling time and decreased density at confluence.[17]

If it is accepted, for purposes of this discussion, that transfection of genomic DNA from human embryo fibroblasts into HeLa cells can, under appropriate conditions, measure DNA sequences involved in senescence, then the results summarized above must be taken into account in any consequent model of aging. In particular, the hypothesis that interspersed repetitive sequences (IRS) may have a role in cellular senescence must be considered. Such a role, although entirely unexpected, would have a number of theoretically attractive features, especially if it is assumed that repetitive sequence elements operate in concert with single copy, polypeptide-encoding growth regulatory genes.

A number of issues and findings complicate the proposition that interspersed repetitive elements play a role in cellular aging. First, there is the influential argument advanced by Orgel and Crick[26] that repetitive sequences are "selfish" in nature and, for the most part, provide no benefit for the eukaryotic cell. Second, there is evidence that few, if any, individual short interspersed sequence elements (SINES)[27] code for an obvious polypeptide or show identity with a putative master gene/consensus sequence;[28,29] such sequence divergence could be interpreted to suggest absence of function. Third, there appears to be substantial variability with respect to the presence of discrete IRS transcripts in different cell types,[30-35] and in several cell lines it has been demonstrated that the specific promoter activities of major repetitive families such as Alu and LINE-1 are absent or very low.[33,36]

These points suggest that if functions are found for repetitive elements with respect to cellular senescence, they may be quite different from those which have been generally anticipated. For example, the absence of conserved open reading frames in IRS does not preclude the functioning of their transcripts as components of small nuclear ribonucleoprotein particles (snRNPs). Likewise, the absence of transcription from repetitive sequence elements may indicate that such sequences frequently exist in the germline genome as inactive "progenes". (Note that if interspersed repetitive sequences mediate growth suppression, then their transcriptional silence in replicative cells may be expected.) It should be stressed that absence of function for most copies of a given repetitive sequence type does not preclude that a majority of copies are important. Moreover, a "selfish" component to the amplification and dispersion of a given repetive sequence family is entirely compatible with individual copies of that family performing a critical function within the host cell.

To further elaborate on these speculations, they will be discussed with respect to sequences belonging to the Alu family, a prototype SINE. Emphasis will be placed on mechanisms by which those sequences might mediate growth suppression.

II. ALU INTERSPERSED REPETITIVE SEQUENCES

The dominant SINE in human and other primate cells is the Alu family. The estimated copy number of DNA sequences with homology to the consensus Alu element is 300,000 to 500,000.[37,38] The number of full-length (ca. 300-bp) copies is probably at least severalfold lower, and the fraction of Alu copies that might retain essential conserved regions required for potential function(s) is unknown. Primate Alu repeats are dimeric in structure, with 3' monomer sequence ca. 30 bp longer than the 5' monomer.[28] It has been proposed that the basic Alu monomer unit is derived from the 7SL RNA gene (see below) by deletion of a 120-bp segment from the central region of the latter and that the final Alu structure in turn is derived from an associated duplication event.[39]

A. STRUCTURAL CONSIDERATIONS

From the point of view of mammalian growth regulation, three features of the Alu family are of particular interest: (1) extensive sequence homology to the 7SL RNA gene; (2) RNA polymerase III (pol III) promoter elements, box A and box B, that reside within the left monomeric unit; and (3) an undecameric sequence, GAGGCNGAGGC, which is identical to the core sequence of papovavirus origins of DNA replication. Theoretically, each of these features may contribute to the potential of Alu sequences to mediate growth suppression.

The significance of the first feature of Alu repeats, homology to the 7SL RNA gene, is based on the function of that gene's transcript. 7SL RNA complexes with six polypeptides to form the signal recognition particle (SRP).[40] Since the SRP is required for translocation of nascent peptides into the endoplasmic reticulum,[41-43] it is essential for transport of proteins that are destined for the cell surface as well as proteins that are to be secreted. The potential link to cellular display of cell surface proteins includes receptors for growth factors, both positive and negative; likewise, the link to secreted proteins includes extracellular growth regulatory proteins. In principle, Alu transcripts could compete with 7SL RNA for binding of proteins involved in SRP function. At least one SRP protein, a 68-kDa polypeptide, has been demonstrated to bind both 7SL RNA and an Alu transcript.[44] Evidence that transfected Alu sequence clusters can modulate a presumed 7SL RNA-dependent function, cell surface expression of a chronic lymphocytic leukemia antigen, has been reported,[45-47] although the mechanism is not known. Clearly, to the extent that Alu sequences have the capacity to modulate 7SL RNA function, they also have the potential to cause pleiotropic alterations in cell growth control. The appearance of growth inhibitory proteins on the cell surface with senescence[48,49] and changes in the responsiveness of senescent cells to serum or extracellular growth factors[50] may be aspects of such pleiotropic alterations.

The potential significance of the GAGGCNGAGGC sequence in Alu repeats is less obvious than the 7SL RNA homology, but it is equally provocative. This sequence occurs within the origins of replication of both simian virus 40 (SV40) and polyoma viruses and is essential for SV40 T antigen binding.[51] Based on the GAGGCNGAGGC sequence identity, Jelinek et al.[52] have proposed that Alu elements may function as mammalian replication origins. Several groups have attempted to confirm this prediction, but they have met with equivocal results. Weak replication activity has been observed *in vivo* in COS cells,[53] whereas relatively efficient *in vitro* replication in COS cell extracts has been reported;[54] unfortunately, the reproducibility of this latter finding has been questioned.[55] Since no origin activity has been found in the absence of T antigen, the significance of these COS cell assays with respect to normal cell replication has remained unclear. Evidence that Alu sequence elements are frequently located in close proximity to cellular origins appears to be less controversial. This evidence derives from observations that Alu-like elements are replicated early in S phase[56] and that procedures designed to purify origins appear to yield mixtures of repetitive sequence elements which are enriched for Alu repeats.[57,58]

One interpretation of the above studies is that Alu repeats are not themselves complete replication origins but rather are partial origins or ancillary, origin-associated components. In keeping with the hypotheses presented in this chapter, we speculate that an important (although not necessarily exclusive) function of Alu sequences may be as negative feedback elements. Unmasking of intra-Alu GAGGCNGAGGC sequences in genomic DNA (e.g., by early replication, loss of specific repressor proteins, change in chromatin structure, and/ or demethylation) may allow them to compete for proteins required for DNA replication. Alternatively, transcript of Alu sequences may yield Alu RNPs that bind to and inhibit replication proteins. A precedent for inhibition of DNA replication by small RNPs can be found in the arrest of host DNA synthesis apparently caused by the vesicular stomatitis virus leader transcript.[59]

The third aspect of Alu sequences that has potential significance vis-à-vis control of DNA replication is the presence of internal RNA pol III control signals.[60-64] There is increasing evidence that pol III transcription may be linked to cellular growth regulation. When cells are stimulated to exit the quiescent state, an increase in pol III transcription of multiple small RNA genes occurs.[65-67] Furthermore, elevated rates of pol III-dependent transcription of the mouse Alu B2 repeat have been found in several transformed cell lines.[68,69] Based on data that the immortalizing adenovirus E1a gene stimulates transcription of adenovirus VA1 and VA2 genes by increasing the level of transcription factor IIIC (TFIIIC), it can be argued that this factor may be limiting for overall transcription of pol III genes.[70,71]

Internal Alu pol III control sequences could potentially cause growth suppression by the mechanisms analogous to those proposed for the GAGGCNGAGGC sequence, i.e., via interaction of pol III transcription factor(s) with either unmasked chromosomal Alu DNA binding sites or Alu RNPs. Pol III-transcribed Alu RNPs, unlike presumably ubiquitous Alu sequences in heterogeneous nuclear RNAs (hnRNAs),[72] should have the potential to form complexes with the nuclear phosphoprotein SS-B/La antigen.[64] Of interest in this regard is the observation by Gottesfeld et al.[73] that a transcription factor with properties similar to those of TFIIIC can be precipitated in an RNP complex by anti-SS-B/La antibody. Pelham and Brown[74] have proposed that pol III-dependent transcription of the *Xenopus* 5S RNA gene is regulated in part through binding and titration of TFIIIA by 5S RNA. If it is confirmed that TFIIIC also binds to pol III transcripts and that pol III transcription in mammalian cells can be regulated by snRNPs, the Pelham-Brown model will be both supported and generalized.

B. MECHANISMS OF ACTIVATION

Implicit in these speculations concerning Alu-mediated growth suppression is the requirement that the activity of these repetitive sequences can increase or decrease, with consequent effects on the processes of quiescence, differentiation, and senescence. Such variations in activity may reflect changes in chromatin structure,[75] transcription,[34] or RNA processing,[76] and they include regulation by cis or trans mechanisms. Activation may occur globally, over a region of DNA corresponding to a functional locus, or at the level of single copies. Rather than discussing all of these mechanisms, emphasis in this section will be placed on the possibilities that cis-acting stochastic derepression of Alu sequences may occur and that such events may in turn influence the rate of cellular senescence.

Alterations in DNA methylation may play an important role in malignant transformation and senescence by predisposing critical growth regulatory genes to stochastic depression.[77,78] Although most attention has been focused on oncogenes, growth suppressing genes could likewise be activated. Consistent with this idea, Wilson and Jones[79] reported a progressive decline in overall methylation levels during the *in vitro* replicative life span of nonimmortalized cells. Shmookler Reis and Goldstein[80] also observed drift in overall methylation levels, but they found that this varied between clonal isolates of human fibroblasts. Virtually all studies on growth control and DNA methylation have focused on RNA polymerase II (pol II)-transcribed genes, since it has generally been assumed that pol II genes are responsible for regulating cellular growth. There is, to our knowledge, no information relevant to the proposal that pol III-dependent Alu transcription may be activated by demethylation of specific pol III control regions. Alu sequences are, however, known to be heavily methylated, and (perhaps suggesting that such modification is important to cell function) methylation appears to be maintained at a relatively constant level in different tissues.[81]

A second mechanism by which activation of Alu sequences may occur concerns the possibility that pol III-dependent transcription may be responsive to cis-acting transcriptional enhancer sequences. This suggestion is supported by the recent observation that the expression of the pol III-transcribed U6 small nuclear RNA (snRNA) gene is stimulated by a regulatory element with enhancer-like properties.[82] Repetitive sequences, including Alu, have been shown to reside in extrachromosomal DNA, and during the *in vitro* life span of mammalian cells, significant translocation of these sequences between high molecular weight chromosomal and extrachromosomal compartments may take place.[83-86] A translocation event that juxtaposes an enhancer and an IRS element may lead to transcriptional activation of the latter. Since enhancers can stimulate expression of nearby promoters in a relatively position- and orientation-independent manner, and since Alu sequences and cellular enhancer elements are both frequent within the genome, the probability of this type of activation event occurring may be relatively high.

The third potential mechanism for regulating Alu expression relates to repressor factors. While some Alu sequence copies are strongly transcribed *in vitro*,[60-64] in immortalized HeLa cells *in vivo* their pol III-dependent transcriptional activity appears to be extremely low.[36] One obvious possible reason for this discrepancy between *in vitro* and *in vivo* expression levels is that these elements may be transcriptionally repressed *in vivo* by specific nuclear proteins that are lost during preparation of the *in vitro* extract. To the extent that Alu copies are silenced *in vivo*, deletion of the repressor binding site from an Alu copy may lead to activation. The frequency with which such activation events occur would be influenced by the details of the repressor-binding site interaction. One intriguing observation with respect to primate Alu sequences is their dimeric structure. If a functional repressor binding site was duplicated by this event, then the rate of random activation would be considerably reduced with a corresponding influence on the rate of senescence.

III. CONCLUSIONS

We have set out the hypothesis that interspersed repetitive sequences may play a role in mediating cellular senescence. Mechanisms by which SINEs might cause growth suppression have been outlined and potential regulatory events leading to stochastic activation of these elements proposed. Although, for purposes of clarity, only the Alu SINE family has been discussed, it is clear that analogous mechanisms may exist for other repetitive sequence families. Indeed, the dominant family involved in negative growth regulation may vary from species to species, and this may have important implications for the characteristic longevity of each species.

What remains uncertain with respect to the hypothesis presented here is why such sequences may be permitted to undergo extensive proliferation in higher organisms if they are so detrimental to cell survival. Two possibilities seem to merit discussion. The first relates to the processes of embryological development. Where detailed embryonic cell lineages have been worked out, such as in the *Caenorhabditis elegans* system,[87] it is clear that there are a surprising number of pathways that end in atrophy of stage-specific structures. This form of involution may well involve molecular mechanisms related to cellular senescence. The second possibility concerns the enormous numbers of cells that must be controlled with respect to growth, especially in large, long-lived animals. Highly redundant mechanisms may have evolved to ensure that cells do not escape control to form tumors, and it is probable that cellular senescence comprises one of those mechanisms.[88] To the extent that SINEs and long interspersed sequence elements (LINEs) are interspersed throughout the genome, the probability of their deletion as tumor suppression elements is exceedingly low.

These speculations on aging and interspersed repetitive elements could be extended to include other interesting possibilities that have been touched on only briefly by reference or have been omitted as being beyond the scope of this chapter. With respect to the possibilities that have been discussed, we have not attempted to address how different growth regulatory mechanisms, especially those involving proto-oncogenes, might be utilized or might interact in various developmental contexts. Finally, it should be pointed out that the hypotheses discussed here provide the basis for numerous predictions that are readily amenable to experimental testing.

REFERENCES

1. **Martin, G. M.,** Syndromes of accelerated aging, *Natl. Cancer Inst. Monogr.,* 60, 241, 1982.
2. **Norwood, T. H. and Smith, J. R.,** The cultured fibroblast-like cell as a model for the study of aging, in *Handbook of the Biology of Aging,* Finch, C. E. and Schneider, E. L., Eds., Van Nostrand Reinhold, New York, 1985, 291.
3. **Rohme, D.,** Evidence for a relationship between longevity of mammalian species and life spans of normal fibroblasts *in vitro* and erythrocytes *in vivo, Proc. Natl. Acad. Sci. U.S.A.,* 78, 5009, 1981.
4. **Schneider, E. L. and Smith, J. R.,** The relationship of *in vitro* studies to *in vivo* human aging, *Int. Rev. Cytol.,* 69, 261, 1981.
5. **Sager, R., Tanaka, K., Lau, C. C., Ebina, Y., and Anisowicz, A.,** Resistance of human cells to tumorigenesis induced by cloned transforming genes, *Proc. Natl. Acad. Sci. U.S.A.,* 80, 7601, 1983.
6. **Geiser, A. G., Der, C. J., Marshall, C. J., and Stanbridge, E. J.,** Suppression of tumorigenicity with continued expression of the c-Ha-*ras* oncogene in EJ bladder carcinoma-human fibroblast hybrid cells, *Proc. Natl. Acad. Sci. U.S.A.,* 83, 5209, 1986.
7. **Stevenson, M. and Volsky, D. J.,** Activated v-*myc* and v-*ras* oncogenes do not transform normal human lymphocytes, *Mol. Cell. Biol.,* 6, 3410, 1986.
8. **Small, M. B., Gluzman, Y., and Ozer, H. L.,** Enhanced transformation of human fibroblasts by origin-defective simian virus 40, *Nature,* 296, 671, 1982.
9. **Stein, G. H.,** SV40-transformed human fibroblasts: evidence for cellular aging in precrisis cells, *J. Cell. Physiol.,* 125, 36, 1985.
10. **Chang, P. L., Gunby, J. L., Tomkins, D. J., Mak, I., Rosa, N. E., and Mak, S.,** Transformation of human cultured fibroblasts with plasmids carrying dominant selection markers and immortalizing potential, *Exp. Cell Res.,* 167, 407, 1986.
11. **Muggleton-Harris, A. L. and DeSimone, D. W.,** Replicative potentials of various fusion products between WI-38 and SV40 transformed WI-38 cells and their components, *Som. Cell Genet.,* 6, 689, 1980.
12. **Pereira-Smith, O. M. and Smith, J. R.,** Expression of SV40 T antigen in finite life-span hybrids of normal and SV40-transformed fibroblasts, *Som. Cell Genet.,* 7, 411, 1981.
13. **Pereira-Smith, O. M. and Smith, J. R.,** Evidence for the recessive nature of cellular immortality, *Science,* 221, 964, 1983.
14. **Holliday, R., Huschtscha, L. I., Tarrant, G. M., and Kirkwood, T. B. L.,** Testing the commitment theory of cellular aging, *Science,* 198, 366, 1977.
15. **Smith, J. R. and Whitney, R. G.,** Intraclonal variation in proliferative potential of human diploid fibroblasts: stochastic mechanism for cellular aging, *Science,* 207, 82, 1980.

16. **Jones, R. B., Whitney, R. G., and Smith, J. R.,** Intramitotic variation in proliferative potential: stochastic events in cellular aging, *Mech. Ageing Dev.*, 29, 143, 1985.

17. **Padmanabhan, R., Howard, T. H., and Howard, B. H.,** Specific growth inhibitory sequences in genomic DNA from quiescent human embryo fibroblasts, *Mol. Cell. Biol.*, 7, 1894, 1987.

18. **Norwood, T. H., Pendergrass, W. R., Sprague, C. A., and Martin, G. M.,** Dominance of the senescent phenotype in heterokaryons between replicative and post-replicative human fibroblast-like cells, *Proc. Natl. Acad. Sci. U.S.A.*, 71, 2231, 1974.

19. **Stein, G. H. and Yanishevsky, R. M.,** Entry into S phase is inhibited in two immortal cell lines fused to senescent human diploid cells, *Exp. Cell Res.*, 120, 155, 1979.

20. **Rabinovitch, P. S. and Norwood, T. H.,** Comparative heterokaryon study of cellular senescence and the serum-deprived state, *Exp. Cell Res.*, 130, 101, 1980.

21. **Yanishevsky, R. M. and Stein, G. H.,** Ongoing DNA synthesis continues in young human diploid cells (HDC) fused to senescent HDC, but entry into S phase is inhibited, *Exp. Cell Res.*, 126, 469, 1980.

22. **Stein, G. H. and Yanishevsky, R. M.,** Quiescent human diploid cells can inhibit entry into S phase in replicative nuclei in heterodikaryons, *Proc. Natl. Acad. Sci. U.S.A.*, 78, 3025, 1981.

23. **Pendergrass, W. R., Saulewicz, A. C., Burmer, G. C., Rabinovitch, P. S., Norwood, T. H., and Martin, G. M.,** Evidence that a critical threshold of DNA polymerase-alpha activity may be required for the initiation of DNA synthesis in mammalian cell heterokaryons, *J. Cell. Physiol.*, 113, 141, 1982.

24. **Stein, G. H., Yanishevsky, R. M., Gordon, L., and Beeson, M.,** Carcinogen-transformed human cells are inhibited from entry into S phase by fusion to senescent cells but cells transformed by DNA tumor viruses overcome the inhibition, *Proc. Natl. Acad. Sci. U.S.A.*, 79, 5287, 1982.

25. **Burmer, G. C., Rabinovitch, P. S., and Norwood, T. H.,** Evidence for differences in the mechanism of cell cycle arrest between senescent and serum-deprived human fibroblasts: heterokaryon and metabolic inhibitor studies, *J. Cell. Physiol.*, 118, 97, 1984.

26. **Orgel, L. E. and Crick, F. H. C.,** Selfish DNA: the ultimate parasite, *Nature*, 284, 604, 1980.

27. **Singer, M. F.,** SINEs and LINEs: highly repeated short and long interspersed sequences in mammalian genomes, *Cell*, 28, 433, 1982.

28. **Deininger, P. L., Jolly, D. J., Rubin, C. M., Friedmann, T., and Schmid, C. W.,** Base sequence studies of 300 nucleotide renatured repeated human DNA clones, *J. Mol. Biol.*, 151, 17, 1981.

29. **Kariya, Y., Kato, K., Hayashizaki, Y., Himeno, S., Tarui, S., and Matsubara, K.,** Revision of consensus of human Alu repeats — a review, *Gene*, 53, 1, 1987.

30. **Young, P. R., Scott, R. W., Hamer, D. H., and Tilghman, S. M.,** Construction and expression *in vivo* of an internally deleted mouse α-fetoprotein gene: presence of a transribed Alu-like repeat within the first intervening sequences, *Nucleic Acids Res.*, 10, 3099, 1982.

31. **Allan, M. and Paul, J.,** Transcription *in vivo* of an Alu family member upstream from the human epsilon-globin gene, *Nucleic Acids Res.*, 12, 1193, 1984.

32. **Ryskov, A. P., Ivanov, P. L., Tokarskaya, O. N., Kramerov, D. A., Grigoryan, M. S., and Georgiev, G. P.,** Major transcripts containing B1 and B2 repetitive sequences in cytoplasmic poly(A)$^+$ RNA from mouse tissues, *FEBS Lett.*, 182, 73, 1985.

33. **Skowronski, J. and Singer, M. F.,** Expression of a cytoplasmic LINE-1 transcript is regulated in a human teratocarcinoma cell line, *Proc. Natl. Acad. Sci. U.S.A.*, 82, 6050, 1985.

34. **Sun, L. H. and Frankel, F. R.** The induction of Alu-sequence transcripts by glucocorticoid in rat liver cells, *J. Steroid Biochem.*, 25, 201, 1986.

35. **Watson, J. B. and Sutcliffe, J. G.,** Primate brain-specific cytoplasmic transcript of the Alu repeat family, *Mol. Cell. Biol.*, 7, 3324, 1987.

36. **Paulson, K. E. and Schmid, C. W.,** Transcriptional inactivity of Alu repeats in HeLa cells, *Nucleic Acids Res.*, 14, 6145, 1986.

37. **Houck, C. M., Rinehart, F. P., and Schmid, C. W.,** A ubiquitous family of repeated DNA sequences in the human genome, *J. Mol. Biol.*, 132, 289, 1979.

38. **Hwu, H. R., Roberts, J. W., Davidson, E. H., and Britten, R. J.,** Insertion and/or deletion of many repeated DNA sequences in human and higher ape evolution, *Proc. Natl. Acad. Sci. U.S.A.*, 83, 3875, 1986.

39. **Ullu, E. and Tschudi, C.,** Alu sequences are processed 7SL RNA genes, *Nature*, 312, 171, 1984.

40. **Walter, P. and Blobel, G.,** Signal recognition particle contains a 7S RNA essential for protein translocation across the endoplasmic reticulum, *Nature*, 299, 691, 1982.

41. **Walter, P., Ibrahimi, I., and Blobel, G.,** Translocation of proteins across the endoplasmic reticulum. I. Signal recognition protein (SRP) binds to *in-vitro*-assembled polysomes synthesizing secretory protein, *J. Cell. Biol.*, 91, 545, 1981.

42. **Walter, P. and Blobel, G.,** Translocation of proteins across the endoplasmic reticulum. II. Signal recognition protein (SRP) mediates the selective binding to microsomal membranes of *in-vitro*-assembled polysomes synthesizing secretory protein, *J. Cell Biol.*, 91, 551, 1981.

43. **Walter, P. and Blobel, G.**, Translocation of proteins across the endoplasmic reticulum. III. Signal recognition protein (SRP) causes signal sequence-dependent and site-specific arrest of chain elongation that is released by microsomal membranes, *J. Cell Biol.*, 91, 557, 1981.

44. **Andrews, P. G. and Kole, R.**, Alu RNA transcribed *in vitro* binds the 68-kDa subunit of the signal recognition particle, *J. Biol. Chem.*, 262, 2908, 1987.

45. **Stanners, C. P., Lam, T., Chamberlain, J. W., Steward, S. S., and Price, G. B.**, Cloning of a functional gene responsible for the expression of a cell surface antigen correlated with human chronic lymphocytic leukemia, *Cell*, 27, 211, 1981.

46. **Beitel, L. K., Chamberlain, J. W., Benchimol, S., Lam, T., Price, G. B., and Stanners, C. P.**, Studies on HSAG, a middle repetitive family of genetic elements which elicit a leukemia-related cellular surface antigen, *Nucleic Acids Res.*, 14, 3391, 1986.

47. **Chamberlain, J. W., Henderson, G., Chang, M. W. M., Lam, T., Dignard, D., Ling, V., Price, G. B., and Stanners, C. P.**, The structure of HSAG-1, a middle repetive genetic element which elicits a leukemia-related cellular surface antigen, *Nucleic Acids Res.*, 14, 3409, 1986.

48. **Pereira-Smith, O. M., Fisher, S. F., and Smith, J. R.**, Senescent and quiescent cell inhibitors of DNA synthesis. Membrane-associated proteins, *Exp. Cell Res.*, 160, 297, 1985.

49. **Stein, G. H. and Atkins, L.**, Membrane-associated inhibitor of DNA synthesis in senescent human diploid fibroblasts: characterization and comparison to quiescent cell inhibitor, *Proc. Natl. Acad. Sci. U.S.A.*, 83, 9030, 1986.

50. **Phillips, P. D., Kaji, K., and Cristofalo, V. J.**, Progressive loss of the proliferative response of senescing WI-38 cells to platelet-derived growth factor, epidermal growth factor, insulin, transferrin, and dexamethasone, *J. Gerontol.*, 39, 11, 1984.

51. **DeLucia, A. L., Lewton, B. A., Tijan, R., and Tegtmeyer, P.**, Topography of simian virus 40 A protein-DNA complexes: arrangement of pentanucleotide interaction sites at the origin of replication, *J. Virol.*, 46, 143, 1983.

52. **Jelinek, W. R., Toomey, T. P., Leinwand, L., Duncan, C. H., Biro, P. A., Choudary, P. V., Weissman, S. M, Rubin, C. M., Houck, C. M., Deininger, P. L., and Schmid, C. W.**, Ubiquitous, interspersed repeated sequences in mammalian genomes, *Proc. Natl. Acad. Sci. U.S.A.*, 77, 1398, 1980.

53. **Johnson, E. M. and Jelinek, W. R.**, Replication of a plasmid bearing a human Alu-family repeat in monkey COS-7 cells, *Proc. Natl. Acad. Sci. U.S.A.*, 83, 4660, 1986.

54. **Ariga, H.**, Replication of cloned DNA containing the Alu family sequence during cell extract-promoting simian virus 40 DNA synthesis, *Mol. Cell. Biol.*, 4, 1476, 1984.

55. **Li, J. J. and Kelly, T. J.**, Simian virus 40 DNA replication *in vitro*: specificity of initiation and evidence for bidirectional replication, *Mol. Cell. Biol.*, 5, 1238, 1985.

56. **Holmquist, G. P. and Caston, L. A.**, Replication time of interspersed repetitive DNA sequences in hamsters, *Biochim. Biophys. Acta*, 868, 164, 1986.

57. **Anachkova, B., Russev, G., and Altmann, H.**, Identification of the short dispersed repetitive DNA sequences isolated from the zones of initiation of DNA synthesis in human cells as Alu-elements, *Biochem. Biophys. Res. Commun.*, 128, 101, 1985.

58. **Zannis-Hadjopoulos, M., Kaufmann, G., Wang, S. S., Lechner, R. L., Karawya, E., Hesse, J., and Martin, R. G.**, Properties of some monkey DNA sequences obtained by a procedure that enriches for DNA replication origins, *Mol. Cell. Biol.*, 5, 1621, 1985.

59. **Remenick, J., Kenney, M. K., and McGowan, J. J.**, Inhibition of adenovirus DNA replication by vesicular stomatitis virus leader RNA, *J. Virol.*, 62, 1286, 1988.

60. **Duncan, C., Biro, P. A., Choudary, P. V., Elder, J. T., Wang, R. R., Forget, B. G., de Riel, J. K., and Weissman, S. M.**, RNA polymerase III transcriptional units are interspersed among human non-alpha-globin genes, *Proc. Natl. Acad. Sci. U.S.A.*, 76, 5095, 1979.

61. **Duncan, C. H., Jagadeeswaran, P., Wang, R. R., and Weissman, S. M.**, Structural analysis of templates and RNA polymerase III transcripts of Alu family sequences interspersed among the human beta-like globin genes, *Gene*, 13, 185, 1981.

62. **Elder, J. T., Pan, J., Duncan, C. H., and Weissman, S. M.**, Transcriptional analysis of interspersed repetitive polymerase III transcription units in human DNA, *Nucleic Acids Res.*, 9, 1171, 1981.

63. **Pan, J., Elder, J. T., Duncan, C. H., and Weissman, S. M.**, Structural analysis of interspersed repetitive polymerase III transcription units in human DNA, *Nucleic Acids Res.*, 9, 1151, 1981.

64. **Shen, C. K. J. and Maniatis, T.**, The organization, structure, and *in vitro* transcription of Alu family RNA polymerase III transcription units in the human α-like globin gene cluster: precipitation of *in vitro* transcripts by Lupus and anti-La antibodies, *J. Mol. Appl. Genet.*, 1, 343, 1982.

65. **Edwards, D. R., Parfett, L. J., and Denhardt, D. T.**, Transcriptional regulation of two serum-induced RNAs in mouse fibroblasts: equivalence of one species to repetitive elements, *Mol. Cell. Biol.*, 5, 3280, 1985.

66. **Tiercy, J. M. and Weil, R.**, Serum-induced stimulation of snRNA synthesis in mouse 3T3 fibroblasts, *Experientia*, 41, 82, 1985.

67. **Lania, L., Pannuti, A., La Mantia, G., and Basilico, C.,** The transcription of B2 repeated sequences is regulated during the transition from quiescent to proliferative state in cultured rodent cells, *FEBS Lett.,* 219, 400, 1987.
68. **Singh, K., Carey, M., Saragosti, S., and Botchan, M.,** Expression of enhanced levels of small RNA polymerase III transcripts encoded by the B2 repeats in simian virus 40-transformed mouse cells, *Nature,* 314, 553, 1985.
69. **Carey, M. F., Singh, K., Botchan, M., and Cozzarelli, N. R.,** Induction of specific transcription by RNA polymerase III in transformed cells, *Mol. Cell. Biol.,* 6, 3068, 1986.
70. **Hoeffler, W. K. and Roeder, R. G.,** Enhancement of RNA polymerase III transcription by the E1A gene product of adenovirus, *Cell,* 41, 955, 1985.
71. **Yoshinaga, S., Dean, N., Han, M., and Berk, A. J.,** Adenovirus stimulation of transcription by RNA polymerase III: evidence for an E1A-dependent increase in transcription factor IIIC concentration, *EMBO J.,* 5, 343, 1986.
72. **Calabretta, B., Robberson, D. L., Maizel, A. L., and Saunders, G. F.,** mRNA in human cells contains sequences complementary to the Alu family of repeated DNA, *Proc. Nat. Acad. Sci. U.S.A.,* 78, 6003, 1981.
73. **Gottesfeld, J. M., Andrews, D. L., and Hoch, S. O.,** Association of an RNA polymerase III transcription factor with a ribonucleoprotein complex recognized by autoimmune sera, *Nucleic Acids Res.,* 12, 3185, 1984.
74. **Pelham, H. R. B. and Brown, D. D.,** A specific transcription factor that can bind either the 5S RNA gene or 5S RNA, *Proc. Natl. Acad. Sci. U.S.A.,* 77, 4170, 1980.
75. **Puvion-Dutilleul, F., Puvion, E., Icard-Liepkalns, C., and Macieira-Coelho, A.,** Chromatin structure, DNA synthesis and transcription through the lifespan of human embryonic lung fibroblasts, *Exp. Cell Res.,* 151, 283, 1984.
76. **Adeniyi-Jones, S. and Zasloff, M.,** Transcription, processing and nuclear transport of a B1 Alu RNA species complementary to an intron of the murine alpha-fetoprotein gene, *Nature,* 317, 81, 1985.
77. **Doerfler, W.,** DNA methylation and gene activity, *Annu. Rev. Biochem.,* 52, 93, 1983.
78. **Holliday, R.,** Strong effects of 5-azacytidine on the *in vitro* lifespan of human diploid fibroblasts, *Exp. Cell Res.,* 166, 543, 1986.
79. **Wilson, V. L. and Jones, P. A.,** DNA methylation decreases in aging but not in immortal cells, *Science,* 220, 1055, 1983.
80. **Shmookler Reis, R. J. and Goldstein, S.,** Variability of DNA methylation patterns during serial passage of human diploid fibroblasts, *Proc. Natl. Acad. Sci. U.S.A.,* 79, 3949, 1982.
81. **Gama-Sosa, M. A., Wang, R. Y. H., Kuo, K. C., Gehrke, C. W., and Ehrlich, M.,** The 5-methyl-cytosine content of highly repeated sequences in human DNA, *Nucleic Acids Res.,* 11, 3087, 1983.
82. **Bark, C., Weller, P., Zabielski, J., Janson, L., and Pettersson, U.,** A distant enhancer element is required for polymerase III transcription of a U6 RNA gene, *Nature,* 328, 356, 1987.
83. **Krolewski, J. J., Bertelsen, A. H., Humayun, M. Z., and Rush, M. G.,** Members of the Alu family of interspersed, repetitive DNA sequences are in the small circular DNA population of monkey cells grown in culture, *J. Mol. Biol.,* 154, 399, 1982.
84. **Kunisada, T., Yamagishi, H., Ogita, Z., Kirakawa, T., and Mitsui, Y.,** Appearance of extrachromosomal circular DNAs during *in vivo* and *in vitro* ageing of mammalian cells, *Mech. Ageing Dev.,* 29, 89, 1985.
85. **Riabowol, K., Shmookler Reis, R. J., and Goldstein, S.,** Interspersed repetitive and tandemly repetitive sequences are differentially represented in extrachromosomal covalently closed circular DNA of human diploid of fibroblasts, *Nucleic Acids Res.,* 13, 5563, 1985.
86. **Yamagishi, H., Kunisada, T., and Takeda, T.,** Amplification of extra-chromosomal small circular DNAs in a murine model of accelerated senescence. A brief note, *Mech. Ageing Dev.,* 29, 101, 1985.
86. **Sulston, J. E., Schierenberg, E., White, J. G., and Thomson, J. N.,** The embryonic cell lineage of the nematode *Caenorhabditis elegans, Dev. Biol.,* 100, 64, 1983.
87. **O'Brien, W., Stenman, G., and Sager, R.,** Suppression of tumor growth by senescence in virally transformed human fibroblasts, *Proc. Natl. Acad. Sci. U.S.A.,* 83, 8659, 1986.

Chapter 13

DNA METHYLATION, MAINTENANCE CpG-METHYLASE, AND SENESCENCE

Robert J. Shmookler Reis, Elena Moerman, and Samuel Goldstein

TABLE OF CONTENTS

ABSTRACT

Evidence is reviewed and new data presented indicating progressive clonal drift of gene-region DNA methylation patterns in human diploid fibroblasts. Losses of methylation generally exceed gains, but these events appear to be random and, at times, to produce methylation increases locally or even globally. A similar trend toward DNA hypomethylation evidently prevails in rodent tissues *in vivo*. Although senescence-associated gene derepressions have been reported both *in vitro* and *in vivo*, the mechanisms are unknown. The loss of methylcytosines from cultured cells cannot be due to insufficient methylase, since its activities increased, relative to the extent of cell cycling, at late passage in a fibroblast strain shown to undergo progressive DNA hypomethylation during its limited replicative life span *in vitro*. Marked elevation was seen in "maintenance" CpG-methylase activity (conversion of hemimethylated sites to full methylation, as in newly replicated DNA) and also in methylase activities which are inappropriate to the maintenance function: *de novo* methylation, defined on unmethylated DNA, and activity in nondividing cells for both hemimethylated and unmethylated DNA substrates. Quantitative changes in methylase enzyme activity also could not account for the independent occurrence of drift in DNA methylation patterns, observed at different gene loci in several clonal lineages. Such observations implicate factors acting locally — e.g., accessibility and/or activity of methylase within specific nuclear chromatin domains.

I. INTRODUCTION

Postsynthetic modification of DNA is limited in vertebrates to cytosine methylation,[1] primarily in CpG dinucleotides, the majority of which are methylated in most mammalian tissues.[1-3] This modification is mediated by a CpG-specific cytosine methylase which has greatly enhanced activity for hemimethylated substrate DNA,[4] thus enabling transmission of methylation patterns to daughter cells.[2,3,5]

The initial expression of many tissue-specific genes during development is correlated with the loss of CpG methylations in and around those loci, especially in the 5' region of transcriptional initiation.[2,5,6] It remains controversial whether such hypomethylation is a cause or an effect of gene expression[5-7] and whether it occurs passively via failure to methylate newly replicated DNA or by active demethylation. However, even if methylation/demethylation is secondary to gene repression/activation, it appears to reinforce or "imprint" that state.[8] Since methylase activity is itself inhibited by the presence of RNA transcripts,[4] such protein-RNA-DNA interactions would create a bistable ("flip-flop") control circuit.

A few tissue-specific genes and a wide assortment of genes which are expressed in most or all cell types (the so-called "housekeeping" genes) are accompanied by GC-rich "islands", usually at their 5' ends. Although CpGs occur in these islands at 10 to 100 times their genomic frequency, they tend to be unmethylated, except on the inactive X chromosome.[8,9]

Gene transcription can be suppressed in transfected DNA by artificial methylation, specifically of the 5' initiation region, as demonstrated for the adenine phosphoribosyltransferase (APRT)[10] and γ-globin genes.[7] Moreover, a number of repressed genes can be activated by treatment of cells with 5-azacytidine or 5-azadeoxycytidine, nucleoside analogs which inhibit methylase activity following their incorporation into DNA.[11] Such activatable genes include those on the inactive X chromosome,[12,13] the metallothionein-I gene in lymphoid cell lines,[14] endogenous viral genes,[15] and tissue-specific genes under developmental regulation such as globin genes in previously uninduced erythroid cells,[11] immunoglobulin genes in B cells,[16] and the phosphoenolpyruvate carboxykinase gene in fetal liver prior to its normal expression.[6]

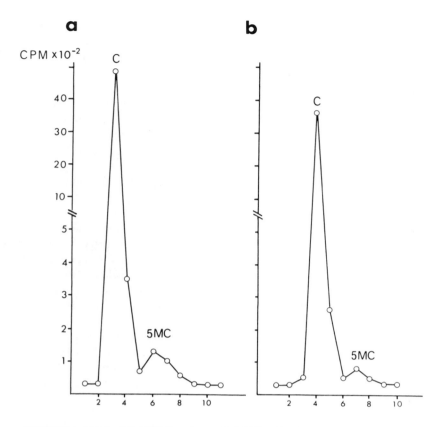

FIGURE 1. C and mC in DNA from human diploid fibroblasts (strain DS) were determined after incorporation of tritiated cytidine (^{3}H-CdR) during three mean population doublings (MPD), commencing at 15 MPD (a) and 52 MPD (b). The CdR (NEN) was uniformly labeled with tritium. DNA samples, hydrolyzed 30 min in 90% formic acid at 175°C, were analyzed by two-dimensional thin-layer chromatography, as described.[17] Radioactivity is shown in fractions (2.5-mm-wide strips) of increasing mobility in both solvents (left to right in the figure), spanning the cytosine (C) and methylcytosine (MC) spots. The fraction of mC (mC/[C + mC]) was 0.045 at early passage (a) and 0.022 at late passage (b). (From Shmookler Reis, R. J. and Goldstein, S., *Proc. Natl. Acad. Sci. U.S.A.*, 79, 3949, 1982. With permission.)

II. METHYLATION DRIFT IN CULTURED CELLS

There are several lines of evidence indicating that diploid human fibroblasts undergo substantial alteration in their DNA methylation patterns during *in vitro* culture. The total methylcytosine (mC) content of normal fibroblast DNA, as measured by quantitation of hydrolyzed DNA on thin-layer chromatograms, was significantly reduced in one such strain as it approached the end of its replicative life span (Figure 1).[17] This observation was subsequently confirmed and extended to normal cultured cells (having limited replicative life spans) from two other species.[18]

A more extensive survey of human fibroblast strains, however, suggests that this decrease is not uniform in all cultures of diploid cells, but it may reflect the net difference between hypo- and hypermethylations of DNA sites, which can occasionally drift up rather than down.[17] Six strains from normal donors were analyzed for methylation of –CCGG– sites, early and late in their life spans *in vitro*, by comparison of DNA fragment lengths after digestion with restriction endonuclease HpaII or MspI. (HpaII cleaves –C–C–G–G– but not

–C–mC–G–G–, whereas MspI cuts both –C–C–G–G– and –C–mC–G–G–.) The percentage of –CCGG– sites methylated — i.e., sites resistant to HpaII but cleaved by MspI — declined in four strains at late passage (from 54 to 48%, 58 to 48%, 55 to 51%, and 52 to 44%), increased in one strain (from 59 to 64%), and remained low but constant in another (45%).[17]

This uneven but prevailing trend toward methylation loss was also evident, on a finer scale, when individual –CCGG– sites were examined by specific probe hybridizations to HpaII- or MspI-cut DNA from fibroblast clones. Methylation losses exceeded gains by at least 2:1, but both events were surprisingly common,[19,20] generating restriction fragment variants at a frequency of ca. 10^{-2} per allele per mean population doubling. More recent studies indicate that clones of human fibroblasts vary in the stability of their DNA methylation patterns, but *not* coordinately at all gene loci.[20,21] As an example, a clonal lineage which underwent frequent hypomethylation in the gene region for the alpha subunit of human chorionic gonadotropin (α-hCG) is shown in Figure 2A; compare the subclones (lanes e to i) to the parental clone (lanes c and d). A second clone, however, maintained essentially full methylation around α-hCG in six subclones and sub-subclones (Figure 2B). Significantly, these trends were reversed in the c-H-*ras* gene region; here, the lineage which had shown progressive α-hCG hypomethylation in subclones (Figure 2A) held a constant HpaII pattern in all subclones, while the other clone (Figure 2B), which had a fully methylated α-hCG region, contained unmethylated sites around both c-H-*ras* alleles, with both hypo- and hypermethylation events apparent in single subclones.[21]

An example of interclonal variation in c-H-*ras* methylation pattern is given in Figure 3, where each pair of lanes represents DNA from a single isolated HDF clone at early to mid-passage and late-passage levels. It should be noted that some clones undergo hypomethylation at late passage (clones 1 and 2 in Figure 3), while others (clones 3 and 4) evidence hypermethylation.

Because gene expression can be suppressed by the acquisition of *de novo* methylations, especially in the region of transcriptional initiation, and induced in some instances by the loss of CpG methylations from such regions, it was initially inferred that DNA methylation must function as a normal negative control mechanism for gene expression.[3,5,7,11] Such a regulatory role for DNA methylation (in particular during development) requires its faithful inheritance,[2,3,5] and, indeed, studies of plasmid DNAs transfected into permanent mouse cell lines have indicated rather faithful heritability of methylated CpG sites[22] following an initial period of instability and extensive hypomethylation.[23,24] In this context, it was surprising to find that inheritance of DNA methylation patterns was rather unfaithful in normal cultured human fibroblasts.[17,19] These observations may thus have important implications for the stability of differentiated gene expression in these cells.

While these apparently contradictory findings are difficult to reconcile, several points should be considered. First, the introduction of exogenous, transfected DNA into permanent cells[23-25] may not reflect the status of endogenous gene loci in normal cells. Thus, it must be considered that genomic DNA methylation may normally be transmitted with substantial infidelity. It nevertheless could be a heritable determinant of gene expression, because tissue-specific differential expression reflects the predominant pattern of RNA and protein synthesis in a given tissue, and may not be seen in all cells comprising that tissue. A variety of stochastic influences which might alter cell-specific gene expression are expected to accumulate during the life span of an individual cell or of cultured cells, e.g., DNA breakage and rearrangement, viral integration, error-prone repair of thymine dimers and DNA adducts, and other sources of somatic mutation. In addition to these factors, cells in culture may be vulnerable to other dysdifferentiative influences due to the artificial nature of their hormonal/growth factor milieu. On the other hand, a regulatory role for DNA methylation is far from established, and (as summarized in Section I) the evidence may be more cautiously interpreted as supporting a "mutual reinforcement" between gene transcription and hypomethylation.[8]

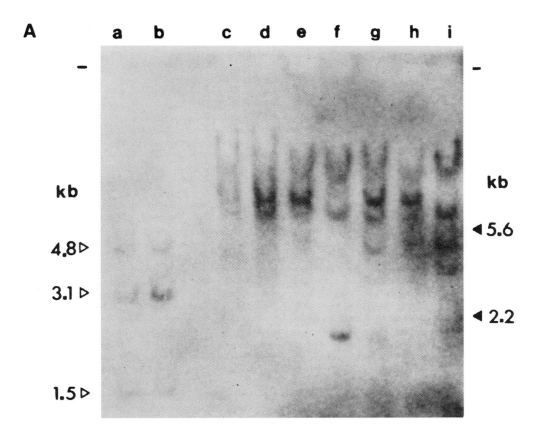

FIGURE 2. DNA methylation patterns in two clonal lineages for the α-hCG gene region. (A) Clone E and subclones. DNA samples (3 μg) of clone E and five of its subclones were digested with MspI (lanes a and b) or HpaII (lanes c to i) at 12 units/μg, electrophoresed in 1.5% agarose, and blotted onto a nitrocellulose filter; autoradiographs were prepared following hybridization to the α-hCG cDNA probe. Lane a: clone E at ca. 50 MPD; lane d: clone E at ca. 65 MPD; lane e: subclone E3 at ca. 60 MPD; lane f: subclone E7 at ca. 60 MPD; lane g: subclone E13 at ca. 60 MPD; lane h: subclone E17 at ca. 60 MPD; and lane i: subclone E20 at ca. 60 MPD. Sizes in right margin correspond to PM2/HindIII markers. (B) Maintenance of methylation patterns in the α-hCG gene region of clone A, its subclones and sub-subclones. DNA samples were analyzed as in A (above) following digestion with HpaII: clone A at middle passage (ca. 50 MPD, lane a) and late passage (ca. 65 MPD, lane b); subclones A_1, A_2, and A_3 of clone A, at 50 MPD (lanes c to e); A_3 subclone at 65 MPD (lane f) and three A_3 sub-subclones at 65 MPD (lanes g to i). (From Goldstein, S. and Shmookler Reis, R. J., *Nucleic Acids Res.*, 13, 7055, 1985. With permission.)

III. ANALYSIS OF METHYLASE ACTIVITY ACROSS THE CELL CYCLE INDICATES THAT METHYLASE IS NOT LIMITING IN SENESCENT CULTURES

Why do methylation patterns change during cell culture? One answer may lie in the nature of the methylation "inheritance": methyl groups must be added to cytosines in DNA postsynthetically by an enzyme activity termed CpG-methylase. The most reactive substrate for this enzyme is hemimethylated DNA, containing at any given site mC–G on one strand and C–G on the complementary strand. Although full or symmetric methylation (on both DNA strands) appears to be the predominant form,[5,8] hemimethylated DNA must arise transiently at the replication fork, where CpG-methylase may function on its preferred substrate, allowing the methylation pattern to be inherited with some degree of fidelity. Any sites that remain hemimethylated at the next replication cycle would lose this methylation

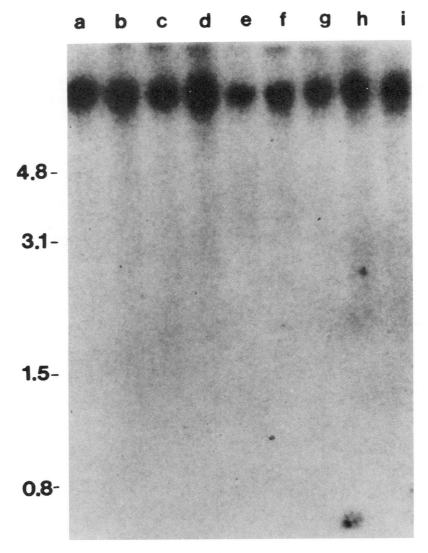

FIGURE 2B.

for all progeny cells in one of the daughter lineages. This might occur due to inadequate methylase levels and/or shielding of specific regions from methylation by other protein-DNA interactions.[8] On the other hand, CpG-methylase also shows some activity (perhaps 1 to 5%, as assessed *in vitro*) toward fully unmethylated sites,[4,5] and it thus has the potential to introduce *de novo* methylations ("hypermethylations"). The frequencies of these two events, hypo- and hypermethylations, may vary between genomic loci,[19-21] between donors of cultured fibroblasts,[17] and between clones within a mass culture.[17,19-21]

In view of the predominant decline in the extent of DNA methylation in late-passage fibroblasts, we examined methylase levels through the cell cycle in synchronized cultures. Our initial studies have utilized the A2 human cell strain, which underwent a "typical" decline in –CCGG– site methylation — from 54 to 48% — toward the end of its replicative life span.[17] The results are shown in Figure 4 for cells at both early passage (panel A) and late passage (panel B). DNA synthesis, monitored by 30-min ^{3}H-TdR incorporation pulses (dashed lines), reached a peak in "young" cells at 18 to 24 h following mitogenic stimulation

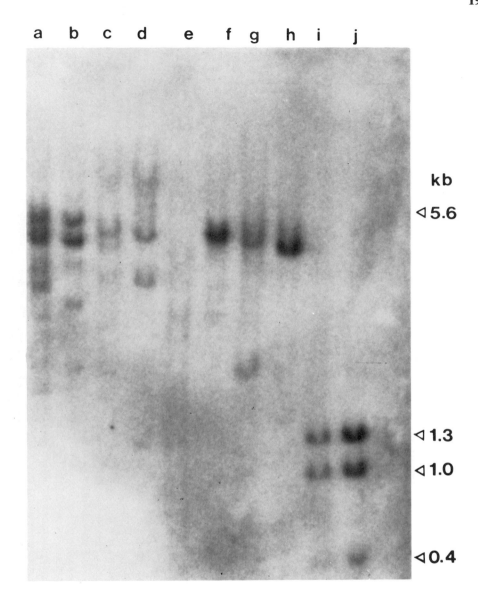

a b c d e f g h i j

kb

◁ 5.6

◁ 1.3

◁ 1.0

◁ 0.4

FIGURE 3. DNA methylation in the c-H-*ras* gene region: clones of HSC172 fibroblasts show variable patterns at early and late passage. DNA samples, digested 2 h at 37°C with HpaII (a to h) or MspI (i, j) at 12 units/μg DNA, were prepared from cells of clones 1, 2, 3, and 4 at early (lanes a, c, e, and g, respectively) and late (lanes b, d, f, and h, respectively) passage. All clones gave identical patterns after MspI digestion, as shown in lanes i and j for clone 4, at early and late passage, respectively. Samples were electrophoresed in 1.5% agarose, transferred to nitrocellulose, and hybridized to a 2.9-kb *ras*-region ^{32}P-DNA probe.

(i.e., splitting from a densely confluent culture and refeeding). The peak of synthesis was slightly later, at 30 h, and it was less pronounced in late-passage, "old" cultures. Maintenance methylase activity (solid lines), assayed on hemimethylated substrate DNA, was strongly cell cycle-dependent, as previously reported,[25] peaking close to the maximal ^{3}H-TdR incorporation in both cultures. In the older cells, however, methylase activity was markedly higher at confluence (0 and CONF in Figure 4), yet it rose relatively little in parallel with DNA synthesis. To some extent, this reflected the poorer degree of synchro-

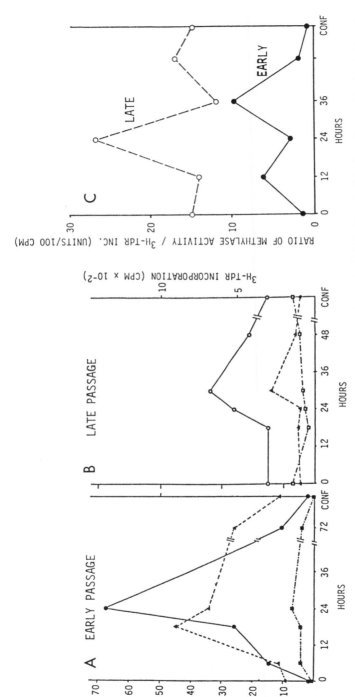

FIGURE 4.   Time course showing the relationship between DNA methylase activity and DNA synthesis following subcultivation of human fibroblasts. Densely confluent fibroblasts at early (panel A) and late passage (panel B) were split at a 1:3 ratio into replicate 850 cm² roller bottles for assay of nuclear methylase activity or into 60-mm petri dishes for tritiated thymidine (³H-TdR) incorporation. DNA synthesis was measured by exposing cells to 1 µCi/ml ³H-TdR in growth medium for 30 min followed by precipitation of aliquots on glass fiber filters (Whatman® GFC) with trichloroacetic acid, and liquid scintillation counting. For DNA methylase determinations,[4] 1 to 5 × 10⁷ cells were harvested, washed, and then disrupted in a Dounce homogenizer to isolate nuclei, which were then frozen at −70°C. After thawing, high-salt extraction, and centrifugation, supernatants were dialyzed and assayed for DNA methylase activity using hemimethylated or unmethylated M13 DNA (provided by K. Zucker and A. Riggs, Duarte, CA) as substrates for maintenance and *de novo* methylases, respectively. These DNAs were prepared from M13 bacteriophage single-stranded DNA as template by use of deoxycytidine triphosphate (dCTP) or 5-methyl-dCTP for *in vitro* second-strand synthesis. Methylase assays with ³H-methyl-S-adenosyl methionine were expressed as units of specific activity (1 unit = 1 pmol of CH₃ incorporated/mg protein/h at 37°C). Symbols, panels A and B: ——●—— = "maintenance" methylase activity, assayed on *in vitro* hemimethylated M13 DNA; --▲-- = *de novo* methylase activity assayed on unmethylated M13 DNA; ---▲--- = ³H-TdR incorporation (acid-precipitated on GFC filters). Symbols, panel C: ——●—— and ----●---- indicate the ratios, at early and late passage, respectively, of maintenance methylase activity to ³H-TdR incorporation.

nization obtainable in senescent cultures, but it may also have indicated a loosening of methylase regulation across the cell cycle. As apparent from Figure 4C, the level of maintenance methylase activity relative to the fraction of cycling cells was actually higher in the older cells, except perhaps at 36 h. *De novo* methylase activity was also assayed using unmethylated substrate DNA (square symbols in Figures 4A and B). This activity was relatively low in early-passage cells, displaying a shallow peak coincident with maintenance methylase. No peak was evident in late-passage cells, for which maximal activity occurred at confluence.

Based on these results and a similar assay for one another cell strain (not shown), late-passage human fibroblasts do not appear to be lacking in maintenance CpG-methylase, but indeed have increased relative methylase levels. Declining methylation of CpG sites (see Figures 1 to 3) in the face of elevated enzyme activity presumably reflects inaccessibility of the appropriate substrate, hemimethylated DNA, at replication forks. The enhanced activity may be compensatory in nature, and it could account for the marked increase observed in *de novo* methylations — especially at confluence (see Figure 4B) and thus presumably not associated with DNA replication. Other contributory factors, such as cell levels of endogenous inhibitors,[4] S-adenosylmethionine, etc., remain to be assessed.

While the levels of methylase and its cofactors could vary between individual clones, as do the extents of overall DNA methylation,[17] such variation alone could not account for the gene-specific manner in which clonal lineages differ in their tendency to lose cytosine methylations. As discussed above, different genes appear to have independent proclivities toward clonal hypomethylation, the explanation for which must lie at the chromatin level — perhaps reflecting interclonal differences in amounts of gene-specific regulatory proteins or in local nucleohistone composition, which influence methylase access to DNA.

IV. IMPLICATIONS FOR REPLICATIVE SENESCENCE AND FOR SENESCENCE *IN VIVO*

An important question arises from these observations: if CpG-methylase enzyme is not lacking in late-passage cells but has restricted access to its substrate at the replication fork, is methylation limiting to senescent-cell proliferation? This question has been indirectly addressed by treatment of cells with 5-azacytidine, a potent methylase inhibitor when incorporated into DNA.[11] The results from two laboratories indicate that such treatment markedly reduces the life span of MRC-5 fibroblasts in culture.[26,27] While a definitive assessment of the role of CpG-methylase in cellular aging cannot be achieved in this way — because many perturbations of cell culture conditions have the potential to reduce the replicative life span — those results are at least consistent with either the inhibition of methylase activity itself, or the consequent reduction in DNA methylation, being responsible for late-passage replicative arrest. Our data argue against the former interpretation, in that methylase activity is not lacking in such late-passage cultures.

Clonal alterations in methylation of specific gene loci are most readily analyzed in cultured cells, where it is possible to identify and harvest pure clones, subclones, and sub-subclones, comprising clonal lineages. Although fibroblasts grown *in vitro* provide a reasonable model for cells which replicate continuously or intermittently *in vivo*, the methylation changes observed in cultured cells may not occur to the same extent in intact tissues, and in any case they may not be observable or interpretable due to the polyclonal nature of tissues. Nevertheless, evidence has been reported of hypomethylation changes consistent with those described above, occurring in tissues of aging mice within a class of long interspersed repeated DNA sequences (LINE1-homologous repeats)[28] and in intracisternal A-particle (IAP) genes.[29]

The relevance of such changes to functional senescence is not known. It might be argued that "epigenetic" events occurring at ca.10^{-2} per generation per gene copy, such as the hypo- and hypermethylations[20,21] or gene amplifications[30] we have assessed, are so much more frequent than true mutations (typically arising spontaneously at 10^{-4} to 10^{-6} per gene) that they are the most likely events to give rise to common clonal aberrations associated with senescence (e.g., atherosclerotic plaque formation, generation of autoimmune lymphocytes, benign polyps, and perhaps also the genesis of malignant tumors).

Nonetheless, hypomethylation of repressed genes is not generally sufficient for their expression; thus far, gene activation has been demonstrated in this manner primarily for cells already on the correct developmental pathway to express those genes, whose precocious appearance is induced by methylation inhibitors.[6,11,16] Other instances include genes on the inactive X chromosome[9,12,13] and metallothionein I in lymphoid cells[14] — both in cells which are expected to be competent for such gene expression. Thus, it is not known to what extent global hypomethylation might induce "leaky" expression of genes characteristic of other differentiated cell types. A clear demonstration that gene hypomethylation may be necessary but not sufficient for expression has been provided by the study of mice inheriting a transgenic c-*myc* gene. Inheritance of this transgene from the father has been absolutely correlated with its hypomethylation in all somatic tissues of carrier progeny, as well as with its expression — which is nevertheless restricted to the myocardium.[31] Such parental "imprinting" has been shown to involve DNA methylation for several marker transgenes, inhibiting their somatic expression upon transmission through the female (or more rarely, the male) germ line. The occurrence of gametic gene methylation, its reversibility in subsequent generations, and whether it is conferred by the female or male parent, differ between lines of transgenic mice and probably depend on the site of integration.[31-34]

There are reports in the literature of senescence-associated gene derepression in rodent tissues,[35,36] but the mechanism is as yet unknown. We have searched for aberrant transcription and translation of hCG subunits α and β in cultured diploid fibroblasts. Complete hCG is normally expressed in the placenta, especially during the first trimester;[37] however, the α-subunit is common to several glycoprotein hormones. By sensitive radioimmunosassay, we have detected β- but not α-hCG in several normal fibroblast strains, including the A2 strain, for which global and α-hCG-specific hypomethylation were observed,[17,19,20] in essential agreement with previous reports.[38,39] We are currently assessing mRNAs corresponding to each hCG chain by Northern blotting and S1 nuclease mapping of transcripts and by kinetic analyses of specific reassociation at very high RNA initial concentration × time (Rot) values.

V. CONCLUSIONS

Evidence accumulated thus far indicates that methylation of DNA is inherited with some infidelity and thus generates clonal lineages *in vitro* with progressive drift in the patterns of gene methylation. The prevailing trend in cultured fibroblasts is toward hypomethylation, and this appears also to be the net effect of constituent clonal changes occurring during *in vivo* senescence. The observation that methylation heritability is both clone- and gene-specific implicates factors acting locally in chromatin, such as other DNA-protein interactions that limit methylase access to its substrate.

Despite evidence indicating that inhibition of CpG-methylase abbreviates the replicative life span of cells in culture, direct measurement of maintenance methylase levels indicates no late-passage decline in this activity. We find instead a substantial increase across the cell cycle in the relative *de novo* methylase activity of senescent cells. Moreover, both maintenance and *de novo* methylase activities are markedly elevated in late- relative to early-passage cells at confluence. Beyond this, the underlying mechanims remains to be investigated, and of course other causal factors must be considered. Nevertheless, the *prima facie* case for a role of methylation drift in senescence, at least *in vitro*, is sufficiently intriguing to warrant further studies, focusing in particular on factors which limit the accessibility of DNA to methylase.

ACKNOWLEDGMENTS

We thank R. Doan for help in preparation of the manuscript, Drs. K. Zucker and A. Riggs for generously providing hemimethylated substrate for methylase assay and for discussions, and Dr. A. Bolden for helpful advice. This work was supported by grants from the National Institutes of Health (AG-03314, AG-03787) and the Veterans Administration.

REFERENCES

1. **Vanyushin, B. F., Tkackeva, S. G., and Belozersky, A. N.,** Rare bases in animal DNA, *Nature,* 225, 948, 1970.
2. **Bird, A. P. and Taggart, M. H.,** Variable patterns of total DNA and rDNA methylated in animals, *Nucleic Acids Res.,* 8, 1485, 1980.
3. **Razin, A. and Riggs., A. D.,** DNA methylation and gene function, *Science,* 210, 604, 1980.
4. **Bolden, A., Ward, C., Siedlecki, J. A., and Weissbach, A.,** DNA methylation: inhibition of *de novo* and maintenance methylation *in vitro* by RNA and synthetic polynucleotides, *J. Biol. Chem.,* 259, 12437, 1984.
5. **Riggs, A. D. and Jones, P. A.,** 5-Methylcytosine, gene regulation, and cancer, *Adv. Cancer Res.,* 40, 1, 1983.
6. **Benvenisty, N., Mencher, D., Meyuhas, O., Razin, A., and Reshef, L.,** Sequential changes in DNA methylation patterns of the rat phosphoenolpyruvate carboxykinase gene during development, *Proc. Natl. Acad. Sci. U.S.A.,* 82, 267, 1985.
7. **Busslinger, M., Hurst, J., and Flavell, R. A.,** DNA methylation and the regulation of globin gene expression, *Cell,* 34, 197, 1983.
8. **Bird, A. P.,** CpG-rich islands and the function of DNA methylation, *Nature,* 321, 209, 1986.
9. **Wolf, S. F., Jolly, D. J., Lunmen, K. D., Friedmann, T., and Migeon, B. R.,** Methylation of the hypoxanthine phosphoribosyl transferase locus on the human X-chromosome: implications for X-chromosome inactivation, *Proc. Natl. Acad. Sci. U.S.A.,* 81, 2806, 1984.
10. **Keshet, I., Yisraeli, J., and Cedar, H.,** Effect of regional DNA methylation on gene expression, *Proc. Natl. Acad. Sci. U.S.A.,* 82, 2560, 1985.
11. **Christman, J. K.,** DNA methylation in Friend erythroleukemia cells: the effects of chemically induced differentiation and of treatment with inhibitors of DNA methylation, *Curr. Top. Microbiol. Immunol.,* 108, 49, 1984.
12. **Veniola, L., Gartler, S. M., Wassman, E. R., Yen, P., Mohandas, T., and Shapiro, L. J.,** Transformation with DNA from 5-azacytidine-reactivated X chromosomes, *Proc. Natl. Acad. Sci. U.S.A.,* 79, 2352, 1983.
13. **Gartler, S. M. and Riggs, A. D.,** Mammalian X-chromosome inactivation, *Annu. Rev. Genet.,* 17, 155, 1983.
14. **Compere, S. J. and Palmiter, R. D.,** DNA methylation controls the inducibility of the mouse metallothionein-1 gene in lymphoid cells, *Cell,* 25, 233, 1981.
15. **Niwa, O. and Sugahara, T.,** 5-Azacytidine induction of mouse endogenous type C virus and suppression of DNA methylation, *Proc. Natl. Acad. Sci. U.S.A.,* 78, 6290, 1981.
16. **Yagi, M. and Koshland, M. E.,** Expression of the J chain gene during B cell differentiation is inversely correlated with DNA methylation, *Proc. Natl. Acad. Sci. U.S.A.,* 78, 4907, 1981.
17. **Shmookler Reis, R. J. and Goldstein, S.,** Variability of DNA methylation patterns during serial passage of human diploid fibroblasts, *Proc. Natl. Acad. Sci. U.S.A.,* 79, 3949, 1982.
18. **Wilson, V. L. and Jones, P. A.,** DNA methylation decreases in aging but not in immortal cells, *Science,* 220, 1055, 1983.
19. **Shmookler Reis, R. J. and Goldstein, S.,** Interclonal variation in methylation patterns for expressed and non-expressed genes, *Nucleic Acids Res.,* 10, 4293, 1982.
20. **Goldstein, S. and Shmookler Reis, R. J.,** Methylation patterns in the gene for the alpha subunit of chorionic gonadotropin are inherited with variable fidelity in clonal lineages of human fibroblasts, *Nucleic Acids Res.,* 13, 7055, 1985.

21. **Shmookler Reis, R. J., Finn, G. K., Smith, K., and Goldstein, S.,** Clonal variation in gene methylation: c-H-*ras* and α-hCG regions vary independently in human fibroblast clones, submitted.

22. **Stein, R., Gruenbaum, Y., Pollack, Y., Razin, A., and Cedar, H.,** Clonal inheritance of the pattern of DNA methylation in mouse cells, *Proc. Natl. Acad. Sci. U.S.A.,* 79, 61, 1982.

23. **Pollack, Y., Stein, R., Razin, A., and Cedar, H.,** Methylation of foreign DNA sequences in eukaryotic cells, *Proc. Natl. Acad. Sci. U.S.A.,* 77, 6463, 1980.

24. **Wigler, M., Levy, D., and Perucho, M.,** The somatic replications of DNA methylation, *Cell,* 24, 33, 1981.

25. **Szyf, M., Kaplan, F., Mann, V., Giloh, H., Kedar, E., and Razin, A.,** Cell cycle-dependent regulation of eukaryotic DNA methylase level, *J. Biol. Chem.,* 260, 8653, 1985.

26. **Holliday, R.,** Strong effects of 5-azacytidine on the *in vitro* lifespan of human diploid fibroblasts, *Exp. Cell Res.,* 166, 543, 1986.

27. **Fairweather, D. S., Fox, M., and Margison, G. P.,** The *in vitro* lifespan of MRC-5 cells is shortened by 5-azacytidine-induced demethylation, *Exp. Cell Res.,* 168, 153, 1987.

28. **Mays-Hoopes, L. L., Brown, A., and Huang, R. C. C.,** Methylation and rearrangement of mouse intracisternal A particle genes in development, aging, and myeloma, *Mol. Cell. Biol.,* 3, 1371, 1983.

29. **Mays-Hoopes, L., Chao, W., Butcher, H. C., and Huang, R. C. C.,** Decreased methylation of the major mouse long interspersed repeated DNA during aging and in myeloma cells, *Dev. Genet.,* 7, 65, 1987.

30. **Srivastava, A., Norris, J. S., Shmookler Reis, R. J., and Goldstein, S.,** c-Ha-*ras*-1 proto-oncogene amplification and overexpression during the limited replicative lifespan of normal human fibroblasts, *J. Biol. Chem.,* 260, 6404, 1985.

31. **Swain, J. L., Stewart, T. A., and Leder, P.,** Parental legacy determines methylation and expression of an autosomal transgene: a molecular mechanism for parental imprinting, *Cell,* 50, 719, 1987.

32. **Reik, W., Collick, A., Norris, M. L., Barton, S. C., and Surani, M. A.,** Genomic imprinting determines methylation of parental alleles in transgenic mice, *Nature,* 328, 248, 1987.

33. **Sapienza, C., Peterson, A. C., Rossant, J., and Balling, R.,** Degree of methylation of transgenes is dependent on gamete of origin, *Nature,* 328, 251, 1987.

34. **Hadchouel, M. Farza, H., Simon, D., Tiollais, P., and Pourcel, C.,** Maternal inhibition of hepatitis B surface antigen gene expression in transgenic mice correlates with *de novo* methylation, *Nature,* 329, 454, 1987.

35. **Wareham, K. A., Lyon, M. F., Glenister, P. H., and Williams, E. D.,** Age related reactivation of an X-linked gene, *Nature,* 327, 725, 1987.

36. **Ono, T., Dean, R. G., Chattopadhay, S. K., and Cutler, R. G.,** Dysdifferentiative nature of aging: age-dependent expressions of MuLV and globin genes in thymus, liver and brain in the AKR mouse strain, *Gerontology,* 31, 362, 1985.

37. **Pierce, J. G. and Parsons, T. F.,** Glycoprotein hormones: structure and function, *Annu. Rev. Biochem.,* 50, 465, 1981.

38. **Rosen, S. W., Weintraub, B. D., and Aaronson, S. A.,** Nonrandom ectopic protein production by malignant cells: direct evidence *in vitro, J. Clin. Endocrinol. Metab.,* 50, 834, 1980.

39. **Milsted, A., Day, D. L., and Cox, R. P.,** Glycopeptide hormone production by cultured human diploid fibroblasts, *J. Cell. Physiol.,* 113, 420, 1982.

Chapter 14

FINITE PROLIFERATIVE CAPACITY OF SYRIAN HAMSTER FETAL AND ADULT FIBROBLASTS *IN VITRO*: A MODEL SYSTEM FOR THE ANALYSIS OF THE CELLULAR BASIS OF AGING

Sarah A. Bruce

TABLE OF CONTENTS

I. INTRODUCTION

Aging is the time-dependent decline in homeostatic efficiency of an individual organism and is characterized by a progressive decline in a variety of physiologic functions, leading to a progressive decline in adaptability.[1] Within renewable tissues there are age-associated alterations in cellular proliferation, a process which plays an important role in tissue homeostasis. Excluding focal increases associated with neoplasia, these alterations in cellular proliferation are generally, but not always, reductions.[1-3] Among rapidly renewing cell types such as gut epithelium and skin epidermis, there is a reduced turnover with increased age.[3-6] Among tissues such as connective tissue, which turn over very slowly except in response to stress, age-related reductions are seen in both deep blister healing and punch biopsy healing in humans,[5] as well as in both open and incision wound healing in a variety of animals and man.[7] Thus, age-related reduced cell proliferation, while unlikely to be the cause of death per se, may contribute to reduced tissue function and reduced organismal homeostasis.[3]

The contribution of reduced proliferative capacity to aging has been studied extensively in cells cultured *in vitro*. The limited *in vitro* proliferative capacity of mammalian cells derived from normal tissue was first described by Swim and Parker[8] and thereafter extensively investigated by Hayflick and Moorhead, who showed that these cells are diploid.[9,10] Hayflick has further proposed that this finite proliferative capacity is an expression, at the cellular level, of aging *in vivo*, and he has termed the phenomenon *in vitro* cellular senescence. In the more than 25 years since Hayflick's original observations, the finite proliferative capacity of normal diploid cells (both fibroblastic and nonfibroblastic) of various species has been studied extensively.[11-14] Despite these efforts, the mechanism of this limit on proliferation *in vitro* and its exact relationship to aging *in vivo* are still not clearly understood.

II. THE SYRIAN HAMSTER AS AN ANIMAL AGING MODEL

Most research on the finite proliferative capacity of normal fibroblasts has focused on human cells, in particular, 3- to 5-month fetal lung fibroblasts such as WI-38, IMR-90, MRC-5, TIG-1, etc. These cells generally undergo 50 to 80 population doublings (PD) before they cease proliferating after 10 to 12 months.[10,11,15-18] In order to establish a rodent model system for the analysis of finite proliferative capacity, we have characterized the proliferation and senescence of diploid fibroblast-like (e.g., mesenchymal) cells derived from the golden or Syrian hamster (*Mesocricetus auratus*).[19]

Laboratory stocks of Syrian hamsters are derived from three animals obtained from their natural habitat in Syria in 1930.[20] Since the introduction of this species as a laboratory animal in biomedical research, it has become an established model system in a variety of research disciplines, notably toxicology and carcinogenesis.[21-23] More recently, the Syrian hamster has been used in aging research.[24,25] An aging colony of male Syrian hamsters was established in our laboratory in 1979 from breedings of outbred stock (LVG) obtained from Charles River. Appropriate numbers of male weanlings have been added at intervals to maintain the colony population at 50 to 60 animals. The mean survival time for animals in this colony is 17.5 months (the median and maximum survival times being 18.5 and 36.0 months, respectively). These values are based on 150 natural deaths. These life span data on LVG hamsters are comparable to those for other strains of Syrian hamsters which exhibit intermediate to long mean life spans (16.4 to 23.0 months).[24,26-28] In contrast, the Turkish hamster (*Mesocricetus brandti*) has a mean life span of 26.7 months (males plus females),[29] and male Chinese hamsters (*Cricetulus griseus*) have a median life span of ca. 38 months.[24]

As described below, the proliferation and senescence of Syrian hamster and human cells are very similar. Notably, there is an inverse correlation between *in vivo* donor age and *in*

TABLE 1
Syrian Hamster Cell Culture
Designations and Sources

Culture designation	Source	
	Tissue	Age
E9	Whole embryo	9 d gestation[a]
FC13	Fetal carcass	13 d gestation[a]
FD13	Fetal dermis	13 d gestation[a,b]
ND3	Neonatal dermis	3 d postpartum
AD6	Adult dermis	6 months[c]
AD24	Adult dermis	24 months[c]

[a] Gestation in the Syrian hamster is 15.5 d.
[b] FC13 cells are equivalent to SHE cells used extensively in neoplastic transformation studies (see Heidelberger et al.[23]).
[c] Median life span and maximum life span of male Syrian hamsters (strain LVG) are 18.5 and 36.0 months, respectively. Animals older than 26 to 28 months are extremely rare.

vitro proliferative capacity in the cells from both species. However, the Syrian hamster system has several distinct advantages over the human cell system. First, the senescence of Syrian hamster cells is characterized by a lower maximum cumulative population doubling level (cumPDL) relative to human cells; the onset of phase III in Syrian hamster fetal cells occurs within approximately 1 month of primary culture, compared to many months in human fetal cell cultures. Second, the easy and constant availability for primary culture of embryonic and fetal tissue of specific gestational age as well as a variety of adult tissues permits analysis of changes occurring during both *in vitro* aging (primary culture to senescence) and *in vivo* aging (primary cultures from variously aged donors). Third, as a rodent model system, direct experimental correlation between *in vitro* cellular behavior and *in vivo* cellular development and function are feasible. Fourth, many inbred strains of the Syrian hamster are available, several of which vary significantly in longevity.[26] Lastly, as a system based on a rodent with a maximal life span of approximately 2 years, the Syrian hamster system is amenable to longitudinal studies on many aspects of aging, including the study of *in vitro* senescence of cells derived from individual hamsters at different ages in the hamster's life span.

III. FINITE PROLIFERATIVE CAPACITY OF SYRIAN HAMSTER FIBROBLASTS *IN VITRO*

A. ISOLATION OF PRIMARY CULTURES AND CULTURE CONDITIONS

Primary cultures were generated from Syrian hamster tissue ranging in developmental stage or age from 9 d gestation to 24-month-old adults (Table 1). In this species, 9 to 10 d gestation is the end of embryonic development (organogenesis),[30,31] and 13 d gestation is approximately equivalent to 16 weeks gestation in human development (from which stage IMR-90 cells were derived). Greater than 90% of the cells in embryonic, fetal, and newborn cultures exhibited the normal diploid chromosome complement for the Syrian hamster (2N = 44), while adult cultures were ca. 80 to 90% diploid, with some tetraploid cells.[19,32] All primary cultures of fetal to adult origin were derived by proteolytic digestion of tissue;[19] embryonic cultures were derived by mechanical disruption.[32] All cells were maintained in Dulbecco's modified Eagle's medium, further modified for Syrian hamster cells,[33] supple-

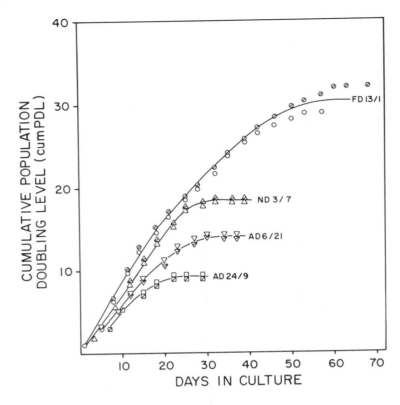

FIGURE 1. *In vitro* proliferation and senescence of Syrian hamster dermal fibro-blastic cells; FD13 (circles), ND3 (triangles), AD6 (inverted triangles), and AD24 (squares). (From Bruce, S. A., Deamond, S. F., and Tso, P. O. P., *Mech. Ageing Dev.*, 34, 151, 1986. With permission.)

mented with non-heat-inactivated fetal bovine serum (20% for E9 cells, 10% for all other cells). Cultures were routinely monitored for mycoplasma contamination by Hoechst staining[34] and/or rRNA:DNA hybridization (Gen-Probe, San Diego, CA). Cells were subcultured twice a week at a constant inoculum previously determined for each cell type to minimize the lag period after subculture and to maintain the cells in exponential growth until the next sub-culture. Cultures were not allowed to attain or incubate at confluence. The efficiency of attachment of freshly subcultured cells (75 to 90%) did not decrease during the proliferative life span of fetal to adult cells, and therefore cumPDL calculations were not corrected for this value. Cultures were judged to be terminally senescent when the harvest number was less than the inoculum number for two consecutive passages.[19]

B. GROWTH PROPERTIES OF FETAL TO ADULT SYRIAN HAMSTER CELLS: INVERSE CORRELATION BETWEEN *IN VIVO* AGE AND *IN VITRO* PROLIFERATION

Nearly all Syrian hamster dermal mesenchymal cell cultures of fetal to aged adult origin exhibited a progressive reduction in proliferative rate following an initial phase of rapid proliferation (Figure 1). The frequency of conversion to a permanent cell line with apparently infinite *in vitro* growth capacity was ca. 3% among replicate flasks.[19] The overall pattern of proliferation and senescence of fetal, neonatal, young adult, and aged adult cells was similar in terms of proliferative changes, as indicated by reductions in ³H-thymidine labeling index, cloning efficiency, and saturation density, as well as an increase in population doubling time and cell volume (Table 2). However, the average maximum cumPDL is characteristic

TABLE 2
**Changes in Growth Characteristics of Syrian
Hamster Mesenchymal Cell Cultures of Fetal to
Aged Adult Origin during *In Vitro* Cellular
Senescence**

Growth characteristics	Low PDL[a]	High PDL[a]
³H-Thymidine labeling index (%)	100	<10
Cloning efficiency (%)	3.4	<0.5
Saturation density (cells/cm²)	1×10^5	2×104
Population doubling time (h)	18	>100
Cell volume (μ³)	800—1400	>6500

[a] PDL = population doubling level.

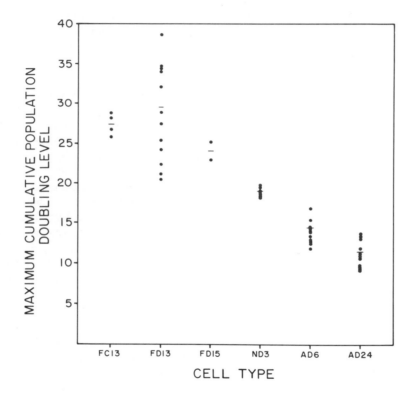

FIGURE 2. Inverse correlation between *in vitro* proliferative capacity of SH dermal fibroblastic cells (maximum cumPDL) and the age of the donor tissue. By *t*-test analysis, differences in the mean maximum cumPDL values are significant at greater than the 99% confidence level (FD13 vs. ND3, p <0.01; ND3 vs. AD6, p <0.001; AD6 vs. AD24, p <0.01). (From Bruce, S. A., Deamond, S. F., and Ts'o, P. O. P., *Mech. Ageing Dev.*, 34, 151, 1986. With permission.)

for each cell type: FD13 cells, 28.6 PDL; ND3 cells, 18.7 PDL; AD6 cells, 13.8 PDL; and AD24 cells, 11.1 PDL (Figure 2). A similar reduction in *in vitro* proliferation relative to increased *in vivo* age is evident in a longitudinal study currently in progress in our laboratory. Thus, the *in vitro* proliferative capacity of Syrian hamster mesenchymal cells is inversely related to the *in vivo* age of the donor tissue in a manner similar to that described for human fibroblasts.[35,36]

C. MORPHOLOGICAL CHANGES ASSOCIATED WITH LOSS OF PROLIFERATIVE CAPACITY IN SYRIAN HAMSTER FIBROBLASTS

Distinct morphological changes were observed as Syrian hamster fetal, neonatal, and adult fibroblastic cells progressed through their *in vitro* life spans. Primary cultures were morphologically homogeneous, containing small, highly proliferative cells with a low cytoplasmic/nuclear ratio. However, within a few passages the cultures became heterogeneous, and there was a continuum of morphological changes until senescence. We have arbitrarily subdivided this continuum into four stages characterized by four cell types (I to IV), distinguished by increasing size and decreasing contact sensitivity, mobility, and proliferative activity (as determined by ^{3}H-thymidine autoradiography) (Figure 3). Primary cultures are composed mostly of type I cells, with some type II cells; terminally senescent cultures are composed nearly exclusively of type IV cells. Intermediate passage cultures exhibit various combinations of all four cell types.[19] Clonal analysis (Figure 4) has shown that the larger, nonproliferative cells (types III and IV) are derived from the smaller, highly proliferative cells (types I and II) in a sequence of morphological differentiation that parallels the loss of proliferative capacity of the mass culture. Similar morphologically distinct cell types have been described in SH lung fibroblast cultures[37,38] and in rat lung and skin fibroblast cultures.[39,40]

IV. MECHANISM OF FINITE PROLIFERATIVE CAPACITY

Theories on the mechanism of the finite proliferative capacity of normal cells in culture can be subdivided into two groups which parallel theories of *in vivo* aging: stochastic (error or damage theory) and deterministic (program theory). Error theory suggests that the loss of proliferative capacity and/or cell death results from a random accumulation, beyond a tolerable limit, of errors or damage in cellular macromolecules.[41] Program theory states that cells in culture are guided by an internal program and, with a given probability at each division, commit to a nonproliferative and, possibly, terminally differentiated state.[42-45] However, these two types of theories are not necessarily mutually exclusive.[46] Aging is most likely the result of a combination of stochastic events (internally generated errors, externally induced damage, external influences on the probability of commitment, etc.) superimposed on a genetic program directing the entire continuum of the life of an individual organism from conception to old age.

One of the major research goals in our laboratory is to analyze the deterministic aspects of the finite proliferative capacity of normal fibroblasts in culture. Specifically, we want to determine the relationship, if any, between the *in vitro* behavior of fetal, newborn, and adult dermal fibroblasts and the normal *in vivo* program of proliferation, differentiation, maturation, and aging of dermal fibroblasts. If the limited *in vitro* proliferative capacity of normal fibroblasts is a manifestation of the *in vivo* program of development, function, and aging of these cells, possibly altered by the artifactual conditions of cell culture, one would expect to find less differentiated stem or progenitor cells and to be able to follow their differentiation with biochemical markers. Furthermore, one would expect to be able to manipulate this program with exogenous proliferation or differentiation factors.

A. PROGENITOR-LIKE CELLS IN CULTURES OF SYRIAN HAMSTER FIBROBLASTS

Early-passage Syrian hamster fetal fibroblast cultures contain a transient, contact-insensitive (CS$^-$) cellular subpopulation which lacks density-dependent inhibition of cell division, a growth property which is frequently characteristic of neoplastically transformed cells.[47] However, in contrast to the permanent expression of the CS$^-$ phenotype in cultured neoplastic cells, this phenotype is only transiently expressed in early-passage fetal cell cultures.[47,48]

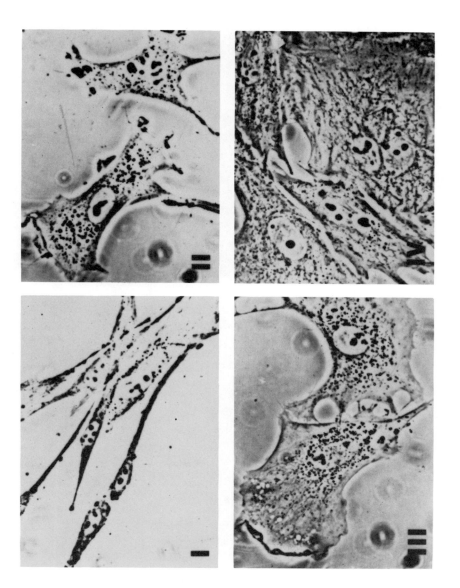

FIGURE 3. Morphology of Syrian hamster dermal fibroblasts. (For description, see text.)

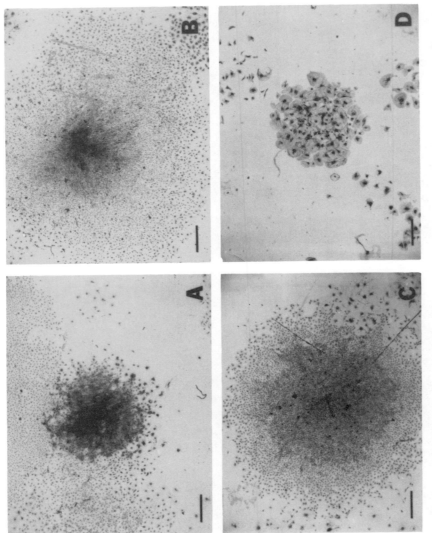

FIGURE 4. Clonal morphology of Syrian hamster dermal fibroblasts. Panel number corresponds to colony type. Type A colonies are composed exclusively of type I cells, are very rare, and have only been observed in low-passage fetal cultures. Type B colonies, observed at low to middle passage, consist of type I cells centrally and type II and III cells peripherally. Type C colonies, predominant at low to middle passage, consist of type II cells centrally and type III and IV cells peripherally. Type D colonies, consisting of type III and IV cells, become predominant at high and terminal passages. (From Bruce, S. A., Deamond, S. F., and Ts'o, P. O. P., *Mech. Ageing Dev.*, 34, 151, 1986. With permission.)

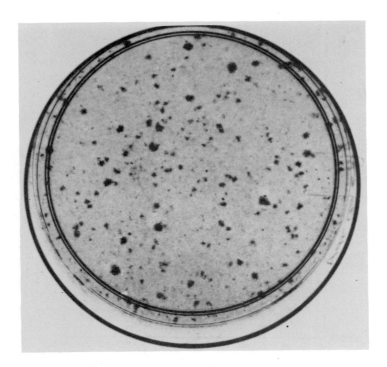

FIGURE 5. Cell mat assay for quantitation of CS^- cells. Cells to be tested (10^4 to 10^6) are superinoculated onto a preformed, lethally irradiated monolayer (cell mat) of contact-sensitive cells, and the plates are incubated 7 to 14 d (with medium changes every third day) and then fixed with methanol and stained with Giemsa.

The proportion of CS^- cells in a culture can be quantitated by measuring the ability of the cells to form colonies on lethally irradiated confluent monolayers (cell mats) of contact-sensitive (CS^+) cells (Figure 5).[47,49] Using this assay, Nakano and colleagues showed that the frequency of CS^- cells decreased to below the limit of resolution of the assay (ca. 10^{-5} to 10^{-6}) by PDL 15 to 20 (Figure 6). Approximately 10 PD after the CS^- cells could no longer be detected in the cell mat assay, the mass culture senesced.[47-49] This loss of CS^- cells was not due to negative selection on a plastic culture substrate, because the same decline was observed when the mass culture was maintained on cell mats.[47] Furthermore, isolation of single colonies from cell mats and reanalysis of their cells on fresh cell mats showed that only 10 to 20% of cells within a cell mat colony were still CS^- after 7 to 14 d of incubation.[47] This suggests that CS^- cells become CS^+ cells as they proliferate.

Assuming this CS^- to CS^+ conversion might be a differentiation-like process, Ueo et al. investigated the effect on the loss of the CS^- cells of tumor-promoting phorbol esters that are known to inhibit many mesenchymal differentiation systems.[50-52] Continuous exposure of mass cultures of fetal fibroblasts to 0.1 μg/ml 12-O-tetradecanoylphorbol-13-acetate (TPA) or phorbol-12,13-didecanoate (PDD), but not the nonpromoting phorbol ester 4αPDD, retarded the loss of the CS^- subpopulation (Figure 6).[48] Such treatment also extended the proliferative life span of fetal cultures by 50 to 100% (see below). In addition, Ueo et al. observed that the frequency of CS^- cells at passage 1 showed a direct relationship to the proliferative capacity of the mass culture (Table 3).[48] Similarly, the frequency of CS^- cells at passage 1 in adult cell cultures with reduced proliferative capacity (Figures 1 and 2) appeared to be lower than that in passage 1 fetal cultures (Table 3).

Thus, the initial frequency and the rate of loss of the CS^- cellular subpopulation appear to be related to the proliferative capacity of the mass culture. Based on these observations,

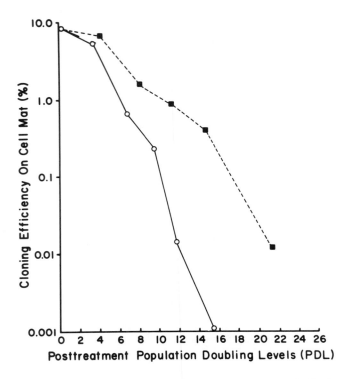

FIGURE 6.   Effect of phorbol ester tumor promoters on the decline in the frequency of CS⁻ cells during serial passage of Syrian hamster fibroblast cultures. Untreated control culture (circles); culture continuously treated with 0.1 μg/ml TPA from passage 1 (squares).

TABLE 3
Relationship between the Frequency of CS⁻ Cells at Passage 1 and the Maximum cumPDL of the Mass Culture

Cell type/ prep no.	% CS⁻ at passage 1	Maximum cumPDL at senescence
F13/U6	22.4	24.9
F13/U2	15.3	21.3
F13/U1	8.8	20.1
F13/U5	7.2	21.5
F13/U4	6.4	16.0
F13/U3	5.8	16.2
AD6/21	0.3	13.2
AD24/9	0.1	9.4

we propose that the phenotypic conversion of CS⁻ cells to CS⁺ cells represents an early stage of fibroblast differentiation, characterized by an altered control of proliferation acquired prior to the appearance of a specific, biochemical phenotype or function which defines the differentiated state. Further, we propose that the CS⁻ to CS⁺ conversion is related to the mechanism of the finite proliferative capacity of normal cells in culture. The isolation and further characterization of these CS⁻ cells is required to verify this hypothesis, to examine the relationship between these CS⁻ cells and the type I cells described above, and to determine

TABLE 4

Modulation of *In Vitro* Proliferative Life Span of Syrian Hamster Fibroblasts

Treatment[a]	Concentration (μg/ml)	Cell type/ prep no.	Maximum cumPDL at senescence (average of replicate flasks)		
			Control	Treated	Percentage change
PDD	0.1	FC13/U5	22.6	40.8	+ 80.5
	0.1	FD13/2	32.7	48.2	+ 47.4
	0.1	ND3/8	19.6	29.3	+ 49.5
	0.1	AD6/21	13.2	16.9	+ 28.0
	0.1	AD24/9	9.4	10.3	+ 9.6
RA	0.05	FC13/46	31.7	9.7	− 69.3
	0.05	FD13/28	39.3	15.1	− 61.7
RA	1.0	FC13/46	31.7	5.5	− 82.6
	1.0	FD13/28	39.3	7.2	− 81.7

[a] PDD = phorbol-12,13-didecanoate; RA = all-trans retinoic acid.

whether these CS⁻/type I cells are examples of the progenitor-like cells proposed in program theory.[44,54]

B. MODULATION OF *IN VITRO* PROLIFERATIVE LIFE SPAN
1. Extension of Proliferative Life Span

Several reports in the literature have shown that supplementation of culture medium with any one of a variety of compounds, including bovine serum albumin,[55] tyrosine,[56] or hydrocortisone,[57] can extend the *in vitro* proliferative capacity of human diploid fibroblasts. Our initial studies in this area were an extension of our observations on the effect of phorbol ester tumor promoters on CS⁻ cells[48] and therefore focused on the effect of tumor promoters on the proliferative life span of Syrian hamster fibroblasts. Continuous exposure of fetal cells from passage 1 with TPA or PDD (0.1 μg/ml) resulted in an extension of the mass culture's *in vitro* proliferative life span (Table 4) but did not result in an increased frequency of conversion to permanent cell lines.[48] As described above, it was separately shown that this concentration of promoter retarded the rate of loss of the subpopulation of CS⁻ cells. The promoter-induced extension of fetal cell proliferation required initiation of treatment at low passage, when the frequency of CS⁻ cells was still relatively high; initiation of treatment at passage 6 had little effect on the proliferative life span of F13 cells.[48] The continuous presence of promoter was also required; removal of promoter after the maximum cumPDL of the control culture had been exceeded resulted in senescence within one to two subcultures.[48] In addition, the magnitude of the promoter-induced extension was reduced in adult cell cultures (Table 4), which have reduced frequencies of CS⁻ cells even at passage 1 (Table 3).[53] These observations link the effect of promoters to the size of the CS⁻ subpopulation and suggest that tumor promoters may alter the probability of the CS⁻ to CS⁺ conversion. The ability to extend the *in vitro* proliferative life span of these cells by treatment with a compound that presumably acts by retarding the loss of a specific subpopulation of cells argues against a time-dependent error and/or damage accumulation mechanism, as well as against a rigidly set clock mechanism. Analysis of isolated populations of CS⁻ cells will clarify the role of these progenitor-like cells in the limited *in vitro* proliferative life span of Syrian hamster fibroblasts.

In contrast to the effect of tumor promoters on fetal and adult fibroblast cultures, similar continuous treatment of 9-d gestation embryonic (E9) cells has no effect on the maximum cumPDL of these cells at senescence, but it does result in an approximately fivefold increase in the frequency of conversion to permanent cell lines. Although most untreated primary E9 cell cultures do exhibit a finite proliferative capacity, more than 65% of replicate E9 cultures treated with PDD or TPA become established cell lines capable of proliferating in the absence of the phorbol ester.[32] E9 cells are also distinct from fetal and adult cells in other aspects; they have a higher serum requirement (20% vs. 10%), a reduced efficiency of attachment to plastic substrate (50 to 60% vs. 80 to 90%), and a ca. fourfold higher frequency of spontaneous conversion to permanent cell lines (ca. 12% vs. ca. 3%). In addition, primary E9 cells differentiate in culture along several distinct mesenchymal lineages (including fibroblastic, adipogenic, and myogenic), and this differentiation capacity is frequently retained by permanent cell lines established from E9 primary cultures.[58]

2. Reduction of Proliferative Life Span

Although the ability to extend *in vitro* proliferative life span argues for a program mechanism for the finite proliferative capacity of normal cells in culture, far stronger evidence would be the ability to induce the expression of the entire cellular program governing *in vitro* proliferation and senescence. *In vitro* cellular senescence, however, is the progression toward a final, generally irreversible nonproliferative state. Therefore, induction of this phenomenon implies induction of a nonproliferative state which is a negative end point. Clearly, this can be accomplished by withdrawal of serum mitogens (e.g., low-serum-arrested, quiescent cells), by maintenance at high cell density (confluent, quiescent cells), or by addition of a cytostatic (but noncytotoxic) compound. Several specific proteins are expressed by both quiescent and senescent human fibroblasts,[59,60] but it is not yet clear whether these proteins are mechanistically related to the cessation of proliferation in senescent cells or if they simply serve as markers of the nonproliferative state characteristic of both senescent and quiescent cells. The combined use of the following three criteria and their reversibility may be useful to distinguish between induction of nonproliferation (i.e., quiescence) and induction of the potentially more complex phenomenon of senescence: (1) a nonproliferative state (negative end point), (2) morphological changes characteristic of Phase III (positive end point), and (3) differential expression of a biochemical marker of fibroblast senescence and/or differentiation (positive end point).

These criteria are being used to analyze the effect of all-trans retinoic acid (RA), a known modulator of differentiation *in vivo* and *in vitro*,[61] on the proliferative life span of Syrian hamster FC13 and FD13 cells. Continuous treatment of F13 cells with noncytotoxic concentrations of RA irreversibly reduces the proliferative life span in a dose-dependent manner (Table 4).[62] Concomitantly, there is a significant increase in the abundance of the type III and IV cells that are characteristic of Phase III in this system. Experiments are in progress to determine the effect of RA on the expression of fibroblast differentiation markers *in vitro*. Similar effects of RA on the proliferation, saturation density, and morphology of human fetal lung fibroblasts have been observed.[63]

In addition to RA, there are several other treatments or culture conditions that appear to reduce the proliferative life span of normal fibroblasts. Several laboratories have reported that 5-azacytidine (5azaC), a DNA-hypomethylating agent, decreases the *in vitro* proliferative life span of human fetal lung fibroblasts.[64-66] Preliminary observations on the effects of 5azaC on Syrian hamster FC13 fibroblasts show a similar dose-dependent effect on *in vitro* life span. The induction of "early" senescence with 5azaC is of particular interest, because there is a general decrease in DNA methylation with age *in vitro*[67,68] and *in vivo*[69] and because 5azaC is known to induce a variety of mesenchymal differentiation lineages *in vitro*.[70]

Growth of fibroblasts on or in extracellular matrix (ECM) or with ECM components also affects *in vitro* proliferation. Both low- and high-PDL human foreskin fibroblasts, incubated in a hydrated collagen gel, shut down proliferation within a few days.[71-73] In addition, treatment of phase II WI-38 cells with heparin, a glycosaminoglycan matrix component, inhibits cell proliferation and stimulates glycosaminoglycan synthesis, particularly that of heparan sulfate and hyaluronic acid.[74] Furthermore, senescence in both TIG-1 and WI-38 human fibroblasts is characterized by a selective increase in the synthesis and accumulation of heparan sulfate.[74,75] We have observed a similar shortening of *in vitro* proliferative life span of Syrian hamster F13 cells continuously passaged on a basement membrane-like ECM or treated with heparan sulfate. These observations suggest that the ECM may play a regulatory role in fibroblast proliferation and senescence *in vitro*.

Identification and further study of treatments and culture conditions that allow us to manipulate coordinately the proliferation and morphology of these cells, as well as the expression of biochemical markers, may help to define the interrelationship between *in vitro* proliferation and senescence and *in vivo* differentiation and function.

C. BIOCHEMICAL MARKERS OF FIBROBLAST DIFFERENTIATION

To determine whether or not the finite proliferative capacity of normal fibroblasts *in vitro* is a manifestation of an intrinsic differentiation program, the proliferation/nonproliferation sequence *in vitro* needs to be defined in terms of biochemical markers that are related to markers of fibroblast developmental and function *in vivo*. The *in vivo* function of loose connective tissue fibroblasts is to produce the ECM (collagen, fibronectin, elastin, other glycoproteins, and proteoglycans). Although the production of these matrix components is not unique to (and therefore diagnostic for) fibroblasts, it is nonetheless reasonable to consider ECM components as markers of fibroblast differentiation. This *in vivo* fibroblast function is also expressed *in vitro*. Normal skin and lung fibroblasts in culture produce and secrete ECM molecules, including collagen types I and III, fibronectin, etc. Secreted ECM molecules are frequently found in a soluble form in the culture medium; however, a pericellular matrix can also be observed.[76-78] In contrast, the production of insoluble, cell-associated ECM is reduced or eliminated in transformed derivatives of normal fibroblasts.[77,79] Furthermore, ECM metabolism *in vitro* appears to reflect *in vivo* metabolism, as suggested by observations that abnormalities in collagen metabolism are generally expressed *in vitro* by skin fibroblasts derived from patients with connective tissue disorders.[80-83]

In vivo developmental and age-related changes in various aspects of collagen metabolism (synthesis, degradation, processing, assembly, etc.) are well known.[84-87] Many reports also describe developmental and/or age-related alterations in collagen metabolism *in vitro*, including increased collagenase activity in human skin fibroblasts from donors of later gestational stage or age,[88] decreased dependence on exogenous ascorbic acid for full hydroxylation in passage 41 vs. passage 26 WI-38 fibroblasts,[89] and variable alterations in collagen synthesis. Collagen synthesis has been reported to decrease with increased PDL in normal adult human skin fibroblasts and IMR-90 human fetal fibroblasts.[90,91] In contrast, collagen synthesis has been observed to increase with increased PDL in rat lung and skin fibroblasts of fetal to adult origin.[39,40] Initial indirect immunofluorescence analysis of Syrian hamster fetal dermal fibroblasts with anti-collagen type I antiserum shows that although both passage 1-2 cells and senescent cells synthesize collagen type I (cytoplasmic signal), senescent cells are characterized by an extensive extracellular matrix not seen in low-passage cells.[92] This latter observation suggests that even if the rate of collagen synthesis does not change, some alteration in processing or assembly may lead to increased extracellular accumulation in senescent cells. An extensive extracellular matrix surrounding senescent cells, similar to the *in vivo* environment of post-mitotic dermal fibroblasts, could act via its connections to the cytoskeletal and, hence, the nuclear matrix[93] to reinforce or even induce the nonproliferative state characteristic of senescent fibroblasts *in vitro* and post-mitotic fibroblasts (fibrocytes) in mature dermis *in vivo*.

The influence of the ECM on cell shape, migration, proliferation, and differentiation *in vivo* and *in vitro* is well documented.[93-96] Thus, understanding changes in the ECM produced by fibroblasts *in vitro* during proliferation and senescence may help to elucidate the mechanism of reduced proliferation characteristic of senescent fibroblasts, as well as the relationship, if any, between *in vitro* fibroblast senescence and *in vivo* fibroblast development, function, and aging.

V. SUMMARY AND CONCLUSION

Normal diploid Syrian hamster dermal fibroblastic cells, regardless of the age of the tissue of origin, exhibit limited *in vitro* proliferative capacity. The proliferative life span of these cells is characterized by reductions in several growth parameters, as well as by a sequence of morphological differentiation. In addition, there is an inverse correlation between *in vitro* proliferative capacity and *in vivo* donor age in this system, as has been demonstrated for human fibroblasts. Primary and low-passage cultures of Syrian hamster fetal fibroblasts contain a transient subpopulation of CS^- cells which are lost during serial subculture. Current data suggest that these cells are lost by conversion to CS^+ cells rather than by negative selection. The initial frequency of CS^- cells and the rate of their loss, which can be retarded by phorbol ester tumor promoters, appear to be directly related to the proliferative capacity of the mass culture. The presence of progenitor-like cells in this system and the ability to modulate both positively and negatively the proliferative life span of these cells suggest a differentiative mechanism for the limited proliferative capacity of cultured Syrian hamster fibroblasts. However, further studies of the CS^- subpopulation and of biochemical markers of fibroblast senescence *in vitro* and differentiation *in vivo* are required to analyze more critically the possible relationship between fibroblast senescence *in vitro* and fibroblast development, function, and aging *in vivo*.

ACKNOWLEDGMENT

This work was supported by the National Institute on Aging (AG 03633) and the Department of Energy (DE-AC02-76-EVO-3280).

REFERENCES

1. **Buetow, D. E.,** Cell number versus age in mammalian tissues and organs, in *Handbook of Cell Biology of Aging,* Cristofalo, V. J., Ed., CRC Press, Boca Raton, FL, 1985, 1.
2. **Walton, J.,** The role of limited cell replicative capacity in pathological age change — a review, *Mech. Ageing Dev.,* 19, 217, 1982.
3. **Bowman, R. D.,** Aging and the cell cycle *in vivo* and *in vitro,* in *Handbook of Cell Biology of Aging,* Cristofalo, V. J., Ed., CRC Press, Boca Raton, FL, 1985, 117.
4. **Lesher, S., Fry, R. J. M., and Kohn, H. I.,** Influence of age on transit time of cells of mouse intestinal epithelium, *Lab. Invest.,* 10, 291, 1961.
5. **Kligman, A. M.,** Perspectives and problems in cutaneous gerontology, *J. Invest. Dermatol.,* 73, 39, 1979.
6. **Leyden, J. J., McGinley, K. J., Grove, G. L., and Kligman, A. M.,** Age-related differences in the rate of desquamation of the skin surface cells, in *Pharmacological Intervention of the Aging Process,* Cristofalo, V. J., Adelman, R. D., and Roberts, J., Eds., Plenum Press, New York, 1977, 297.
7. **Goodson, W. H. and Hunt, T. K.,** Wound healing and aging, *J. Invest. Dermatol.,* 73, 88, 1979.
8. **Swim, H. E. and Parker, R. F.,** Culture characteristics of human fibroblasts propagated serially, *Am. J. Hyg.,* 66, 235, 1957.
9. **Hayflick, L. and Moorhead, P. S.,** The serial cultivation of human diploid cell strains, *Exp. Cell Res.,* 25, 585, 1961.
10. **Hayflick, L.,** The limited *in vitro* lifetime of human diploid cell strains, *Exp. Cell Res.,* 37, 614, 1965.

11. **Smith, J. R. and Pereira-Smith, O. M.,** Lung-derived fibroblast-like human cells in culture, in *Handbook of Cell Biology of Aging,* Cristofalo, V. J., Ed., CRC Press, Boca Raton, FL, 1985, 375.
12. **Schneider, E. L.,** Human skin-derived fibroblast-like cells in culture, in *Handbook of Cell Biology of Aging,* Cristofalo, V. J., Ed., CRC Press, Boca Raton, FL, 1985, 425.
13. **Williams, J. and Dearfield, K. L.,** Non-human fibroblast-like cells in culture, in *Handbook of Cell Biology of Aging,* Cristofalo, V. J., Ed., CRC Press, Boca Raton, FL, 1985, 433.
14. **Hayflick, L.,** Ageing and death of vertebrate cells, in *Lectures on Gerontology,* On the Biology of Ageing, Vol. 1, Viidik, A., Ed., Academic Press, London, 1982, 59.
15. **Nichols, W .W., Murphy, D. G., Cristofalo, V. J., Toji, L. H., Greene, A. E., and Dwight, S. A.,** Characterization of a new human diploid cell strain, IMR-90, *Science,* 196, 60, 1977.
16. **Ohashi, M., Aizawa, S., Ooka, H., Ohsawa, T., Kaji, K., Kondo, H., Kobayashi, T., Noumura, T., Matsuo, M., Mitsui, Y., Murota, S., Yamamoto, K., Ito, H., Shimada, H., and Utako, T.,** A new human diploid cell strain, T1G-1, for the research on cellular aging, *Exp. Gerontol.,* 15, 121, 1980.
17. **Jacobs, J. P., Jones, C. M., and Baille, J. P.,** Characterization of a human diploid cell designated MRC-5, *Nature,* 227, 168, 1970.
18. **Hayflick, L.,** Intracellular determinants of cell aging, *Mech. Ageing Dev.,* 28, 177, 1984.
19. **Bruce, S. A., Deamond, S. F., and Ts'o, P. O. P.,** *In vitro* senescence of Syrian hamster mesenchymal cells of fetal to aged adult origin. Inverse relationship between *in vivo* donor age and *in vitro* proliferative capacity, *Mech. Ageing Dev.,* 34, 151, 1986.
20. **Yerganian, G.,** History and cytogenetics of hamsters, *Prog. Exp. Tumor Res.,* 16, 2, 1972.
21. **Van Hoosier, G. L., Jr. and McPherson, C. W.,** *Laboratory Hamsters,* Academic Press, Orlando, 1987.
22. **Homburger, F., Ed.,** Pathology of the Syrian Hamster, *Prog. Exp. Tumor Res.,* 16, 1, 1972.
23. **Heidelberger, C., Freeman, A. E., Pienta, R. J., Sivak, A., Bertram, J. S., Casto, B. C., Dunkel, V. C., Francis, M. W., Kakunaga, T., Little, J. B., and Schechtman, L. M.,** Cell transformation by chemical agents — a review and analysis of the literature, *Mutat. Res.,* 114, 283, 1983.
24. National Research Council Committee on Animal Models for Research on Aging, *Mammalian Models for Research on Aging,* National Academy Press, Washington, D.C., 1981, 146.
25. **Schmidt, R. E., Eason, R. L., Hubbard, G. B., Young, J. T., and Eisenbrandt, D. L.,** *Pathology of Aging Syrian Hamsters,* CRC Press, Boca Raton, FL, 1983.
26. **Festing, M. F. W.,** *Inbred Strains in Biomedical Research,* Oxford University Press, New York, 1979, 297.
27. **Redman, H. C., Hobb, C. H., and Rebar, A. H.,** Survival distribution of Syrian hamsters (*Mesocricetus auratus,* Sch:SYR) used during 1972—1977, *Prog. Exp. Tumor Res.,* 24, 108, 1979.
28. **Bernfeld, P.,** Longevity of the Syrian hamster, *Prog. Exp. Tumor Res.,* 24, 118, 1979.
29. **Lyman, C. P., O'Brien, R. C., Green, G. C., and Papafrangos, E. D.,** Hibernation and longevity in the Turkish hamster *Mesocricetus brandti, Science,* 212, 668, 1981.
30. **Boyer, C. C.,** Embryology, in *The Golden Hamster,* Hoffman, R. A., Robinson, P. F., and Megalhaes, H., Eds., Iowa State University Press, Ames, 1968, 73.
31. **Bruce, S. A., Gyi, K. K., Nakano, S., Ueo, H., Zajac-Kaye, M., and Ts'o, P. O. P.,** Genetic and development determinants in neoplastic transformation, in *Biochemical Basis of Chemical Carcinogenesis,* Greim, H., Jung, R., Kramer, M., Marquardt, H., and Oesch, F., Eds., Raven Press, New York, 1984, 159.
32. **Okeda, T., Yokogawa, Y., Ueo, H., Bury, M. A., Ts'o, P. O. P., and Bruce, S. A.,** Two classes of continuous cell lines established from Syrian hamster 9 day gestation embryos — preneoplastic and progenitor cells, submitted.
33. **Casto, B. C.,** Enhancement of adenovirus transformation by treatment of hamster cells with ultraviolet light, DNA base analogues, and dibenz(*a,h*) anthracene, *Cancer Res.,* 33, 402, 1973.
34. **Chen, T. R.,** *In situ* detection of mycoplasma contamination in cell cultures by fluorescent Hoechst 33258 stain, *Exp. Cell Res.,* 104, 255, 1977.
35. **Martin, G. M., Sprague, C. A., and Epstein, C. J.,** Replicative life span of cultivated human cells, *Lab. Invest.,* 23, 86, 1970.
36. **Schneider, E. L. and Mitsui, Y.,** The relationship between *in vitro* cellular aging and *in vivo* human age, *Proc. Natl. Acad. Sci. U.S.A.,* 73, 3584, 1976.
37. **Raes, M. and Remacle, J.,** Ageing of hamster embryo fibroblasts as the result of both differentiation and stochastic mechanisms, *Exp. Gerontol.,* 18, 223, 1983.
38. **Raes, M., Geuens, G., de Brabander, M., and Remacle, J.,** Microtubules and microfilaments in ageing hamster embryo fibroblasts *in vitro, Exp. Gerontol.,* 18, 241, 1983.
39. **Kontermann, K. and Bayreuther, K.,** The cellular aging of rat fibroblasts *in vitro* is a differentiation process, *Gerontology,* 25, 261, 1979.
40. **Mollenhauer, J. and Bayreuther, K.,** Donor-age-related changes in the morphology, growth potential and collagen biosynthesis in rat fibroblast subpopulations *in vitro, Differentiation,* 32, 165, 1986.

41. **Orgel, L. E.,** Ageing of clones of mammalian cells, *Nature,* 243, 441, 1973.
42. **Martin, G. M., Sprague, C. A., Norwood, T. H., and Pendergrass, W. R.,** Clonal selection, attenuation and differentiation in an *in vitro* model of hyperplasia, *Am. J. Pathol.,* 74, 137, 1974.
43. **Holliday, R., Huschtscha, L. I., Tarrant, G. M., and Kirkwood, T. B. L.,** Testing the commitment theory of cellular aging, *Science,* 198, 366, 1977.
44. **Holliday, R., Huschtscha, L. I., and Kirkwood, T. B. L.,** Cellular aging: further evidence for the commitment theory, *Science,* 213, 1505, 1981.
45. **Bell, E., Marek, L., Sher, S., Merrill, C., Levinstone, D., and Young, I.,** Do diploid fibroblasts in culture age?, *Int. Rev. Cytol.,* Suppl. 10, 1, 1979.
46. **Goldstein, S. and Shmookler-Reis, R. J.,** Genetic modification during cellular aging, *Mol. Cell. Biochem.,* 64, 15, 1984.
47. **Nakano, S. and Ts'o, P. O. P.,** Cellular differentiation and neoplasia: characterization of subpopulations of cells that have neoplasia-related growth properties in Syrian hamster cell cultures, *Proc. Natl. Acad. Sci. U.S.A.,* 78, 4995, 1981.
48. **Ueo, H., Bruce, S. A., Nakano, S., and Ts'o, P. O. P.,** Tumor promoters retard the loss of a transient subpopulation of cells in low passage Syrian hamster cell cultures — effect on *in vitro* proliferative life span of the culture, submitted.
49. **Nakano, S., Bruce, S. A., Ueo, H., and Ts'o, P. O. P.,** A qualitative and quantitative assay for cells lacking post-confluence inhibition of cell division. Characterization of this phenotype in carcinogen-treated Syrian hamster embyro cells in culture, *Cancer Res.,* 42, 3132, 1982.
50. **Cohen, R., Pacifici, M., Rubinstein, N., Biehl, J., and Holtzer, H.,** Effect of tumour promoter on myogenesis, *Nature,* 266, 538, 1977.
51. **Diamond, L., O'Brien, T. G., and Rovera, G.,** Inhibition of adipose conversion of 3T3 fibroblasts by tumor promoters, *Nature,* 269, 247, 1977.
52. **Pacifici, M. and Holtzer, H.,** Effect of a tumor-promoting agent on chondrogenesis, *Am. J. Anat.,* 150, 207, 1977.
53. **Bruce, S. A., Deamond, S. F., Ueo, H., and Ts'o, P. O. P.,** Age-related difference in promoter-induced extension of *in vitro* life span of Syrian hamster cells, *J. Cell Biol.,* 97, 346a, 1983.
54. **Bell, E., Marek, L. F., Levinstone, D. S., Merrill, C., Sher, S., Young, I. T., and Eden, M.,** Loss of division potential *in vitro:* aging or differentiation?, *Science,* 202, 1158, 1978.
55. **Todaro, G. J. and Green, H.,** Serum albumin supplemented medium for long term cultivation of mammalian fibroblast strains, *Proc. Soc. Exp. Biol. Med.,* 116, 688, 1964.
56. **Litwin, J.,** Human diploid cell response to variations in relative amino acid concentrations in Eagle medium, *Exp. Cell Res.,* 72, 566, 1972.
57. **Cristofalo, V. J. and Stanulis-Praeger, B. M.,** Cellular senescence *in vitro, Adv. Cell Cult.,* 2, 1, 1982.
58. **Yokogawa, Y., Okeda, T., and Bruce, S. A.,** Isolation of a diploid myoblast cell line from Syrian hamster embryos, *J. Cell Biol.,* 101, 170a, 1985.
59. **Wang, E.,** A 57,000-mol-wt protein uniquely present in nonproliferating cells and senescent human fibroblasts, *J. Cell Biol.,* 100, 545, 1985.
60. **Lumpkin, C. K., Jr., McClung, J. K., Pereira-Smith, O. M., and Smith, J. R.,** Existence of high abundance antiproliferative mRNA's in senescent human diploid fibroblasts, *Science,* 232, 393, 1986.
61. **Lotan, R.,** Effects of vitamin A and its analogues (retinoids) on normal and neoplastic cells, *Biochim. Biophys. Acta,* 605, 33, 1980.
62. **Bruce, S. A. and Deamond, S. F.,** Retinoic acid reduces the *in vitro* proliferative life span of Syrian hamster fetal fibroblasts, *J. Cell Biol.,* 105, 270a, 1987.
63. **Stanulis-Praeger, B. M.,** Filopodia number increases with age and quiescence in populations of normal WI-38 cells, and is correlated with drug-induced changes in proliferation in both normal and transformed populations, *Mech. Ageing Dev.,* 33, 221, 1986.
64. **Fairweather, D. S., Fox, M., and Margison, G. P.,** The *in vitro* life span of MRC-5 cells is shortened by 5-azacytidine induced demethylation, *Exp. Cell Res.,* 168, 153, 1986.
65. **Honda, S. and Matsuo, M.,** 5-Azacytidine shortens the *in vitro* life span of human diploid cells, *Cell Biol. Int. Rep.,* 11, 141, 1987.
66. **Holliday, R.,** Strong effects of 5-azacytidine on the *in vitro* life span of human diploid fibroblasts, *Exp. Cell Res.,* 166, 543, 1987.
67. **Wilson, V. L. and Jones, P. A.,** DNA methylation decreases in aging but not in immortal cells, *Science,* 220, 1055, 1983.
68. **Shmookler-Reis, R. J. and Goldstein, S.,** Variability of DNA methylation patterns during serial passage of human diploid fibroblasts, *Proc. Natl. Acad. Sci. U.S.A.,* 79, 3949, 1982.
69. **Wilson, V. L., Smith, R. A., Ma, S., and Cutler, R. G.,** Genomic 5-methyldeoxycytidine decreases with age, *J. Biol. Chem.,* 262, 9948, 1987.
70. **Jones, P. A., Taylor, S. M., and Wilson, V.,** DNA modification, differentiation, and transformation, *J. Exp. Zool.,* 228, 287, 1983.

71. **Bell, E., Ivarsson, B., and Merrill, C.,** Production of a tissue-like structure by contraction of collagen lattices by human fibroblasts of different proliferative potential *in vitro, Proc. Natl. Acad. Sci. U.S.A.,* 76, 1274, 1979.

72. **Sarber, R., Merrill, C., Soranno, T., and Bell, E.,** Regulation of proliferation of fibroblasts of low and high population doubling level grown in collagen lattices, *Mech. Ageing Dev.,* 17, 107, 1981.

73. **Guidry, C. and Grinnell, F.,** Studies on the mechanism of hydrated collagen gel reorganization by human skin fibroblasts, *J. Cell Sci.,* 79, 67, 1985.

74. **Wever, J., Schachtschabel, D. O., Sluke, G., and Wever, G.,** Effect of short- or long-term treatment with exogenous glycosaminoglycans on growth and glycosaminoglycan synthesis of human fibroblasts (WI-38) in culture, *Mech. Ageing Dev.,* 14, 89, 1980.

75. **Matuoka, K. and Mitsui, Y.,** Changes in cell-surface glycosaminoglycans in human diploid fibroblasts during *in vitro* aging, *Mech. Ageing Dev.,* 15, 153, 1981.

76. **Bornstein, P. and Ash, J. F.,** Cell surface-associated structural proteins in connective tissue cells, *Proc. Natl. Acad. Sci. U.S.A.,* 74, 2480, 1977.

77. **Vaheri, A., Kurkinen, M., Lehto, V.-P., Linder, E., and Timpl, R.,** Codistribution of pericellular matrix proteins in cultured fibroblasts and loss in transformation: fibronectin and procollagen, *Proc. Natl. Acad. Sci. U.S.A.,* 75, 4944, 1978.

78. **Hedman, K., Johansson, S., Vartio, T., Kjellén, T., Vaheri, A., and Höök, M.,** Structure of the pericellular matrix: association of heparan and chondroitin sulfates with fibronectin-procollagen fibers, *Cell,* 28, 663, 1982.

79. **Sandmeyer, S., Smith, R., Kiehn, D., and Bornstein, P.,** Correlation of collagen synthesis and procollagen messenger RNA levels with transformation in rat embryo fibroblasts, *Cancer Res.,* 41, 830, 1981.

80. **Aumailley, M., Rieg, T., Dessau, W., Müller, P. K., Timpl, R., and Bricaud, H.,** Biochemical and immunological studies of fibroblasts derived from a patient with Ehlers-Danlos Syndrome type IV demonstrate reduced type III collagen synthesis, *Arch. Dermatol. Res.,* 269, 169, 1980.

81. **Uitto, J., Perejda, A. J., Abergel, R. P., Chu, M.-L., and Ramirez, F.,** Altered steady-state ratio of type I/III procollagen mRNAs correlate with selectively increased type I procollagen biosynthesis in cultured keloid fibroblasts, *Proc. Natl. Acad. Sci. U.S.A.,* 82, 5935, 1985.

82. **Graves, P. N., Weiss, I. K., Perlish, J. S., and Fleischmajer, R.,** Increased procollagen mRNA levels in scleroderma skin fibroblasts, *J. Invest. Dermatol.,* 80, 130, 1983.

83. **Minor, R. R., Sippola-Thiele, M., McKeon, J., Berger, J., and Prockop, D. J.,** Defects in the processing of procollagen to collagen are demonstrable in cultured fibroblasts from patients with the Ehlers-Danlos and osteogenesis imperfecta syndromes, *J. Biol. Chem.,* 261, 10006, 1986.

84. **Tolstoshev, P., Haber, R., Trapnell, B. C., and Crystal, R. G.,** Procollagen messenger RNA levels and activity and collagen synthesis during the fetal development of sheep lung, tendon and skin, *J. Biol. Chem.,* 256, 9672, 1981.

85. **Bochantin, J. and Mays, L. L.,** Age-dependence of collagen tail fiber breaking strength in Sprague-Dawley and Fischer 344 rats, *Exp. Gerontol.,* 16, 101, 1981.

86. **Lavker, R. M., Zheng, P., and Dong, G.,** Aged skin: a study by light, transmission electron, and scanning electron microscopy, *J. Invest. Dermatol.,* 88, 44s, 1987.

87. **Barnes, M. J., Constable, B. J., Morton, L. F., and Royce, P. M.,** Age-related variations in hydroxylation of lysine and proline in collagen, *Biochem. J.,* 139, 461, 1974.

88. **Bauer, E. A., Kronberger, A., Stricklin, G. P., Smith, L. T., and Holbrook, K. A.,** Age-related changes in collagenase expression in cultured embryonic and fetal human skin fibroblasts, *Exp. Cell Res.,* 161, 484, 1985.

89. **Paz, M. A. and Gallop, P. M.,** Collagen synthesized and modified by aging fibroblasts in culture, *In Vitro,* 11, 302, 1975.

90. **McCoy, B. J., Galdun, J., and Cohen, I. K.,** Effects of density and cellular aging on collagen synthesis and growth kinetics in keloid and normal skin fibroblasts, *In Vitro,* 18, 79, 1982.

91. **Hildebran, J. N., Absher, M., and Low, R. B.,** Altered rates of collagen synthesis in *in vitro* aged human lung fibroblasts, *In Vitro,* 19, 307, 1983.

92. **Bruce, S. A., Saboori, A. M., and Deamond, S. F.,** Increased extracellular accumulation of collagen type I matrix in senescent Syrian hamster fetal dermal fibroblasts, in preparation.

93. **Bissel, M. J., Hall, H. G., and Parry, G.,** How does the extracellular matrix direct gene expression?, *J. Theor. Biol.,* 99, 31, 1982.

94. **Hay, E. D., Ed.,** *Cell Biology of Extracellular Matrix,* Plenum Press, New York, 1981.

95. **Trelstad, R. L., Ed.,** *The Role of the Extracellular Matrix in Development,* Alan R. Liss, New York, 1984.

96. **Reddi, A. H., Ed.,** *Extracellular Matrix: Structure and Function,* Alan R. Liss, New York, 1985.

Chapter 15

CELLULAR FACTORS RELATED TO CESSATION OF PROLIFERATION AND DIFFERENTIATION

Audrey L. Muggleton-Harris

TABLE OF CONTENTS

I. INTRODUCTION

To fully comprehend the cellular factors which influence and/or control proliferation and differentiation during the aging process, it is necessary to study the cells' interactions with one another and with their environment. Although the genes of the individual provide the "blueprint" for development, it is the environment in which the genes operate that influences cell behavior and, ultimately, senescence and the aging phenomenon. For instance, at one level the cell cycle is tightly regulated by growth factors, and the basic processes needed for cell replication depend on cytoskeletal proteins and membrane components, but these in turn relate to the extracellular matrix. Intercellular relationships depend on the extracellular matrix, which in turn modulates differentiation, growth, and cell behavior at the organ and whole-animal levels.

It is impossible to relate the *in vivo* aging process to cellular studies without taking this multiphase process into consideration. The isolation and study of cells *in vitro* from donors of known age can provide basic data on the cellular aging phenomenon, and an extensive body of work has capitalized on the human diploid fibroblast model system to study the cellular and molecular mechanisms associated with the cells' senescence and the aging process *in vitro*. Other contributors to this volume have covered these aspects in detail; I will refer to this system only where it has a bearing on the aspects I am discussing in this chapter.

There are specific questions which I would like to address. Do cell populations age *in vivo* as *in vitro*? Are there specific cellular stages of differentiation and development which are more susceptible to the effects of aging? Which cellular factors influence proliferation with regard to specific populations of cells in aging tissue?

To understand the nature of cell proliferation, specific cell markers and techniques are required to detect cells and their progeny. Cloning of cells *in vitro* from a defined source of tissue *in vivo* provides a suitable cellular system. However, the difficult aspect is to relate the aging parameters noted *in vitro* with what may be taking place *in vivo*. The "black box" situation is always in operation, and in many instances only an approximation can be made. Conversely, how is one to study stages in cell differentiation and development which occur *in vivo* at the cellular level without isolating the cells from the aging tissue for analysis? There are tissues and organs which lend themselves more readily to such studies than others. Where replication, growth, differentiation, and organogenesis occur *in vitro* in somatic cells derived from a defined tissue of known age and stage of development, a comparison of cell quiescence, senescence, and the aging phenomenon can then be made with similar cells *in vivo*.

The majority of cells studied *in vitro* that have been derived from somatic tissue have a well-defined population doubling level and growth pattern. The finite life span of normal human diploid cells was first described by Hayflick and Moorhead in 1961.[1] The spontaneous transformation of normal cells is a rare occurrence, and it has been stated that mouse cells have the unique property of almost always transforming *in vitro*.[2] Therefore, in general, cultured somatic tissues have a defined life span unless they have been transformed. The immortality of the germ line can be studied in comparison with embryonic cell lineages at a stage when the aging phenomenon is just beginning. I will attempt to discuss those factors which appear to regulate cell maturation and replication at these different levels, so that a judgment may be made as to whether those factors of importance in regulating cells during embryogenesis are also effective in the aging organism (Figure 1).

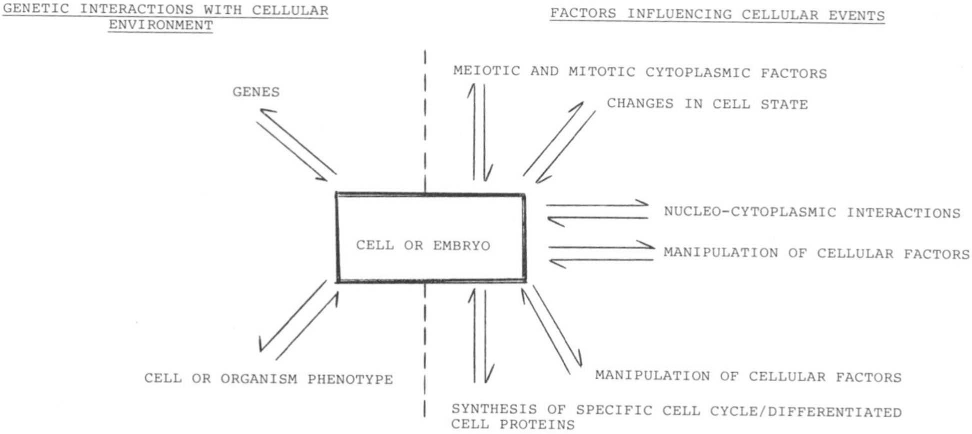

GENETIC INTERACTIONS WITH CELLULAR
ENVIRONMENT

FACTORS INFLUENCING CELLULAR EVENTS

GENES

MEIOTIC AND MITOTIC CYTOPLASMIC FACTORS

CHANGES IN CELL STATE

CELL OR EMBRYO

NUCLEO-CYTOPLASMIC INTERACTIONS

MANIPULATION OF CELLULAR FACTORS

CELL OR ORGANISM PHENOTYPE

MANIPULATION OF CELLULAR FACTORS

SYNTHESIS OF SPECIFIC CELL CYCLE/DIFFERENTIATED
CELL PROTEINS

FIGURE 1.   A diagram outlining the points discussed on the genetic and environmental factors influencing cellular events.

II. THE OOCYTE AND EMBRYO AS A SOURCE OF INFORMATION ON CELLULAR FACTORS RELATED TO CELL MATURATION AND PROLIFERATION

A recent paper has discussed the temporal and regional changes in DNA methylation during early mouse embryo development.[3] Stage- and tissue-specific global demethylation and remethylation occur during embryonic development. The egg genome is undermethylated and the sperm genome methylated. The authors propose that much of the methylation observed in somatic tissues acts to stabilize and reinforce prior events that regulate the activity of specific genes. Fetal germ cell DNA is markedly undermethylated, and it is postulated that the germ lineage is set aside before the occurrence of extensive methylation of DNA in fetal precursor cells. DNA modification mechanisms and gene activity during development may depend on the enzymatic modification of specific bases in repeated DNA sequences.[3] A continual interaction between cytoplasmic enzymes and DNA sequences could initiate developmental programs.[4] A clonal inheritance of the pattern of DNA methylation in mouse cells is based on a cytosine/guanine (CG)-specific methylase that operates on newly replicated hemimethylated DNA. Studies have shown that the DNA methylation pattern can remain fixed for more than 100 cell generations.[5]

A. MEIOTIC AND MITOTIC CYTOPLASMIC FACTORS
The factors which regulate the cleavage of the embryo are of interest because the mitotic promoting factors which are in operation may also act at later stages in development. The lack of such factors would play an important role in senescent and/or quiescent cells (Figures 2 and 3).

A simplistic but relevant model system which has supported the role of cytoplasmic factors controlling mitotic events is that of the two-cell block experienced by some strains of mouse embryos *in vitro*. A small amount of injected cytoplasm from a nonblocking strain is sufficient to initiate a resumption of normal cleavage and development in the blocked embryo.[5] Recently we have shown that these factors are cell cycle specific and are at their most effective in the G_1 and late G_2 stages of the cell cycle.[6] We have noted that the organization and distribution of mitochondria are different in the blocked and nonblocked embryos; however, because of the physiological parameters of both donor cytoplasm and recipient host, we know that the mitochondria are not the active components in the cytoplasm,

Maturation of oocyte

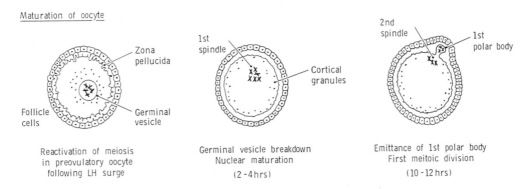

Reactivation of meiosis
in preovulatory oocyte
following LH surge

Germinal vesicle breakdown
Nuclear maturation
(2-4hrs)

Emittance of 1st polar body
First meiotic division
(10-12hrs)

FIGURE 2. An outline of the events leading to oocyte maturation beyond the second meiotic metaphase. The oocyte may remain arrested in meiosis for an extended period of time prior to activation or fertilization.

Following Fertilisation

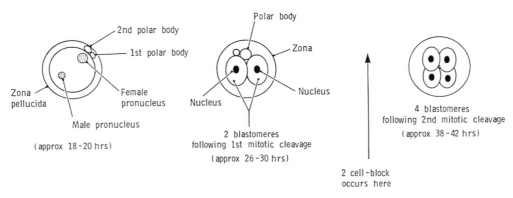

(approx 18-20 hrs)

2 blastomeres
following 1st mitotic cleavage
(approx 26-30 hrs)

4 blastomeres
following 2nd mitotic cleavage
(approx 38-42 hrs)

2 cell-block
occurs here

FIGURE 3. Once the oocyte has been fertilized, the egg proceeds to the first cleavage stage. The *in vitro* two-cell block occurs in the mouse embryo of certain strains and prevents if from undergoing a second cleavage to a four-blastomere embryo.

which rescues the cells. Recent experiments have indicated that when the active factors in the cytoplasm are injected into the blocked two-cell embryo, a redistribution of the mitochondria pattern occurs, and a more homogeneous state similar to nonblocked two-cell embryos is seen.[6a] Mitochondrial reorganization during resumption of arrested meiosis in the mouse oocyte has been noted by Van Blerkom and Runner;[7] the authors state that this localization of mitochondria could be necessary for restricted activities of ATP in the oocyte. With regard to the mitochondria themselves, a recent review[8] has linked fungal senescence to the mitochondrial system. The authors suggest that, in the light of this work, Orgel's error catastrophy theory seems applicable to fungal senescence, and that because of its high mutation rate mitochondrial DNA would show signs of senescence earlier than nuclear DNA.

The relevance of factors which control oocyte maturation[9,10] may make a significant contribution to our understanding of the regulation of cells during the inhibited, quiescent, and/or senescent stages of replication. The activity of cytoplasmic factors and of the surrounding environment on the arrested oocyte allows maturation beyond the second meiotic metaphase. The oocyte may remain arrested in meiosis for an extended period until activated a short while before ovulation occurs. If the oocytes are isolated from the follicles and placed in hormone-free media, they resume meiosis.[11] Oocytes remaining in contact with follicular fluid and granulosa cells continue to be arrested.[12-15] The factors which maintain the oocyte in this arrested state are possibly produced by the granulosa cells. A number of studies have suggested that cyclic adenosine $3',5'$-monophosphate (cAMP) is responsible for maintaining the oocyte in its suspended state.[16-19] A significant decrease in oocyte cAMP occurs just before meiosis resumes.[20,21] Activators of adenylate cyclase and inhibitors of phosphodiesterase lead to an increase in cAMP in the oocyte.[10,22,23] By measuring the level of cAMP in oocytes during meiotic arrest and following the resumption of meiosis, it has been shown that a fall in the cAMP level preceeds germinal vesicle breakdown, which coincides with a rise in the level of cAMP in follicular fluid and cumulus cells. Nuclear maturation in mouse oocytes is inhibited by cAMP.[24] A factor in porcine follicular fluid which blocks mouse oocyte maturation *in vitro* when combined with cAMP[14] has been identified as hypoxanthine.[15] The impact of these factors derived from the follicular fluid may in turn be influenced by hormones which regulate follicular development *in vivo*.[25,26] The rate of *in vitro* maturation is affected by hormones and culture conditions. Follicle-stimulating hormone (FSH) can induce oocytes to resume meiosis *in vitro*, and the sequence of events is the same as it is *in vivo*.[27]

Cytoplasmic factors of inhibited and active oocytes have been previously identified; one is the maturating-promoting factor (MPF) and another is a cytostatic factor (CSF).[9,28-30] Cytoplasmic factors with activities similar to MPF are also present in mitotic mammalian cells; MPF from oocytes and somatic cells is interchangeable and active between species.[31-34] For instance, cytoplasm from maturing mouse oocytes was injected into starfish oocytes, inducing germinal vesicle breakdown.[35] Also, when hybrid cells from immature and maturing oocytes are formed they will proceed to the first metaphase.[36-38] An important role in the presence and effectiveness of cytoplasmic factors is played by cell cycle dynamics. Protein synthesis is required for the initial appearance of MPF but not for its disappearance, and serial cytoplasmic transfers suggest that it is self-amplifying.[40,41] Partially purified MPF has its highest activity when extracted in the presence of phosphatase inhibitors and injected in combination with ATP.[42] The ability of MPF to induce a transition from G_2 to M of the cell cycle is not species specific,[35] and thus MPF may have a role in regulating the cessation of proliferation as well as allowing the transition from one stage of the cell cycle to another.

The nuclear reconstitution *in vitro* and the disassembly of the nucleus in cell-free systems have permitted the dissection and identification of mitotic factors.[43] The disassembly extracts are active on nuclei from a variety of vertebrate cell types, implying that the primary interactions between disassembly factors and nuclei involve proteins rather than DNA and that those interactions in turn regulate the behavior of the nucleic acid.[44]

B. EMBRYONIC CELL LINES

Normal, developing embryonic cells undergo a succession of transitions, and most cells in the adult are committed to express a unique terminal phenotype, although stem cells remain multipotential. To study factors that control the replication and differentiation of different cell lineages at the embryonic level, it is necessary to isolate and follow specific cells and, once having identified those factors which regulate cell behavior, to be able to manipulate them.

Teratomas are tumors composed of various tissues which, although foreign to their site of origin, are composed of well-differentiated somatic cells which have limited growth potential. These benign growths are easily generated in several animal species by transplanting embryos, or parts of embryos, to extrauterine sites.[45] However, malignant teratomas (teratocarcinomas) contain cells which are undifferentiated and malignant. In mice, these can only be produced from mouse embryos which are transplanted to extrauterine sites. The production and characteristics of teratocarcinoma cells have been described in detail.[46]

An analysis of the factors influencing the development of malignant tumors from embryos delineates two categories: embryo-related and host-related. By experimentally manipulating these factors, it has been shown that the strain of mouse used as the recipient is the critical host factor; there are a few nonpermissive inbred strains, such as C57BL/6 and AKR, which are resistant. The incidence of teratocarcinoma can be increased when histocompatible mice are immunized with teratocarcinoma-derived cells. Thus, it appears that a normal or specifically stimulated immune response is beneficial for development of embryo-derived teratocarcinomas. Non-H-2 alloantigens are detected on preimplantation mouse embryos, and maternal non-H-2 alloantigens are detectable at all stages (from a two-cell to a $4\frac{1}{2}$-d blastocyst), whereas paternal antigens first become evident at the six- to eight-cell stage. No convincing evidence of the presence of alloantigens associated with the H-2 haplotype has been found on the preimplantation embryo,[47] as neither murine embryonal carcinoma (EC) cells nor early embryos express major histocompatibility antigens (H-2).[46] The embryo-related factor which has emerged as the most important is the age of the embryo; only embryos of 7 d or less (that is, before the formation of the neural fold) are able to grow into teratocarcinomas.

One assay of the factors which regulate the cells' replication and tumorigenicity properties can be made when EC cells from *in vitro* cell lines or from *in vivo* tumors are injected into early mouse embryos during the blastocyst stage. These embryos are returned to foster mothers and allowed to develop. The participation of the cells in the tissue of the resultant offspring can then be followed using genetic markers such as pigment or enzyme differences. A table compiled by Papaioannou and Rossnat in 1983[48] shows that these cells can contribute to normal differentiated tissues in the experimental mice. It is interesting to note that contributions to the germ line were rare.

Stem cell lines derived from early embryonic material provide a source of cells which are able to spontaneously differentiate *in vitro* or following manipulation of the culture conditions. Multipotential stem cells may be isolated and maintained as cell lines from mouse teratocarcinomas. Numerous studies have been undertaken on such cells to study the mechanisms by which chemical inducers promote differentiation. Retinoic acid (RA) appears to be the most potent and reliable chemical for inducing differentiation of cells from all mutant EC lines tested.[49] The uptake of RA by the cell, its interaction with a cellular RA-binding protein (CRABP) in the cytoplasm, translocation of the complex into the nucleus, and direct intervention with the genetic program would lead to phenotypic alteration by interaction with the chromatin.[50] With regard to the extrapolation of the results with EC cell lines *in vitro* to the *in vivo* situation, this must be carried out with the knowledge that the EC cell lines are obtained in the first instance by selecting for cells from tumors or embryos that can proliferate rapidly. Such cells may have lost the ability to respond to differentiation signals that the cells possessed prior to establishment of the cell line.

There are mouse EC cell lines which have stable karyotypes despite their rapid proliferation (8 to 12 h) pattern. Studies *in vivo* and *in vitro* strongly suggest that the developmental sequence of events which occurs in differentiating populations of EC cells resembles the analogous chain of events occurring in the developing embryo. Morphological and biochemical properties of embryonic tissues provide the means to identify the types of cells present. If a variety of cellular properties are used, a particular cell lineage can be identified, providing that both morphological and biochemical markers are employed. *In situ* staining procedures with monospecific antibodies are particularly useful; antisera to cytoskeletal proteins have been used extensively on such cells. These monoclonal antibodies are commercially available and can identify a particular cell type as being derived from neurofilament, epithelial, muscle, or mesoderm sources. Details on the methods and markers used with these embryonic stem cell lines are described by Rudnicki and McBurney.[51] Mouse mutants have been used to provide models for human diseases and abnormalities, as embryonic stem cell lines (ES) can be injected into blastocysts to create mutants for particular genetic-related disorders. One such approach to produce hypoxanthine phosphoribosyl transferase (HPRT)-deficient mice has recently been successful in our unit.[52] Therefore, directed mutation in the mouse is possible, and the identification and/or creation of mouse models which parallel human disorders and form a significant part of the human aging process could be considered for future studies.

Teratocarcinoma cells can be used both as a model to understand growth control mechanisms in early development and as a source of factors which can delineate subpopulations of cells in well-characterized cell lines derived from the embryonal carcinoma. The initial derivation methods involve the use of feeder layers, although the culturing of EC cells in serum-free medium on gelatinized tissue culture dishes is also possible.[53] Identification and purification of an embryonal carcinoma-derived growth factor (ECDGF) which can regulate the growth of heterologous cell types in culture have shown it to be a powerful mitogen of quiescent cells.[54,55] It would appear from the facts presented that genetic predisposition as well as epigenetic factors can modify developmental processes; the same may hold true for the processes of aging.

III. THE LENS, A MODEL SYSTEM TO STUDY AGING OF CELLS AND TISSUE

Embryonic and adult studies of a unique nature can be undertaken on the lens. Because the developing and adult lenses are enclosed by collagenous capsules, none of the cells are lost throughout the life of the organism. Thus, one has available sufficient cells and tissue of a known age, cell cycle, and stage of differentiation. Also, the aging cell populations of the lens are not exposed to fluctuating and unknown growth factors in the blood supply due to their being avascular. A number of reviews and papers have described the cellular, embryonic, adult, and aging features of the lens. Harding and Dilley discussed the structural proteins of the mammalian lens with relation to changes in development, aging, and cataract.[56] Lens proteins can be used as markers of terminal differentiation, and the molecular aspects of lens differentiation have been well researched.[57] Cell divisions are precisely controlled in specific populations of the developing and adult lens. Differential gene expression is clearly and concisely observed with specific stages of differentiation, and terminal differentiation of lens fiber cells occurs.

Congenital and early developmental cataracts are common ocular abnormalities; 10 to 38% of all blindness in children is caused by such cataracts. Mouse mutants suitable for studies of these cataracts are discussed in a recent review.[58] The aging cataract, which is a major cause of blindness in the majority of individuals over the age of 60, is normally identified as a subcapsular cataract. The essential changes that occur in these ocular abnormalities are[59-61]

1. The proliferation of the central epithelial cells, which have been quiescent throughout the period following birth (G_1 cell cycle)
2. The wrinkling of the collagenous capsule surrounding the lens
3. The multilayering of the proliferating central lens cells
4. The migration and invagination of the abnormally replicating cells into the fiber structure

The basement membrane of the capsule surrounding the lens is composed of collagen, glycoproteins, and proteoglycans.[62] The extracellular matrix glycoprotein concentration appears to influence early lens differentiation *in vivo*.[63] Dramatic changes in growth, morphology, cell interactions, and gene expression occur during development of the embryonic lens. Specific lens proteins, the alpha and beta crystallins, are synthesized by the central

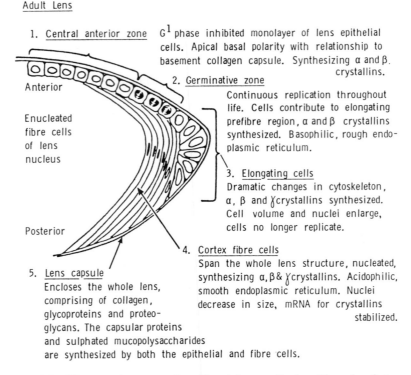

Adult Lens

1. Central anterior zone G^1 phase inhibited monolayer of lens epithelial
cells. Apical basal polarity with relationship to
basement collagen capsule. Synthesizing α and β.
crystallins.

Anterior

2. Germinative zone
Continuous replication throughout
life. Cells contribute to elongating
Enucleated   prefibre region, α and β crystallins
fibre cells   synthesized. Basophilic, rough endo-
of lens   plasmic reticulum.
nucleus

3. Elongating cells
Dramatic changes in cytoskeleton,
α, β and ɣcrystallins synthesized.
Cell volume and nuclei enlarge,
Posterior   cells no longer replicate.

4. Cortex fibre cells
Span the whole lens structure, nucleated,
5. Lens capsule   synthesizing α,β& ɣcrystallins. Acidophilic,
Encloses the whole lens,   smooth endoplasmic reticulum. Nuclei
comprising of collagen,   decrease in size, mRNA for crystallins
glycoproteins and proteo-   stabilized.
glycans. The capsular proteins
and sulphated mucopolysaccharides
are synthesized by both the epithelial and fibre cells.

FIGURE 4.   Diagrammatic representation of the adult mammalian lens. The various features
of cell proliferation, differentiation, and morphology shown underscore the many advantages
of this system.

lens epithelial cells subjacent to the collagen capsule. The fully differentiated and morphologically distinct elongated fiber cells synthesize both of these crystallins as well as the
gamma crystallins.[64-67] The components of the collagen lens capsule are synthesized by both
the epithelial and fiber cells (Figure 4).[68]

The monolayer of lens epithelial cells that is subjacent to the lens capsule has a dorsal-
apical polarity, and after birth the cells are inhibited in the G_1 phase of the cell cycle.
However, the cells retain the ability to replicate and produce more lens epithelial cells if
stimulated by the culture of the whole lens, mass cultures, or individual cells.[69-74]

Cloned murine lens epithelial cells have a finite life span and retain a diploid set of
chromosomes.[75-77] Naturally occurring neoplastic tumors of the lens are not known. However,
the congenital and aging cataractous cells are transformed from a tightly well-regulated
functional population in the normal lens (Figures 5 and 8) to an abnormal pathological state,
examples of which are shown in Figures 6 and 7 and have been described in detail in a
number of publications.[76,78-82]

The morphological and cellular changes that occur with replication, differentiation,
development, and senescence of lens cells *in vitro* are related to the presence and association
of the lens capsule with the cells.[77] The role of the substratum in regulating the morphology
and behavior of lens cells is obviously very important.[74,83-85] The behavior of the replicating
lens cells when they are no longer anchored to their substrata is very similar to a transformed
cell phenotype.[77,85] Multiple layers of the replicating epithelial cells are found in the anterior
of the cataractous lens, and in many instances the cells "invade" and disturb the structure
of the lens fibers. We have noted that the basement membrane of the posterior of the lens
thickens and in many cases breaks down.[76] The cells found in the posterior area have a
morphology similar to those cultured in hydrated collagen gels.[77]

FIGURE 5. Section through a 15-d gestation, noncataractous (DBA/2) mouse lens. The germinative zone (gz) nuclei form a regular pattern. The lens fibers (f) run the length of the lens. Bar graph = 50 μm.

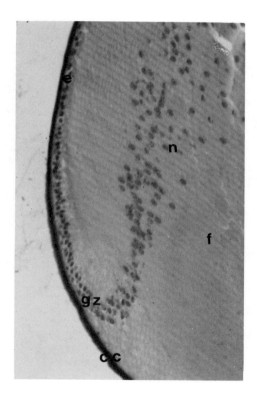

FIGURE 8. A high-power magnification of a section through the adult noncataractous (DBA/2) mouse lens, showing the tightly regulated lens fiber cells (f) which run the full length of the lens. The nuclei (n) form the normal mitotic bow pattern. The epithelial cells (e) are in a monolayer, except at the germinative zone (gz) where they are differentiating into fiber cells. The collagen capsule (cc) encapsulates the lens.

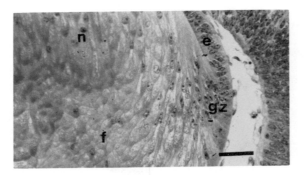

FIGURE 6.   Section through a 15-d gestation, congenital cataractous (CAT) mouse lens. The lens fibers (f) are swollen and
club-shaped. In the interior area of the cortex fibers there are
rounded cells. The nuclei (n) of the fiber cells are becoming
pyknotic; gz = germinative zone. Bar graph = 50 μm.

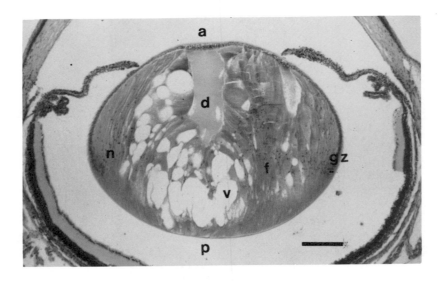

FIGURE 7.   Section through an adult congenital cataractous lens. The lens epithelial cells
in the anterior (a) of the lens have replicated and formed a multilayer. The epithelial cells
closer to the germinative zone (gz) area remain in a monolayer. The inner cortex fibers (f)
are distressed and have large vacuoles (v) between them. The area within the center of the
lens lacks any fiber structure, and the fibers have degenerated (d). The normal pattern of the
nuclei of the lens fibers (n) is disturbed. Bar graph = 240 μm.

The congenital cataractous mouse (CAT), for whom the gene(s) responsible for the
cataractous genotype have recently been identified,[80] and a noncataractous strain (DBA/2)
have provided the material for an informative developmental study.[86] Early embryos were
aggregated from both strains to form chimeras, and a rescue from the genetically dominant
inherited lens abnormalities was achieved in the majority of adult mice. The manner in
which the regulation of the abnormally replicating lens epithelial cells took place has not
been determined; however, the presence of a high proportion of the cataractous lens cell
genotype in the "rescued" lens suggests that it is at the cellular level. An exchange of
cellular components and information would have to have taken place by intercellular communication. Previously we found that regulation of cultured mouse lens epithelial cell rep-

munication. Previously we found that regulation of cultured mouse lens epithelial cell replication depends on the somatic cell hybridization of the two cell types.[74] The chimeric lens consists of separate cells of both genotypes with no *in vivo* hybridization taking place. The receptors on the cell surface of the extracellular matrix may play an important role in regulating the chimeras' lens cells during embryogenesis. It has been shown that gap-junctional communication regulates and coordinates cellular growth.[87-90] Preferential coupling may have taken place in the cells of the chimeric lens and regulated the cells' response to growth factors.[91,92]

The ability to culture lens cells from the epithelial layer of mice of different ages and the availability of cells which can be readily dissected from the lens provide a system which meets the criteria for the relevance of the *in vitro* cellular studies with the *in vivo* situation. Also, "aged" cells (either due to donor age or to remaining in a nonproliferative phase for extended periods, in many instances from birth until death) provide defined cell populations for the study of the cellular and molecular factors concerned with, or responsible for, cessation of replication, senescence, and aging.

There are specific proteins associated with lens cell populations, and studies on the lens proteins of the developing, aging, and cataractous lens have shown that there are changes in their expression.[65,67,93-99] The lens crystallins and their gene families have been reviewed by Piatigorsky,[100] and a number of interesting molecular studies on mRNA levels and selective loss of gene transcripts of the crystallins have provided a quantitative basis for exploring the differential expression of these proteins. Assignment of chromosomal location for the families of lens crystallins has shown them to be dispersed and not clustered on one chromosome.[101] In contrast, the gene locations for specific lens abnormalities appear to be related to specific chromosomes; those associated with congenital cataracts in mice have been reviewed recently,[58] and the Cat^{Fr} mouse congenital cataract has now been assigned to chromosome 10.[80] Gene transfer systems may permit the study of developmentally regulated genes, e.g., crystallins. If specific genes (or nuclei) are introduced into totipotential or multipotential embryonic cells, the genes may be subject to normal regulatory events. The availability of pure DNA probes to the crystallins will permit study of gene activation, inactivation, and regulatory phenomena. Microinjection of foreign genes into the germ line of mice, with site-specific recombination to specific chromosomes associated with a known pathogenic state (e.g., congenital cataract/chromosome 10), may be the way to study gene regulation at the embryonic level.[102] Such technology could be used for other chromosome-related disabilities associated with the aging phenomenon.

Do cell populations age *in vitro* as they do *in vivo*? If one takes the specific synthesis of lens proteins associated with known stages of differentiation as a marker, the development of a "lentoid body" *in vitro* from cultured mouse lens cells is a multiphase process involving cell replication, synthesis of mucosubstances, and a basement collagen membrane. The synthesis of lens proteins confirms that the genes regulating normal differentiation *in vivo* are operating in the *in vitro* system. The behavior of different subpopulations of the cultured lens cells is similar to that observed *in vivo*. When replicating cells cease dividing, they either differentiate to a further stage in their committed pathway or they remain quiescent and with time phase out of culture. When encapsulated by a collagen capsule which is synthesized by the lens cells *in vitro*, a "lentoid body" is formed. The terminally differentiating lens cells *in vitro* elongate and synthesize the gamma crystallins, apparently responding to a positional cue from the collagen capsule. There are subpopulations of the cell monolayer that proliferate, then remain quiescent for extended periods of time before undergoing obvious alterations in nucleocytoplasmic ratio, accumulation of lysosomal vacuoles, etc. (senescence). These cells then gradually phase out of culture in a manner similar to other cultured mamalian cells which have a finite life span.[77]

Recently we have undertaken a collaborative study using Eugenia Wang's monoclonal mouse antibody, statin. This probe was found to stain nuclear proteins of nonproliferating cells in senescent human fibroblast cultures.[103] When statin was used on the various stages of proliferation and differentiation of cultured mouse lens epithelial cells, it readily detected those cells which were no longer replicating. More importantly for these studies, it also identified those cells of the population which were terminally differentiating.[104] The availability of such probes to specific proteins associated with nonproliferation and senescence is exciting, especially when one relates it to a protein called cyclin A. This protein induced entry into M phase and the resumption of meiosis in *Xenopus* oocytes; the rise in cyclin A played a direct role in driving cells into M phase.[105] Perhaps statin also acts in a positive manner in regulating cessation of proliferation and terminal differentiation of cells.

IV. DISCUSSION

My interest in the nucleocytoplasmic factors which control or influence cell replication began in the 1950s when I was working with J. F. Danielli, FRS, and it has continued to this present time. A review on this subject discussing a number of very diverse systems was presented in 1979;[106] special emphasis was placed on work undertaken on the reassembly of cellular components for the study of aging and finite life span. In the concluding remarks I suggested that the cytoplasmic influence on genetic events was obviously of importance, even when considering the clones of cells which make up the developing embryo. Despite the diversity of systems and cells used, ranging from the amoeba to the human fibroblast, the conclusion was that neither the nucleus nor an inherited genetic event alone could be responsible for the aging phenomenon.

In 1976, Leonard Hayflick and I published a paper which reported that when individual reconstituted cells from components (cytoplast and karyoplast) of the human diploid fibroblast were made, they were viable and capable of replicating. Clonal analysis of these cells allowed the role of the nucleus and cytoplasm in the control of replication to be examined by transferring an "aged" nucleus (karyoplast) into a "young" cytoplasm and a "young" nucleus into an "aged" cytoplasm. This study showed that the "aged" karyoplast played a dominant role with regard to the number of proliferations that the cell underwent. However, the cytoplasm played an equally important role in the maintenance of the nonreplicative phase of the cell cycle.[107] When individual mononucleate senescent phase III diploid cells received an additional nucleus from a young diploid WI-38 fibroblast, or a transformed CL-1 fibroblast or HeLa cell, a nucleus from the replicating cell stimulated the quiescent/senescent cell to undergo further mitosis. We observed that the recipient fibroblast cell retained its typical morphology and that the finite life span of the fibroblast cell predominated in the majority (98%) of the experimental cells despite transformed and/or malignant cells' genes being present.[108] Less than 2% of the experimental hybrids derived from the fusion of phase II diploid fibroblasts with transformed fibroblasts were capable of sustained replication beyond that of a phase II fibroblast nucleus fused with a similar senescent cell. In the case of the minority of cells which did transform from the finite life span state, we noted that they also changed their fibroblast-like morphology.[109] The conclusion from these experiments was that the replication of cells could be reprogrammed beyond the normal life span expectancy, but that replication in the majority of cells remained a controlled event. We suggested that either inhibitors in the senescent fibroblast cytoplasm influence the transformed cell's genes or that the transformed nucleus needs specific activating factors which were in the transformed cell's cytoplasm.

Much of the present work on oncogenes requires an understanding of the factors and/or circumstances that control proliferation. It is known that the action of many oncogenic proteins circumvents normal regulatory processes of proliferation, but a full understanding of the mechanics of the process is lacking. Proto-oncogenes in normal cells appear capable of inducing some of the phenotypic changes characteristic of transformed cells. Most, if not all, genetically altered proto-oncogenes contribute to some extent to the multistage development of neoplasia. However, it does appear that for such an oncogene (for example, *ras*) to play a definitive role in tumor progression, it must be activated in cells that have already acquired continuous proliferating properties.[110] Nuclear microinjection of C-H-*ras* DNA induced global DNA synthesis in quiescent human diploid fibroblasts. The proto-oncogene and oncogene forms were equally effective inducers. In contrast, C-H-*ras* DNA, either alone or in combination with the adenovirus EIA (pAdl2E1A) gene, did not cause senescent cells to synthesize DNA.[111] The authors propose that there is an inhibitor present in the senescent cells which controls the action of the oncogene. Perhaps this inhibitor is similar, if not identical, to cytoplasmic inhibitors which were suggested to be present in the cells used in previous studies.[109] It will be of interest to follow the progress toward the specific identification of the proposed inhibitors and a correlation between DNA synthesis and cell replication.

Genetic differences in onco-virus protein expression are evident during embryogenesis as well as in adult animals, and the expression of the protein may be influenced by the state of differentiation of the cell. Scott and Maercklein[112] propose that an initiator of carcinogenesis selectively inhibits stem cell differentiation and that the process involves multiple phases; the development of defects in the control of differentiation and proliferation is mediated at distinct cell cycle stages.

To return to the specific questions posed earlier in this chapter: do cell populations age *in vitro* as *in vivo*? If one takes the lens system, for example, the answer is an unequivical yes. Are there specific cellular stages of differentiation and development susceptible to the effects of age? From the evidence presented, it would appear that the cell cycle stage and the period for which a cell remains quiescent possibly increases the amount, stability, and/or effectivity of intracellular factors in the meiotic oocyte, senescent/quiescent fibroblast, or terminally differentiating lens cell. Which cellular factors influence proliferation with regard to specific populations of cells in aging tissue? Because of the data discussed with regard to the oocyte, embryo, and lens, I would hesitate to identify any factors which act only on aging tissues. However, both the importance of the cells' relationship to one another and the influence of the extracellular environment on proliferation are evident. For example, intracellular Ca^{2+} changes with cell state in both the maturing oocyte and the differentiated somatic cell; extracellular Ca^{2+} promotes cartilage differentiation,[113] and since intracellular Ca^{2+} and cAMP levels are associated, it would appear that cell-to-cell contact is an important factor to consider in aging tissue. A metabolic component could influence or control an age-related functional deficiency. The ATP turnover with increasing population doubling level *in vitro* and the increasing *in vivo* age of human diploid fibroblasts have been shown to decrease.[114] The same age-matched fibroblasts were shown to have a decreased response to stimulus.[115] Earlier examples given with regard to the oocyte also underline the possibility of a compromised energetic state in inhibited cells.[5,6,28]

A discussion of the factors which regulate cells during maturation, embryogenesis, development, differentiation, senescence, and aging has been presented. It will hopefully underscore how important it is not to confine the study of aging to one specific aging system but to view the phenomena of senescence and age as extensions of the developing organism.

REFERENCES

1. **Hayflick, L. and Moorhead, P. S.,** The serial cultivation of human diploid embryonic cell strains, *Exp. Cell Res.*, 25, 585, 1981.
2. **Hayflick, L.,** The cellular basis for biological ageing, in *Handbook of the Biology of Ageing,* Finch, C. and Hayflick, L., Eds., Van Nostrand Reinhold, New York, 1977.

3. **Monk, M., Bonbelik, M., and Lehnert, S.,** Temporal and regional changes in DNA methylation in the embryonic, extraembryonic and germ cell lineages during mouse embryonic development, *Development,* 99, 371, 1987.

4. **Stein, R., Gruerbaum, Y., Polack, Y., Razin, A., and Cedar, H.,** Clonal inheritance of the pattern of DNA methylation in mouse cells, *Proc. Natl. Acad. Sci. U.S.A.,* 799, 61, 1982.

5. **Muggleton-Harris, A. L., Whittingham, D. G., and Wilson, L.,** Cytoplasmic control of preimplantation development in-vitro in the mouse, *Nature,* 299, 460, 1982.

6. **Pratt, H. P. M. and Muggleton-Harris, A. L.,** Cycling cytoplasm factors that promote mitosis in the cultured 2-cell mouse embryo, *Development,* Vol. 104, 1988.

6a. **Muggleton-Harris, A. L. and Brown, J. J. G.,** Cytoplasmic factors influence mitochondrial reorganization and resumption of clearage during culture of early mouse embryos, *Hum. Reprod.,* 3, 1020, 1988.

7. **Van Blerkom, J. and Runner, M. V.,** Mitochondrial reorganisation during resumption of arrested meiosis in the mouse oocyte, *Am. J. Anat.,* 171, 335, 1984.

8. **Benne, R. and Tabak, H. F.,** Senescence comes of age, *Trends Genet.,* June, 147, 1986.

9. **Masui, Y. and Clarke, H.,** Oocyte maturation, *Int. Rev. Cytol.,* 57, 185, 1979.

10. **Sato, E. and Kiode, S. S.,** Biochemical transmitters regulating the arrest and resumption of meiosis in oocytes, *Int. Rev. Cytol.,* 106, 1, 1987.

11. **Edwards, R. G.,** Maturation in-vitro of mouse, sheep, cow, pig, rhesus monkey and human ovarian oocytes, *Nature,* 208, 349, 1965.

12. **Tsafriri, A.,** in *Ovarian Follicular and Corpus Luteum Function,* Channing, C. D., Marsh, J., and Sadler, W. A., Eds., Plenum Press, New York, 1978, 269.

13. **Stone, S. L., Pomerantz, S. H., Schwartz-Kripner, A., and Channing, C. P.,** Inhibitor of oocyte maturation from Porcine follicular fluid: further purification and evidence for reversible action, *Biol. Reprod.,* 19, 585, 1978.

14. **Eppig, J. J. and Downs, S. M.,** Chemical signal that regulates mammalian oocyte maturation, *Biol. Reprod.,* 30, 1, 1984.

15. **Downs, S. M., Coleman, D. L., Ward-Bailey, P. F., and Eppig, J. J.,** Hypoxanthine is the principle inhibitor of murine oocyte maturation in a low molecular weight fraction of porcine fluid, *Proc. Natl. Acad. Sci. U.S.A.,* 82, 454, 1985.

16. **Cho, W. K., Stern, S., and Biggers, J. D.,** Inhibitory effect of dibutyryl cAMP on mouse oocyte maturation in-vitro, *J. Exp. Zool.,* 187, 383, 1974.

17. **Hillensjö, T., Channing, C. P., Pomerantz, S. H., and Schwartz-Kripner, A.,** Intrafollicular control of oocyte maturation in the pig, *In Vitro,* 15, 32, 1979.

18. **Dekel, N. and Beers, W. H.,** Development of the rat oocyte in-vitro. Inhibition and induction of maturation in the presence or absence of the cumulus oophorus, *Dev. Biol.,* 75, 247, 1980.

19. **Gilula, N. B., Epstein, M. L., and Beers, W. H.,** Cell to cell communication. A study of the cumulus-oocyte complex, *J. Cell Biol.,* 78, 58, 1978.

20. **Schuetz, R. M., Montgomery, R. R., and Belanoff, J. R.,** Regulation of mouse oocyte meiotic maturation. Implication of a decrease in oocyte cAMP and protein dephosphorylation in commitment to resume meiosis, *Dev. Biol.,* 97, 264, 1983.

21. **Schuetz, R. M., Montgomery, R. R., Ward-Bailey, P. F., and Eppig, J. J.,** Regulation of oocyte maturation in the mouse: possible roles of intercellular communication, cAMP and testosterone, *Dev. Biol.,* 95, 294, 1983.

22. **Hubbard, C. J. and Terranova, P. F.,** Inhibitory action of cyclic guanosine 5^1-phosphoric acid (GMP) on oocyte maturation: dependence on an intact cumulus, *Biol. Reprod.,* 26, 628, 1982.

23. **Powers, R. D. and Paleos, G. A.,** Combined effects of calcium and dibutyryl cyclic AMP on germinal vesicle breakdown in the mouse oocyte, *J. Reprod. Fertil.,* 66, 1, 1982.

24. **Wasserman, W. J., Richter, J. D., and Smith, L. D.,** Protein synthesis and maturation-promoting factor during progesterone-induced maturation in *Xenopus* oocytes, *Dev. Biol.,* 89, 152, 1982.

25. **Bar-Ami, S., Nimrod, A., Brodie, A. M. H., and Tsafriri, A.,** Role of FSH and oestradiol β-17 in the development of meiotic competence in rat oocytes, *J. Steroid Biochem.,* 19, 965, 1983.

26. **Tsafriri, A., Dekel, N., and Bar-Ami, S.,** The role of oocyte maturation inhibitor in follicular regulation of oocyte maturation, *J. Reprod. Fertil.,* 64, 641, 1982.

27. **Salustri, A. and Siracusa, G.,** Metabolic coupling, cumulus expansion and meiotic resumption in mouse cumuli oophori cultures in-vitro in the presence of FSH or dc AMP, or stimulated in-vivo by hCG, *J. Reprod. Fertil.,* 68, 335, 1983.

28. **Sorenson, R. A., Cyert, M. S., and Pederson, R. A.,** Active maturation-promoting factor is present in mature mouse oocytes, *J. Cell Biol.,* 100, 1637, 1985.

29. **Matler, J. L. and Krebs, E. G.,** Regulation of amphibian oocyte maturation, *Curr. Top. Cell. Regul.,* 16, 271, 1980.

30. **Sorensen, R. A. and Wasserman, M. P.,** Relationship between growth and meiotic maturation-promoting factor is present in mature mouse oocytes, *Dev. Biol.,* 50, 531, 1971.

31. **Sunkaru, P. S., Wright, D. A., and Rao, P. N.,** Mitotic factors from mammalian cells induce germinal vesicle breakdown and chromosome condensation in amphibian oocytes, *Proc. Natl. Acad. Sci. U.S.A.,* 76, 2799, 1979.

32. **Nelkin, B., Nichols, C., and Vogelstein, B.,** Protein factor(s) from mitotic CHO cells induce meiotic maturation in *Xenopus laevis* oocytes, *FEBS Lett.,* 109, 223, 1980.

33. **Kishomoto, T. I., Kuriyama, R., Konda, H., and Kanatani, H.,** Generality of the action of various maturation-promoting factors, *Exp. Cell Res.,* 137, 121, 1982.

34. **Miake-Lye, R. and Kirschner, M. W.,** Induction of early mitotic events in a cell-free system, *Cell,* 41, 165, 1985.

35. **Kishomoto, T., Yamazaki, K., Sato, Y., Kato, S., Koide, S., and Kantani, H.,** Induction of starfish oocyte maturation by maturation-promoting factor of mouse and surf clam oocytes, *J. Exp. Zool.,* 231, 293, 1984.

36. **Balakier, H.,** Induction of maturation in small oocytes from sexually immature mice by fusion with meiotic or mitotic cells, *Exp. Cell Res.,* 112, 137, 1978.

37. **Fulka, J., Jr.,** Maturation-inhibiting activity in growing mouse oocytes, *Cell Differ.,* 17, 45, 1985.

38. **Fulka, J., Jr.,** Nuclear maturation in pig and rabbit oocytes after interspecific fusion, *Exp. Cell Res.,* 146, 212, 1983.

39. **Gerhart, J. C., Wu, M., and Kirschner, M.,** Cell cycle dynamics of an M-phase cytoplasmic factor in *Xenopus laevis* oocyte and eggs, *J. Cell Biol.,* 98, 1247, 1984.

40. **Reynhart, J. K. and Smith, L. D.,** Studies on the appearance and nature of a maturation-promoting factor in the cytoplasm of amphibian oocytes exposed to progesterone, *Dev. Biol.,* 38, 394, 1974.

41. **Richter, J. D. and Smith, L. D.,** Reversible inhibition of *Xenopus* oocyte specific proteins, *Nature,* 309, 378, 1984.

42. **Wu, M. and Gearhart, J. C.,** Partial purification and characterisation of the maturation-promoting factor from eggs of *Xenopus laevis, Dev. Biol.,* 79, 465, 1980.

43. **Newport, J. and Spann, T.,** Nuclear reconstitution in-vitro: stages of assembly around protein-free DNA, *Cell,* 48(2), 205, 1987.

44. **Fisher, P. A.,** Disassembly and reassembly of nuclei in cell free systems, *Cell,* 48(2), 175, 1987.

45. **Solter, D., Skreb, N., and Damjanov, I.,** Extrauterine growth of mouse egg cylinders results in malignant teratoma, *Nature,* 227, 503, 1970.

46. **Damjanov, I., Damjanov, A., and Solter, D.,** Production of teratocarcinomas from embryos transplanted to extra-uterine sites, in *Teratocarcinomas and Embryonic Stem Cells, a Practical Approach,* Robertson, E. J., Ed., IRL Press, Oxford, 1987, 1.

47. **Muggleton-Harris, A. L. and Johnson, M. H.,** The nature and distribution of serologically detectable alloantigens on the preimplantation mouse embryo, *J. Embryol. Exp. Morphol.,* 35(1), 59, 1976.

48. **Papaioannou, V. and Rossant, J.,** EC-embryo chimaeras. Table 3, in *Teratocarcinoma Stem Cells,* Cold Spring Harbor Conferences on Cell Proliferation, Vol. 10, Silver, L. M., Martin, G. R., and Strickland, S., Eds., Cold Spring Harbor Laboratory, Cold Spring Harbor, NY, 1983, 734.

49. **Strickland, S. and Mahdavi, V.,** The induction of differentiation in teratocarcinoma cells by retinoic acid, *Cell,* 15, 393, 1978.

50. **Jelten, A. M.,** Possible role of retinoic acid binding protein in retinoid stimulation of embryonal carcinoma cell differentiation, *Nature,* 278, 180, 1979.

51. **Rudnicki, M. A. and McBurney, M. W.,** Cell culture methods and induction of differentiation of embryonal carcinoma cell lines, in *Teratocarcinomas and Embryonic Stem Cells, a Practical Approach,* Robertson, E. J., Ed., IRL Press, Oxford, 1987, 19.

52. **Hooper, M., Hardy, K., Handyside, A., Hunter, S., and Monk, M.,** HPRT-deficient (Lesch-Nyhan) mouse embryos derived from germline colonization by cultured cells, *Nature,* 326, 292, 1987.

53. **Engstrom, W., Rees, A. R., and Heath, J. K.,** Proliferation of a human embryonal carcinoma-derived cell line in serum free medium: inter-relationship between growth factor, *J. Cell Sci.,* 73, 361, 1985.

54. **Isacke, C. M. and Deller, M. J.,** Teratocarcinoma cells exhibit growth cooperatively in-vitro, *J. Cell. Physiol.,* 117, 407, 1983.

55. **Heath, J. K. and Isacke, C. M.,** PC13 embryonal carcinoma-derived growth factor, *EMBO J.,* 3, 2957, 1984.

56. **Harding, J. J. and Dilley, K. J.,** Animal models for inherited cataracts: a review, *Exp. Eye Res.,* 3(5), 765, 1984.

57. **Piatigorsky, J.,** Lens differentiation in vertebrates. A review of cellular and molecular features, *Differentiation,* 19, 134, 1981.

58. **Muggleton-Harris, A. L.,** Mouse mutants: model systems to study congenital cataract, *Int. Rev. Cytol.,* 104, 25, 1986.

59. **Bartholemew, R. S., Clayton, R. M., Cuthbert, J., Phillips, C. I., McReid, J., Seth-Truman, D. E. S., Wilson, C., and Yim, S. H.,** in *Ageing of the Lens,* Regnault, F., Hockwin, O., and Courtois, Y., Eds., Elsevier/North-Holland, Amsterdam, 1980, 241.

60. **Iwata, S. and Kinoshita, J. H.**, Mechanism of development of hereditary cataract in mice, *Invest. Ophthalmol.*, 10, 504, 1971.

61. **Spector, A.**, Report on Nat. Eye Inst. cataract workshop, *Invest. Ophthalmol.*, 13, 325, 1974.

62. **Kefalides, N. A., Alper, R., and Clark, C. C.**, Biochemistry of basement membranes, *Int. Rev. Cytol.*, 61, 162, 1979.

63. **Hendrix, R. W. and Zwaan, J.**, Changes in the glycoprotein concentration of extracellular matrix between lens and optic vesicle associated with early differentiation, *Differentiation*, 2, 337, 1974.

64. **Papaconstaniou, J.**, Molecular aspects of lens cell differentiation, *Science*, 156, 338, 1967.

65. **Harding, J. J. and Dilley, K. H.**, Structural proteins of the mammalian lens: a review with emphasis on changes in development, ageing and cataract, *Exp. Eye Res.*, 22, 581, 1976.

66. **McAvoy, J. W.**, Cell division, cell elongation and distribution of alpha, beta and gamma crystallins in rat lens, *J. Embryol. Exp. Morphol.*, 44, 149, 1978.

67. **Piatigorsky, J.**, Lens differentiation in vertebrates: a review of cellular and molecular features, *Differentiation*, 19, 134, 1981.

68. **Rafferty, N. S. and Goosens, W.**, Growth and ageing of lens capsule, *Growth*, 42, 375, 1978.

69. **Muggleton-Harris, A. L.**, Ageing effects at the cellular level studied by transferring nuclei from organ cultured lens cells, *Exp. Gerontol.*, 7, 219, 1972.

70. **Muggleton-Harris, A. L., Lipman, R. D., and Kearns, J.**, In-vitro characteristics of normal and cataractous lens epithelial cells, *Exp. Eye Res.*, 32, 563, 1986.

71. **Courtois, Y., Counis, M. F., Laurent, M., Simonneau, L., and Trenton, J.**, In-vitro cultivation of bovine, chick and human epithelial lens cells in ageing studies, *Int. Top. Gerontol.*, 12, 2, 1978.

72. **Rink, H.**, Lens epithelial cells in-vitro, *Int. Top. Gerontol.*, 12, 24, 1978.

73. **Russell, P., Fukui, H., Tsunematsu, H., and Kinoshita, J. H.**, Tissue culture of lens epithelial cells from normal and Nakano mice, *Invest. Ophthalmol. Visual Sci.*, 16, 243, 1977.

74. **Creighton, M. O., Mousa, G. Y., Miller, G. C., Blair, D. G., and Trevithick, J. R.**, Differentiation of rat lens epithelial cells in tissue culture. Some characteristics in the process, including in-vitro models for pathogenic processes in cataractogenesis, *Vision Res.*, 21, 25, 1981.

75. **Lipman, R. D. and Muggleton-Harris, A. L.**, Modification of the cataractous phenotype by somatic cell hybridisation, *Som. Cell Genet.*, 8, 791, 1982.

76. **Muggleton-Harris, A. L. and Higbee, N.**, In-vitro and in-vivo characterization of the Lop mutant lens, *Exp. Eye Res.*, 44(6), 805, 1987.

77. **Muggleton-Harris, A. L. and Higbee, N.**, Factors modulating mouse lens epithelial cell morphology with differentiation and development of a lentoid structure in-vitro, *Development*, 99, 25, 1987.

78. **Zwaan, J. and Williams, R. M.**, Morphogenesis of the eye lens in a mouse strain with hereditary cataracts, *J. Exp. Zool.*, 169, 407, 1968.

79. **Zwaan, J. and Williams, R. M.**, Cataracts and abnormal proliferation of the eye lens epithelium in mice carrying the Cat^{Fr} gene, *Exp. Eye Res.*, 8, 161, 1969.

80. **Muggleton-Harris, A. L., Festing, M. F. W., and Hall, M.**, A gene location for the inheritance of the Cataract Fraser (Cat^{Fr}) mouse congenital cataract, *Genet. Res.*, 49(3), 235, 1987.

81. **Frazer, F. C. and Schabtech, G.**, "Shrivelled": a hereditary degeneration of the lens in the house mouse, *Genet. Res.*, 3, 383, 1962.

82. **Oda, S., Watanabe, K., Fujisawa, H., and Kamegama, Y.**, Impaired development of the lens fibres in genetic microthalmia eye lens obsolescence *Elo* of the mouse, *Exp. Eye Res.*, 31, 673, 1980.

83. **Iwig, M. and Glaessar, D.**, On the role of the substratum in cell shape variations of bovine lens cells, *Ophthalmic Res.*, 11, 298, 1979.

84. **Eldsdale, T. and Bard, J.**, Collagen substrata for studies on cell behaviour, *J. Cell Biol.*, 154, 626, 1972.

85. **Greenburg, G. and Hay, E. D.**, Cytodifferentiation and tissue phenotype change during transformation of embryonic lens epithelium to mesenchyme-like cells in-vitro, *Dev. Biol.*, 115, 363, 1986.

86. **Muggleton-Harris, A. L., Hardy, K., and Higbee, N.**, Rescue of developmental lens abnormalities in chimaeras of non-cataractous and congenital cataractous mice, *Development*, 99, 473, 1987.

87. **Pitts, J. D. and Finbow, M. E.**, Junctional permeability and its consequences, in *Intercellular Junctions and Synopses*, Feldman, J., Gelula, N. B., and Pitts, J. D., Eds., Chapman & Hall, London, 1977, 63.

88. **Lowenstein, W. R.**, Junctional intercellular communication and the control of growth, *Biochim. Biophys. Acta*, 560, 1, 1979.

89. **Sheridan, J.**, Cell coupling and cell communication during embryogenesis, in *Cell Surface in Animal Embryogenesis and Development*, Cell Surface Reviews I, Poete, G. and Nicholson, G. L., Eds., Elsevier/North-Holland, Amsterdam, 1977, 409.

90. **Wolpert, L.**, Gap junctions, channels for communications in development, in *Intercellular Junctions and Synopses*, Feldman, J., Gilula, N. B., and Pitts, J. D., Eds., Chapman & Hall, London, 1978, 83.

91. **Rothstein, H., Worgul, B., and Weinsieder, A.**, Regulation of lens morphogenesis by pituitary-development insulin-like mitogens, in *Cellular Communication during Ocular Development*, Sheffield, J. B. and Hilfer, S. R., Eds., Springer-Verlag, New York, 1982, 111.

92. **Goodall, H.,** Bridging the junction gap, *Nature,* 317, 286, 1985.

93. **Bloemendal, H.,** Lens proteins as markers of terminal differentiation, *Science,* 197, 127, 1977.

94. **Zwaan, J.,** Immunofluorescent studies on aphakia, a mutation of a gene involved in the control of lens differentiation in the mouse embryo, *Dev. Biol.,* 44, 306, 1975.

95. **Moustafapour, M. K. and Reddy, V. N.,** Studies on lens proteins. II. Soluble lens proteins and membrane proteins in normal and cataractous lenses, *Doc. Ophthalmol.,* 18, 193, 178.

96. **Garber, A. T., Goring, D., and Gold, R. J. M.,** Characterization of abnormal proteins in the soluble lens proteins of Cat^{Fr} mice, *J. Biol. Chem.,* 259(16), 10376, 1984.

97. **Garber, A. T., Winkler, C., Shinohara, T., King, C. R., Inna, G., Piatigorsky, J., and Gold, R. J. M.,** Selective loss of a family of gene transcripts in a hereditary murine cataract, *Science,* 227, 74, 1985.

98. **Dreyfus, J. C., Banroques, J., Poenaru, L., Skala, H., and Vibert, M.,** Ageing of proteins and lens, in *Lens Ageing and Development of Senile Cataracts,* Vol. 13, Hockin, O. S., Ed., S. Karger, Basel, 1978, 147.

99. **Zigler, J. S., Jr., Carper, D. A., and Kinoshita, J. H.,** Changes in lens crystallins during cataract development in the Philly mouse, *Ophthalmic Res.,* 13, 237, 1981.

100. **Piatigorsky, J.,** Lens crystallins and their gene families, *Cell,* 38, 620, 1984.

101. **Weber, N., Church, R. L., Piatigorsky, J., Petrash, M., and Lalley, P. A.,** Assignment of the mouse alpha A-crystallin structural gene to chromosome 17, *Curr. Eye Res.,* 4(12), 1263, 1985.

102. **Westphal, H.,** Transgenic mice, *Bioessays,* 6(2), 73, 1987.

103. **Wang, E. and Lin, S. L.,** Appearance of Statin, a protein marker for non-proliferating and senescent cells following serum-stimulated cell cycle entry, *Exp. Cell Res.,* 167, 135, 1986.

104. **Muggleton-Harris, A. L. and Wang, E.,** Statin expression associated with terminally differentiating and post replicative lens epithelial cells, *Exp. Cell Res.,* in press, 1989.

105. **Swenson, K. I., Farrell, K. M., and Ruderman, J. V.,** The clam embryo protein cyclin A induces entry into M phase and the resumption of meiosis in *Xenopus* oocytes, *Cell,* 47, 861, 1986.

106. **Muggleton-Harris, A. L.,** The reassembly of cellular components for the study of ageing and finite life-span, *Int. Rev. Cytol. Suppl.,* 9, 279, 1979.

107. **Muggleton-Harris, A. L. and Hayflick, L.,** Cellular ageing studies by the reconstruction of replicating cells from nuclei and cytoplasms isolated from normal human diploid fibroblasts, *Exp. Cell Res.,* 103, 321, 1976.

108. **Muggleton-Harris, A. L. and Palumbo, M.,** Nucleocytoplasmic interactions in experimental binucleates from normal and transformal components, *Som. Cell Genet.,* 5, 397, 1979.

109. **Muggleton-Harris, A. L. and DeSimmone, D. W.,** Replicative potentials of various fusion products between W1-38 and SV40 transformed W1-38 cells and their components, *Som. Cell Genet.,* 6, 689, 1980.

110. **Barbacid, M.,** Mutagens, oncogenes and cancer, *Trends Genet.,* July, 188, 1986.

111. **Lumpkin, C. K., Knepper, J. E., Butel, J. S., Smith, J. R., and Pereira-Smith, O. M.,** Mitogenic effects of the proto-oncogene and ongene forms of C-H-*ras* DNA in human dipoid fibroblasts, *Mol. Cell. Biol.,* 6(8), 2990, 1986.

112. **Scott, R. E. and Maercklein, P. B.,** An initiator of carcinogenesis selectively and stably inhibits stem-cell differentiation. A concept that initiation of carcinogenesis involved multiple phases, *Proc. Natl. Acad. Sci. U.S.A.,* 82, 2995, 1985.

113. **Bee, J. A. and Jeffries, R.,** The relationship between calcium levels and limb-bud chondrogenesis in-vitro, *Development,* 100(1), 73, 1987.

114. **Muggleton-Harris, A. L. and DeFuria, R.,** Age dependent metabolic changes in cultured human fibroblasts. In-vitro, *Cell. Dev. Biol.,* 21(5), 271, 1985.

115. **Muggleton-Harris, A. L., Reisert, P. S., and Berghoff, R. L.,** In-vitro characterization of response to stimulus (wounding) with regard to the ageing in human skin fibroblasts, *Mech. Ageing Dev.,* 19, 37, 1982.

Chapter 16

CONTROL OF MUSCLE DIFFERENTIATION BY MITOGENS: A RATIONALE FOR MULTIPLE RESTRICTION POINTS IN THE CELL CYCLE

Burton Wice, Brian Lathrop, and Luis Glaser

TABLE OF CONTENTS

I. INTRODUCTION

Cell growth and differentiation are usually mutually exclusive events. A good illustration of this phenomenon can be seen during the development of skeletal muscle, where myoblasts differentiate following growth arrest early in the G_1 phase of the cell cycle.[1] However, all quiescent myoblasts do not differentiate. For example, muscle satellite cells remain quiescent and nondifferentiated until they are induced to proliferate and then differentiate following injury to muscle fibers.[2-4] Thus, myoblasts, when quiescent, must be able to regulate their state of differentiation. The molecular mechanism by which this is accomplished has remained obscure.

We have examined in detail the role of mitogens in controlling the differentiation of a muscle-like cell line, BC_3H1. Under appropriate conditions (for example, low serum), these cells express high levels of muscle-specific proteins — the muscle isoform of creatine phosphokinase (M-CPK),[5-8] vascular smooth muscle α-actin (α-actin),[9,10] and the acetylcholine receptor;[5,11,12] however, they do not fuse to form myotubes.[5,7] Of particular importance is the observation that upon the readdition of high concentrations of serum, differentiated BC_3H1 cells can reenter the cell cycle, lose their differentiated phenotype, and resume logarithmic growth.[6-15] In addition, the differentiation state of quiescent BC_3H1 cells can be regulated by a single well-characterized polypeptide growth factor, bovine pituitary-derived fibroblast growth factor (FGF),[8,10,13,14] or by sodium orthovanadate,[10] an inhibitor of tyrosine phosphate phosphatases. These cells, therefore, provide an excellent system with which to study the control of cell growth and differentiation.

We will summarize in this chapter our observations on the control of muscle differentiation by mitogens, and we conclude from these observations that FGF and, potentially, other mitogenic molecules can control cell differentiation in the absence of cell growth. In the discussion, we will focus on the important biological consequences that are derived from these observations.

II. RESULTS

A. DIFFERENTIATION OF BC$_3$H1 CELLS

Figure 1 illustrates the inverse relationship between the growth and differentiation of BC_3H1 cells. When logarithmically growing cells are transferred from medium containing 20% serum to medium containing 1% serum, the cells continue to grow exponentially for approximately 24 h, and they cease growing within the next 24 h (Figure 1A). Concomitant with this cessation of growth is the induction of M-CPK synthesis (Figure 1B), α-actin synthesis (Figure 1C), and cell surface expression of functional acetylcholine receptors (not shown).[11] Simultaneously, there is a decrease in the rate of synthesis of the β- and γ-actins, both of which are characteristic of somatic cells (Figures 1D, E).

B. DIFFERENTIATION OF BC$_3$H1 CELLS IS REVERSED BY SERUM

Normally, the commitment of fused myoblasts to the differentiation program is irreversible, and this precludes a detailed study of the role of mitogens in the control of their differentiation program. However, this can be studied in BC_3H1 cells, since they differentiate in the absence of cell fusion.[5] The readdition of 20% serum to subconfluent, quiescent, differentiating BC_3H1 cells results in reinitiation of growth (at a rate similar to that of cells maintained in 20% serum).[7] As illustrated in Figures 1 and 2, serum addition also results in the inhibition of M-CPK and α-actin synthesis and causes an increase in the rate of synthesis of β- and γ- nonmuscle actins. It is important to note that the increase in the rate of synthesis of β- and γ-isoactin synthesis induced by serum is maximal about 6 h after serum addition and is a general response to serum, since it is also observed with α-actin synthesis (Figure

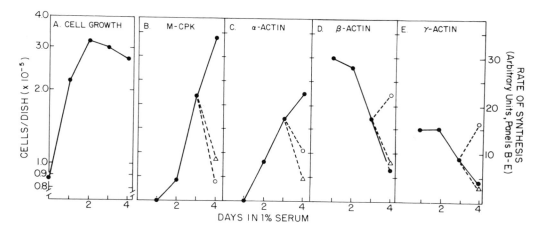

FIGURE 1. Relationship between cell growth and differentiation in BC₃H1 cells. Logarithmically growing BC₃H1 cells in 35-mm dishes were transferred from Dulbecco's modified minimal essential medium (D-MEM) containing 20% fetal calf serum (FCS) to D-MEM containing 1% FCS on day 0. Cells were harvested at the times indicated. ●, cells in 1% FCS; ○, cells to which 20% FCS was added on day 3; △, cells to which FGF was added on day 3. Cell number (panel A) was determined using a Coulter Counter®. The rate of M-CPK synthesis (panel B) was measured essentially as described.[8,13] Briefly, cells were labeled for 1.5 h with ³⁵S-methionine and CPK was immunoprecipitated from cell homogenates. The muscle and brain isozymes were separated by sodium dodecyl sulfate (SDS) gel electrophoresis. After fluorography of the gels, the amount of ³⁵S-CPK was quantitated by densitometry and normalized to the total rate of protein synthesis. The rate of isoactin synthesis (panels C to E) was determined as described.[10] Briefly, cells were labeled for 1 h with ³⁵S-methionine and the actins were purified from cell homogenates using DNAse I affinity chromatography. The ³⁵S-actins were then subjected to isoelectric focusing to separate the α-, β-, and γ-isoforms or to SDS gel electrophoresis to determine the amount of total ³⁵S-actin. The amount of ³⁵S-actin in the gels was quantitated as for M-CPK.

2). However, the rate of α-actin synthesis decreases after longer exposure to 20% serum, while the rate of synthesis of β- and γ-actin remains elevated (Figure 2).

C. FGF REVERSES THE DIFFERENTIATION OF BC₃H1 CELLS IN THE ABSENCE OF CELL GROWTH

The previous results suggest that there is a strong correlation between the reinitiation of cell growth and the dedifferentiation of BC₃H1 cells. However, since serum is an ill-defined mixture of compounds, the possibility exists that these observations are due to (a) nonmitogenic component(s) of the serum. Therefore, we examined the effects of a single, well-characterized polypeptide growth factor,* FGF, on differentiated BC₃H1 cells.

As can be seen in Figures 1 and 2, FGF, like 20% serum, inhibits the synthesis of α-actin and M-CPK. However, there are some important differences between the effects of FGF and serum. First, although serum addition causes differentiated BC₃H1 cells to reinitiate growth, treatment with FGF does not result in any increase in cell number, nor does FGF stimulate the rate of incorporation of ³H-thymidine into DNA (not shown).[8,10,13] Therefore, the inhibition of the muscle phenotype can occur in the absence of cell growth. Secondly, unlike serum, FGF does not induce the synthesis of the nonmuscle actins (Figures 1 and 2); in fact, their rate of synthesis continues to decrease as in untreated, differentiating control cells. We can conclude from this result that the inhibition of the synthesis of muscle-specific

* Acidic brain fibroblast growth factor and basic pituitary-derived fibroblast growth factor are different polypeptides.[16] Recent observations, however, have shown significant sequence homology between these two growth factors, which may be important for their biological function.[17,18] Furthermore, both mitogens can interact with the same receptors in BHK-21 cells,[19] and as shown here and in previous publications, both can repress the synthesis of M-CPK in BC₃H1 cells.[8,10,13]

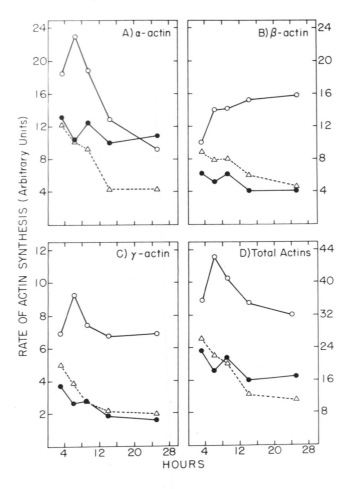

FIGURE 2.   Effects of mitogens on isoactin expression. The experimental protocol was the same as that in Figure 1 and shows the rate of isoactin synthesis between days 3 and 4 at the hourly intervals indicated. Cells were labeled with ^{35}S-methionine for 1 h, and the time shown was halfway through the pulse period. ●, control; ○, 20% FCS; △, FGF. (From Wice, B., Milbrandt, J., and Glaser, L., *J. Biol. Chem.*, 262, 1810, 1987. With permission.)

proteins can take place independently from the induction of the synthesis of proteins characteristic of growing cells.

D. FGF INDUCES THE TRANSIT OF DIFFERENTIATED BC$_3$H1 CELLS FROM G$_{1d}$ to G$_{1q}$ IN THE CELL CYCLE

Differentiated BC$_3$H1 cells appear to be arrested early in the G$_1$ portion of the cell cycle, at a restriction point usually referred to as G$_0$, but which we will refer to as G$_{1d}$.* Addition of 20% serum to such cells results in their reentry into the cell cycle, and they reach the S phase of the cell cycle approximately 12 to 15 h later (Figure 3). If BC$_3$H1 cells that have been treated with FGF (i.e., quiescent but nondifferentiated) are still at this restriction point,

---

*   We refer to the restriction point in the G$_1$ portion of the cell cycle that is permissive for differentiation as G$_{1d}$, which is identical to G$_0$, and to that restriction point where cells are quiescent but not differentiated as G$_{1q}$.

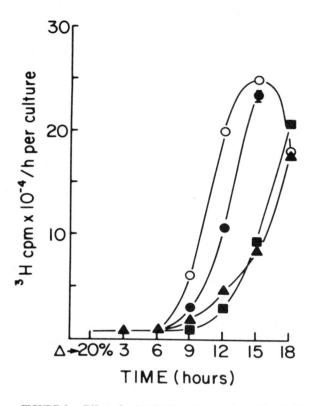

FIGURE 3.   Effect of acidic FGF on the duration of G_1. Acidic FGF was added directly to differentiated cultures at 5 ng/ml. Control cultures received no FGF. After incubation for the times indicated, cells were stimulated to enter S phase by refeeding with serum. The entry into S phase was determined by the increased rate of incorporation of ^{3}H-thymidine into a trichloroacetic acid-precipitable form. ▲, control cultures; ■, no pretreatment; ●, 4-h pretreatment; ○, 12-h pretreatment. Similar results were obtained using basic FGF. (Reproduced from *The Journal of Cell Biology*, 1985, 101, p. 2196 by copyright permission of the Rockefeller University Press.)

then they should also enter S phase 12 to 15 h after the addition of 20% serum. As can be seen in Figure 3, differentiated BC$_3$H1 cells that have been preincubated with FGF for 4 or 12 h and are then stimulated by serum incorporate ^{3}H-thymidine 4 or 6 h, respectively, earlier than control cells. Similar results are obtained by counting ^{3}H-thymidine-labeled nuclei following autoradiography as an index of entry into S phase (not shown).[13] These results suggest that addition of FGF causes differentiated cells to immediately reenter the cell cycle and advance 6 h from G_{1d} to a new restriction point, which we will identify as G_{1q}, at which cells can remain quiescent but not differentiated. If, after 24 h, the cells are not exposed to fresh FGF, they return to G_{1d} and again synthesize high levels of M-CPK.[13] Furthermore, the cells reenter G_{1d} without incorporating ^{3}H-thymidine into DNA, indicating that movement between G_{1d} and G_{1q} is reversible and can take place without transit through the entire cell cycle.

Since FGF causes an immediate reentry into the cell cycle, serum or FGF-treated cells advance simultaneously through G_1 for 6 h, at which time FGF-treated cells enter G_{1q} and serum-treated cells continue to progress through the cell cycle. The serum-induced increase in β- and γ-actin synthesis is maximal 6 h following serum addition, and FGF treatment of

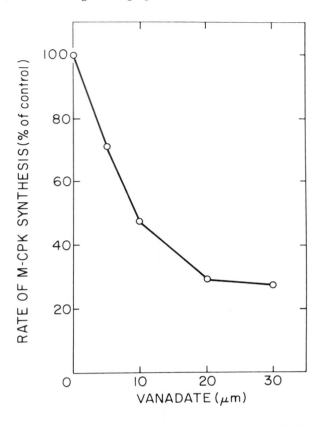

FIGURE 4.   Effects of vanadate on M-CPK synthesis in BC₃H1 cells. Logarithmically growing cells were switched from medium containing 20% FCS to medium containing 1% FCS and allowed to differentiate for 3 d. Cells were then treated with the indicated concentration of vanadate for 24 h. M-CPK synthesis was determined as described in Figure 1. (From Wice, B., Milbrandt, J., and Glaser, L., *J. Biol. Chem.*, 262, 1810, 1987. With permission.)

differentiated cells does not result in increased β- or γ-actin synthesis (Figure 2). We can conclude, therefore, that the induction of β- and γ-actin synthesis is not a result of movement into the cell cycle, but rather is a response to component(s) of serum which are not involved in the transit of cells from G_{1d} to G_{1q}.

E. VANADATE, AN INHIBITOR OF TYROSINE PHOSPHATE PHOSPHATASES, MIMICS THE EFFECTS OF FGF

Since the receptors of some mitogenic polypeptides (for example, epidermal growth factor,[20] platelet-derived growth factor,[21,22] and FGF[23]) are tyrosine-specific protein kinases, we examined the effects of sodium orthovanadate (an inhibitor of tyrosine phosphate phosphatases)[24-27] on the differentiation of BC₃H1 cells. As can be seen in Figure 4, addition of vanadate to differentiating cells results in a decreased rate of synthesis of M-CPK in a concentration-dependent manner, with maximal inhibition observed with 20 μ*M* vanadate. Vanadate also decreased the rate of synthesis of α-actin (not shown).[10] The degrees of inhibition of M-CPK and α-actin by FGF and 20 μ*M* vanadate were similar (compare Figures 1 and 4, for example). Vanadate had no effect on total protein synthesis (not shown) and, like FGF, did not induce the synthesis of β- or γ-actin (not shown).[10] Although vanadate is known to have several effects in addition to inhibiting tyrosine phosphate phosphatases,[28-35] these other effects do not appear likely to explain our results (see Section III).

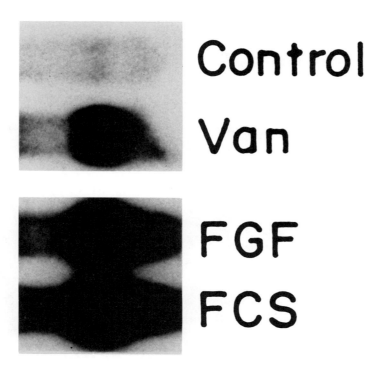

FIGURE 5.   Induction of c-*fos* mRNA by treatment with vanadate, FGF, or FCS. Total cellular RNA was prepared from differentiated BC_3H1 cells 3 d after the cells were switched to 1% FCS as previously described.[10] Some dishes were treated with 20 μM vanadate (VAN), FGF, or 20% FCS for 30 min before cells were harvested. RNA was separated by size using 1.4% agarose, 2.2 M formaldehyde gels and then transferred to Gene Screen Plus. The c-*fos* transcripts were detected by hybridization to a v-*fos* DNA probe.[39] (From Wice, B., Milbrandt, J., and Glaser, L., *J. Biol. Chem.*, 262, 1810, 1987. With permission.)

Since vanadate and FGF have similar effects on differentiated BC_3H1 cells, it is of interest to determine if vanadate, like FGF, causes cells to move from G_{1d} to G_{1q} of the cell cycle. However, vanadate, unlike FGF, prevents the serum-induced movement of quiescent BC_3H1 cells into S phase (not shown);[10] therefore, we cannot determine how far into G_1 vanadate-treated cells have progressed using the methods we employed with cells exposed to FGF. Recent observations have shown that the concentration of the mRNA of the proto-oncogene c-*fos* is transiently increased when cells exit G_0 and enter G_1,[36-38] providing a method to assess the exit of cells from G_{1d}. We therefore looked for an increase in c-*fos* transcripts as an indication of movement into the cell cycle. As can be seen in Figure 5, treatment of differentiated BC_3H1 cells with FGF or 20% serum resulted in an increased level of c-*fos* transcripts in these cells. Vanadate treatment of differentiated BC_3H1 cells caused a similar increase in the level of c-*fos* mRNA (Figure 5), suggesting that vanadate, like FGF and serum, also moves cells into the G_1 portion of the cell cycle. However, we do not know if vanadate-treated cells advance to G_{1q} or if they become quiescent at a different restriction point in G_1. It should also be noted that the time course of c-*fos* mRNA expression is similar regardless of whether FGF, serum, or vanadate is added to differentiated cells (maximal expression at ca. 30 min and back to near basal levels by 60 min; not shown). This is consistent with the notion of a rapid and simultaneous reentry into the cell cycle following the addition of serum, FGF, or vanadate to differentiated cells.

F.   TRANSCRIPTIONAL AND TRANSLATIONAL REGULATION OF BC_3H1 CELL DIFFERENTIATION

Logarithmically growing BC_3H1 cells contain low levels of M-CPK and α-actin mRNA

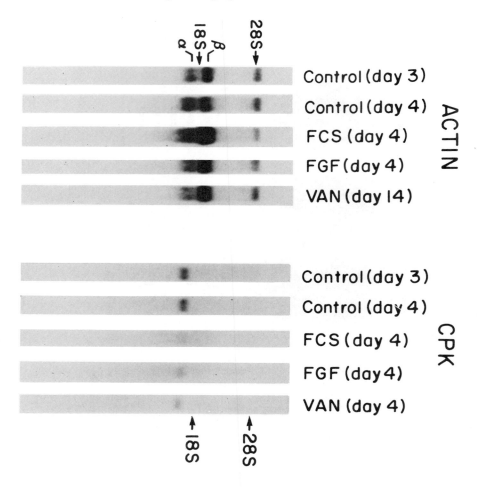

FIGURE 6.   Northern analysis of M-CPK and actin mRNA levels. BC₃H1 cells were plated in 150-mm tissue culture dishes (1.1×10^6 cells/dish) and were switched to DMEM containing 1% FCS 48 h later (day 0). Total cellular RNA was prepared on day 3 or 4 as indicated. Some dishes received 20% FCS, FGF, or vanadate (VAN) for the final 24 h. The α- and β- plus γ-actin mRNAs are 1500 and 2100 nucleotides in length, respectively. CPK mRNA is ca. 1500 nucleotides in length. Each lane represents 6 μg of total cell RNA. (Reproduced in part from Wice, B., Milbrandt, J., and Glaser, L., *J. Biol. Chem.*, 262, 1810, 1987. With permission.)

and high levels of β- plus γ-actin mRNA, whereas differentiated BC₃H1 cells express high levels of M-CPK and α-actin mRNA and low levels of β- plus γ-actin mRNA.[14,40] The synthesis of M-CPK and the isoactins appears to correlate very well with the level of their respective mRNAs. As can be seen in Figure 6, there is a dramatic decrease in the level of M-CPK mRNA 24 h after the addition of FGF, serum, or vanadate to differentiated BC₃H1 cells. This decrease in the amount of CPK mRNA is similar to the decreased rate of M-CPK synthesis observed under the same conditions (compare Figures 1 and 6), suggesting that M-CPK expression is almost totally regulated at the level of transcription. Other laboratories have obtained similar results.[14,41]

The regulation of isoactin synthesis is dramatically different from the regulation of M-CPK synthesis. Within 6 h after the addition of 20% serum to differentiated cells there is a twofold increase in the levels of α- and β- plus γ-isoactin mRNAs, and these higher levels are essentially maintained throughout the next 18 h (Figure 6, Table 1). Thus, the serum-induced increase in the rate of β- and γ-actin synthesis, as well as the initial (6 h) increase

TABLE 1
Level of Actin mRNA in BC$_3$H1 Cells

Time (h)	Addition	mRNA	
		α-actin	β- plus γ-actin
6	None	2199	1911
	FCS	4456	4250
	FGF	2891	2501
24	None	3865	2291
	FCS	3435	4735
	FGF	2330	1916

Note: BC$_3$H1 cells were allowed to differentiate for 3 d, after which they were incubated with the indicated additions for either 6 or 24 h. RNA was isolated as described,[10] and levels of α-actin mRNA and β- plus γ-actin mRNA were quantitated from Northern blots hybridized to an α^{32}P-actin probe by cutting out the appropriate regions and determining the ^{32}P content in a scintillation counter. All results, expressed as cpm, are the averages of duplicate dishes. (From Wice, B., Milbrandt, J., and Glaser, L., *J. Biol. Chem.*, 262, 1810, 1987. With permission.)

in the rate of α-actin synthesis, can be explained by increases in their respective mRNA levels. However, the serum-induced decrease in the rate of α-actin synthesis (Figures 1 and 2) is not associated with a decreased level of α-actin mRNA (Figure 6, Table 1).

The level of α-actin mRNA 24 h after vanadate or FGF treatment is similar to the level in control cells (Figure 6, Table 1); however, the rate of α-actin synthesis is greatly reduced only in FGF- and vanadate-treated cells (compare Figure 6 with Figures 1 and 2). The rate of β- and γ-actin synthesis decreases in control as well as FGF- or vanadate-treated cells (Figure 1), but the level of β- plus γ-actin mRNA remains elevated in FGF- or vanadate-treated cells (Figure 6). We conclude that translational control is at least in part responsible for the observed reduction in α-actin synthesis following the addition of FGF, serum, or vanadate to differentiated BC$_3$H1 cells, whereas both transcriptional control (in the case of serum) and translational control (in the case of FGF or vanadate) are responsible for the change in the rate of synthesis of β- and γ-actin.

III. DISCUSSION

In a number of cells, mitogenic polypeptides act as "competence factors": they cause cells to move from G$_0$ partially through G$_1$. Additional compounds, which have been referred to as progression factors, are then required in order for cells to complete their transit through G$_1$ and to enter S phase.[13,42-44] However, the biological implications of having two or more independently regulated restriction points for the control of the movement of cells through G$_1$ have not been carefully considered.

The results described in this chapter can be summarized as shown in Figure 7, which schematically represents the cell cycle and the two restriction points that we call G$_{1d}$ and G$_{1q}$. Exponentially growing BC$_3$H1 cells that are transferred to low serum exit the cell cycle and enter G$_{1d}$, where they express high levels of muscle-specific proteins. The subsequent addition of FGF to these cells causes them to reenter the cell cycle and advance 4 to 6 h in

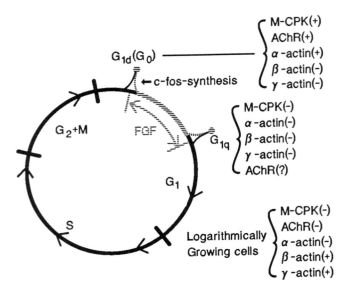

FIGURE 7. The cell cycle and control of differentiation. The figure diagrams the observations presented in the text and, in a strict sense, applies only to BC$_3$H1 cells, although the principles may be generally applicable. BC$_3$H1 cells can remain stably quiescent in two positions in the cell cycle: G$_{1d}$, which is differentiation competent, and G$_{1q}$ which is differentiation negative. The expression of the various proteins associated with differentiation or with logarithmic growth is indicated. Movement between G$_{1q}$ and G$_{1d}$ is fully reversible. In the presence of FGF, cells move from G$_{1d}$ to G$_{1q}$, and in its absence, the converse takes place.

G$_1$ to a new restriction point that we refer to as G$_{1q}$, where the cells remain quiescent but not differentiated. The cells can move between G$_{1q}$ and G$_{1d}$ without having to transit the cell cycle, or they can undergo one or more cycles of cell proliferation before entering G$_{1d}$ in order to differentiate. Since the synthesis of high levels of β- and γ- nonmuscle actin, which is indicative of cells during exponential growth, is not induced by FGF treatment, the induction of this phenotype takes place independently of the cells' movement from G$_{1d}$ to G$_{1q}$.

As we have shown here as well as in a previous publication,[10] the inhibition of the synthesis of one muscle-specific protein (α-actin) is regulated at the level of translation, whereas under identical conditions the synthesis of another protein characteristic of differentiated muscle (M-CPK) appears to be regulated at the level of transcription. In view of our results, as well as those recently reported by Endo and Nadal-Ginard,[45] who have also shown that the synthesis of several muscle-specific proteins can be controlled at the level of translation, it seems likely that several independent mechanisms are at work to bring about the coordinate regulation of the synthesis of proteins associated with the muscle-specific phenotype.

The mechanism for signal transduction is not yet understood in detail for any mitogenic polypeptide. We have used vanadate in an attempt to determine the nature of the intracellular signals that follow binding of FGF to its receptor. Vanadate is known to inhibit tyrosine phosphate phosphatase, both *in vitro*[24-26] and *in vivo*,[27] at concentrations similar to those that inhibit the expression of the muscle phenotype of BC$_3$H1 cells. However, other known effects of vanadate include (1) the activation of adenylate cyclase,[30] (2) mimicking the action of insulin,[31,32] and (3) inhibition of (Na$^+$,K$^+$)-ATPase *in vitro*.[28,29] Neither forskolin, an activator of adenylate cyclase,[46-51] nor insulin has any effect on the rate of α-actin or M-

CPK synthesis following its addition to differentiated BC_3H1 cells (not shown), suggesting that adenylate cyclase and insulin-like effects are not responsible for the effects of vanadate addition to BC_3H1 cells. Furthermore, although vanadate inhibits (Na^+,K^+)-ATPase *in vitro*,[28,29] it has little, if any, effect on this activity *in vivo*,[29,52-54] even at concentrations of vanadate much higher than those sufficient (5 to 20 μM) to exert its effect as a mitogen[33-35] or to repress the muscle-like phenotype of BC_3H1 cells. Therefore, it appears that vanadate is exerting its effects on differentiated BC_3H1 cells by its ability to inhibit tyrosine phosphate phosphatases. This is consistent with the results of Huang and Huang, who have shown that the FGF receptor acts as a protein-specific tyrosine kinase.[23] Although we do not know if vanadate treatment causes BC_3H1 cells to growth arrest at G_{1q}, the failure of cells to differentiate after exit from G_{1d} appears to be at least causally related to the activation of one or more tyrosine-specific protein kinases.

The existence of two restriction points, G_{1q} and G_{1d}, is compatible with the long-term survival of quiescent cells, allowing them to easily move from a differentiation incompetent state to a differentiation competent state. In the whole animal, not all quiescent cells differentiate, nor does it seem reasonable that they should do so. Satellite cells, which regenerate skeletal muscle, remain quiescent but do not express the muscle phenotype. One could hypothesize that the presence of FGF-like molecules may generate this situation and that the presence of other, as yet unidentified molecules is required to stimulate these cells to grow. Differentiation might then be initiated by the removal of these FGF-like molecules. In this regard, it should be noted that transforming growth factor β can also inhibit the synthesis of M-CPK in differentiated BC_3H1 cells without inducing cell growth.[41] Therefore, molecules structurally unrelated to FGF but which have some overlapping effects with FGF (see Folkman and Klagsbrun[55] for review) could also be involved in the regulation of muscle differentiation. The growth and differentiation of "nonfusing" muscle cells (e.g., vascular smooth muscle) could be regulated in a similar fashion, except that reversible movement between G_{1d} and G_{1q} is possible, since differentiated cells do not irreversibly withdraw from the cell cycle.[56]

Movement of the cells through the cell cycle is associated with changes in the pattern of protein synthesis, and the expression of different proteins, often of unknown function, is associated with particular phases of the cell cycle. The observations presented here suggest a cautionary approach to the interpretation that such proteins are necessarily related in a causal way to transit through the cell cycle. These proteins may be equally well related to the differentiated/undifferentiated phenotype of the cells, and their function may not be related to movement through the cell cycle. For example, one mitogen-induced cell protein has been identified as the ADP/ATP exchange protein of mitochondria,[57] a protein certainly not causally related to transit through the cell cycle but probably related to the increased energy demand of growing vs. nongrowing cells. Actin has also been implicated as a marker for cell movement from G_0 through G_1; its synthesis was transiently increased 4 to 6 h after serum stimulation of quiescent, serum-deprived Swiss mouse 3T3 cells.[58] However, our results clearly demonstrate that, at least in BC_3H1 cells, actin synthesis is controlled by nonmitogenic component(s) of serum and not by movement from G_0 into the cell cycle.

ACKNOWLEDGMENT

This work was supported by NIH grant Rol GM38285 and by a grant from Monsanto Chemical Company. Research by the authors was carried out in the Department of Biological Chemistry, Washington University School of Medicine, St. Louis, MO 63110.

REFERENCES

1. **Buckingham, M. E.,** Muscle protein synthesis and its control during the differentiation of skeletal muscle cells *in vitro, Int. Rev. Biochem.* 15, 269, 1977.
2. **Stockdale, F. E. and Holtzer, H.,** DNA synthesis and myogenesis, *Exp. Cell Res.,* 24, 508, 1961.
3. **Moss, F. P. and Leblond, C. P.,** Satellite cells as the source of nuclei in muscles of growing rats, *Anat. Rec.,* 170, 421, 1971.
4. **Goldspink, G.,** Development of muscle, in *Differentiation and Growth of Cells in Vertebrate Tissues,* Goldspink, G., Ed., Chapman and Hall, London, 1974, chap. 3.
5. **Schubert, D., Harris, A. J., Devine, C. E., and Heinemann, S.,** Characterizaiton of a unique muscle cell line, *J. Cell Biol.,* 61, 398, 1974.
6. **Olson, E. N., Caldwell, K. L., Gordon, J. I., and Glaser, L.,** Regulation of creatine phosphokinase expression during differentiation of BC$_3$H1 cells, *J. Biol. Chem.,* 258, 2644, 1983.
7. **Munson, R., Jr., Caldwell, K. L., and Glaser, L.,** Multiple controls for the synthesis of muscle-specific proteins in BC$_3$H1 cells, *J. Cell Biol.,* 92, 350, 1982.
8. **Lathrop, B., Olson, E., and Glaser, L.,** Control by fibroblast growth factor of differentiation in the BC$_3$H1 muscle cell line, *J. Cell Biol.,* 100, 1540, 1985.
9. **Strauch, A. R. and Rubenstein, P. A.,** Induction of vascular smooth muscle α-isoactin expression in BC$_3$H1 cells, *J. Biol. Chem.,* 259, 3152, 1984.
10. **Wice, B., Milbrandt, J., and Glaser, L.,** Control of muscle differentiation in BC$_3$H1 cells by fibroblast growth factor and vanadate, *J. Biol. Chem.,* 262, 1810, 1987.
11. **Olson, E. N., Glaser, L., Merlie, J. P., Sebanne, R., and Lindstrom, J.,** Regulation of surface expression of acetylcholine receptors in response to serum and cell growth in the BC$_3$H1 muscle cell line, *J. Biol. Chem.,* 258, 13946, 1983.
12. **Patrick, J., McMillan, J., Wolfson, H., and O'Brien, J. C.,** Acetylcholine receptor metabolism in a non-fusing muscle cell line, *J. Biol. Chem.,* 252, 2143, 1977.
13. **Lathrop, B., Thomas, K., and Glaser, L.,** Control of myogenic differentiation by fibroblast growth factor is mediated by position in the G$_1$ phase of the cell cycle, *J. Cell Biol.,* 101, 2194, 1985.
14. **Spizz, G., Roman, D., Strauss, A., and Olson, E. N.,** Serum and fibroblast growth factor inhibit myogenic differentiation through a mechanism dependent on protein synthesis and independent of cell proliferation, *J. Biol. Chem.,* 261, 9483, 1986.
15. **Olson, E. N., Glaser, L., Merlie, J. P., and Lindstrom, J.,** Expression of acetylcholine receptor α-subunit mRNA during differentiation of the BC$_3$H1 muscle cell line, *J. Biol. Chem.,* 259, 3330, 1984.
16. **Bohlen, P., Esch, F., Baird, A., and Gospodarowicz, D.,** Acidic fibroblast growth factor (FGF) from bovine brain: amino-terminal sequence and comparison with basic FGF, *EMBO J.,* 4, 1951, 1985.
17. **Esch, F., Baird, A., Ling, N., Ueno, N., Hill, F., Denoroy, L., Klepper, R., Gospodarowicz, D., Bohlen, P., and Guillemin, R.,** Primary structure of bovine pituitary basic fibroblast growth factor (FGF) and comparison with the amino-terminal sequence of bovine brain acidic FGF, *Proc. Natl. Acad. Sci. U.S.A.,* 82, 6507, 1985.
18. **Thomas, K. A. and Gimenez-Gallego, G.,** Fibroblast growth factors: broad spectrum mitogens with potent angiogenic activity, *Trends Biochem. Sci.,* 11, 81, 1986.
19. **Neufeld, G. and Gospodarowicz, D.,** Basic and acidic fibroblast growth factors interact with the same cell surface receptors, *J. Biol. Chem.,* 261, 5631, 1986.
20. **Ushiro, H. and Cohen, S.,** Identification of phosphotyrosine as a product of epidermal growth factor-activated protein kinase in A-431 cell membranes, *J. Biol. Chem.,* 255, 8363, 1980.
21. **Ek, B., Westermark, B., Wasteson, Å., and Heldin, C. H.,** Stimulation of tyrosine-specific phosphorylation by platelet derived growth factor, *Nature,* 295, 419, 1982.
22. **Nishimura, J., Huang, J. S., and Deuel, T. F.,** Platelet derived growth factor stimulates tyrosine specific protein kinase activity in Swiss mouse 3T3 cell membranes, *Proc. Natl. Acad. Sci. U.S.A.,* 79, 4303, 1982.
23. **Huang, S. S. and Huang, J. S.,** Association of bovine brain-derived growth factor receptor with protein tyrosine kinase activity, *J. Biol. Chem.,* 261, 9568, 1986.
24. **Swarup, C., Speeg, K. V., Jr., Cohen, S., and Garbers, D. L.,** Phosphotyrosyl-protein phosphatase of TCRC-2 cells, *J. Biol. Chem.,* 257, 7298, 1982.
25. **Leis, J. F. and Kaplan, N. O.,** An acid phosphatase in the plasma membranes of human astrocytoma showing marked specificity toward phosphotyrosine protein, *Proc. Natl. Acad. Sci. U.S.A.,* 79, 6507, 1982.
26. **Nelson, R. L. and Branton, P. E.,** Identification, purification, and characterization of phosphotyrosine-specific protein phosphatases from cultured chicken embryo fibroblasts, *Mol. Cell. Biol.,* 4, 1003, 1984.
27. **Klarlund, J. K.,** Transformation of cells by an inhibitor of phosphatases acting on phosphotyrosine in proteins, *Cell,* 41, 707, 1985.

28. **Cantley, L. C., Jr., Cantley, L. G., and Josephson, L.,** A characterization of vanadate interactions with (Na,K)-ATPase, *J. Biol. Chem.,* 253, 7361, 1978.

29. **Cantley, L. C., Jr., Resh, M. D., and Guidotti, G.,** Vanadate inhibits the red cell (Na$^+$,K$^+$)-ATPase from the cytoplasmic side, *Nature,* 272, 552, 1978.

30. **Schwabe, U., Puchstein, C., Hannemann, H., and Sochtig, E.,** Activation of adenylate cyclase by vanadate, *Nature,* 277, 143, 1979.

31. **Tamura, S., Brown, T. A., Dubler, R. E., and Larner, J.,** Insulin-like effect of vanadate on adipocyte glycogen synthase and on phosphorylation of 95,000 dalton subunit of insulin receptor, *Biochem. Biophys. Res. Commun.,* 113, 80, 1983.

32. **Tamura, S., Brown, T. A., Whipple, J. H., Fujita-Yamaguchi, Y., Dubler, R. E., Cheng, K., and Larner, J.,** A novel mechanism for the insulin-like effect of vanadate on glycogen synthase in rat adipocytes, *J. Biol. Chem.,* 259, 6650, 1984.

33. **Hori, C. and Oka, T.,** Vanadate enhances the stimulatory action of insulin on DNA synthesis in cultured mouse mammary gland, *Biochim. Biophys. Acta,* 610, 235, 1980.

34. **Carpenter, G.,** Vanadate, epidermal growth factor and the stimulation of DNA synthesis, *Biochem. Biophys. Res. Commun.,* 102, 1115, 1981.

35. **Smith, J. B.,** Vanadium ions stimulate DNA synthesis in Swiss mouse 3T3 and 3T6 cells, *Proc. Natl. Acad. Sci. U.S.A.,* 80, 6162, 1983.

36. **Greenberg, M. E. and Ziff, E. G.,** Stimulation of 3T3 cells induces transcription of the c-fos proto-oncogene, *Nature,* 311, 433, 1984.

37. **Müller, R., Bravo, R., Burckhardt, J., and Curran, T.,** Induction of c-fos gene and protein by growth factors precedes activation of c-myc, *Nature,* 312, 716, 1984.

38. **Kruijer, W., Cooper, J. A., Hunter, T., and Verma, I. M.,** Platelet-derived growth factor induces rapid but transient expression of the c-fos gene and protein, *Nature,* 312, 711, 1984.

39. **Milbrandt, J.,** Nerve growth factor rapidly induces c-fos mRNA in PC12 rat pheochromocytoma cells, *Proc. Natl. Acad. Sci. U.S.A.,* 83, 4789, 1986.

40. **Strauch, A. R., Offord, J. D., Chalkley, R., and Rubenstein, P. A.,** Characterization of actin mRNA levels during BC$_3$H1 cell differentiation, *J. Biol. Chem.,* 261, 849, 1986.

41. **Olson, E. N., Sternberg, E., Hu, J. S., Spizz, G., and Wilcox, C.,** Regulation of myogenic differentiation by type β transforming growth factor, *J. Cell Biol.,* 103, 1799, 1986.

42. **Pledger, W. J., Stiles, C. D., Antoniades, H. N., and Scher, C. D.,** Induction of DNA synthesis in Balb/c3T3 cells by serum components: reevaluation of the commitment process, *Proc. Natl. Acad. Sci. U.S.A.,* 74, 4481, 1977.

43. **Pledger, W. J., Stiles, C. D., Antoniades, H. N., and Scher, C. D.,** An ordered sequence of events is required before Balb/c-3T3 cells become committed to DNA synthesis, *Proc. Natl. Acad. Sci. U.S.A.,* 75, 2839, 1978.

44. **Stiles, C. D., Capone, G. T., Scher, C. D., Antoniades, H. N., Van Wyk, J. J., and Pledger, W. J.,** Dual control of cell growth by somatomedins and platelet-derived growth factor, *Proc. Natl. Acad. Sci. U.S.A.,* 76, 1279, 1979.

45. **Endo, T. and Nadal-Ginard, B.,** Three types of muscle-specific gene expression in fusion blocked rat skeletal muscle cells: translational control in EGTA-treated cells, *Cell,* 49, 515, 1987.

46. **Seamon, K. B., Padgett, W., and Daly, J. W.,** Forskolin: unique diterpene activator of adenylate cyclase in membranes and in intact cells, *Proc. Natl. Acad. Sci. U.S.A.,* 78, 3363, 1981.

47. **Fradkin, J. E., Cook, G. H., Kilhoffer, M. C., and Wolff, J.,** Forskolin stimulation of thyroid adenylate cyclase and cyclic 3′,5′-adenosine monophosphate accumulation, *Endocrinology,* 111, 849, 1982.

48. **Insel, P. A., Stengel, D., Ferry, N., and Hanoune, J.,** Regulation of adenylate cyclase of human platelet membranes by forskolin, *J. Biol. Chem.,* 257, 7485, 1982.

49. **Litosh, I., Hudson, T. H., Mills, I., Li, S.-Y., and Fain, J. N.,** Forskolin as an activator of cyclic AMP accumulation and lipolysis in rat adipocytes, *Mol. Pharmacol.,* 22, 109, 1982.

50. **Takeda, J., Adachi, K., Halprin, K. M., Itami, S., Levine, V., and Woodyard, C.,** Forskolin activates adenylate cyclase activity and inhibits mitosis *in vitro* in pig epidermis, *J. Invest. Dermatol.,* 81, 236, 1983.

51. **Urner, F., Hermann, W. L., Baulieu, E. E., and Schorderet-Slatkine, S.,** Inhibition of denuded mouse oocyte meiotic maturation by forskolin, an activator of adenylate cyclase, *Endocrinology,* 113, 1170, 1983.

52. **Cassel, D., Zhuang, Y.-X., and Glaser, L.,** Vanadate stimulates Na$^+$/H$^+$ exchange activity in A431 cells, *Biochem. Biophys. Res. Commun.,* 118, 675, 1984.

53. **Werdan, K., Bauriedel, G., Bozsik, M., Krawietz, W., and Erdmann, E.,** Effects of vanadate in cultured rat heart muscle cells: vanadate transport, intracellular binding and vanadate-induced changes in beating and in active cation flux, *Biochim. Biophys. Acta,* 597, 364, 1980.

54. **Dubyak, G. R. and Kleinzeller, A.,** The insulin-mimetic effects of vanadate in isolated rate adipocytes, *J. Biol. Chem.,* 255, 5306, 1980.

55. **Folkman, J. and Klagsbrun, M.,** Angiogenic factors, *Science,* 235, 442, 1987.
56. **Chamley-Campbell, J., Campbell, G. R., and Ross, R.,** The smooth muscle cell in culture, *Physiol. Rev.,* 59, 1, 1979.
57. **Battini, R., Ferrari, S., Kaczmarek, L., Calabretta, B., Chen, S. T., and Baserga, R.,** Molecular cloning of a cDNA for a human ADP/ATP carrier which is growth-related, *J. Biol. Chem.,* 262, 4355, 1987.
58. **Riddle, V. G. H., Dubrow, R., and Pardee, A. B.,** Changes in the synthesis of actin and other cell proteins after stimulation of serum-arrested cells, *Proc. Natl. Acad. Sci. U.S.A.,* 76, 1298, 1979.

Section V. Summary

SUMMARY

Huber R. Warner

Recent work in the area of control of cell proliferation, as evidenced by the papers in this volume, is providing new insights into the molecular events involved in traverse of the cell cycle. Although papers on both positive and negative control have been included here, the role of negative control has been and remains of particular interest to experimental gerontologists. However real the cell cycle itself may be, it is at least a useful concept for studying cellular events related to proliferation. Cristofalo et al. (Chapter 1) have provided evidence that representatives from three classes of growth factors must be present to completely stimulate resting cells to enter the G_1 phase of the cell cycle. Once senescent cells enter the G_1 phase they are apparently competent to express most, if not all, G_1-specific genes (Rittling and Baserga, Chapter 2). However, for reasons which are still not clear, these cells are unable to complete the traverse of the G_1 phase and enter S phase and synthesize DNA. Although the decreased specific activity of DNA polymerase alpha in senescent cells could play a role in this (Busbee et al., Chapter 6), it seems unlikely that the failure to enter S phase can be blamed entirely on this deficit. Nevertheless, it is intriguing that cyclin, a protein appearing toward the end of the G_1 phase but preceding S phase, is a DNA polymerase delta auxiliary protein (Mathews, Chapter 7).

The metabolism and role of proliferin or mitogen-regulated protein (MRP) have been discussed by Parfett and Denhardt (Chapter 3). Although this protein is undetectable in primary cell cultures of mouse fibroblasts, it appears in both senescent cells and immortal cell lines which develop from the primary culture. One possible explanation is that this protein is necessary in small amounts for normal cell proliferation but is induced in larger amounts in immortal lines or in cells where DNA replication is blocked. This could be analogous to the induction of other enzymes in a pathway when the pathway is inhibited at any step. The relationship of this protein to the activator of DNA synthesis (ADR) in lymphocytes stimulated by IL-2 and various mitogens is unclear (Fresa and Cohen, Chapter 4). However, it is interesting to note that ADR, like MRP, is found in senescent cells at normal levels, and the defect appears to be in the ability of the nuclei to respond. In Chapter 5, Miller has demonstrated that one important defect may occur very early in the induction cascade, i.e., the increase in intracellular calcium levels, but the molecular basis for this defect is not clear.

Intriguing, but still mysterious, is (are) the role(s) of one or more proteins which correlate with the entry of cells into the resting and/or senescent state (E. Wang, Chapter 11; Pereira-Smith et al., Chapter 8; Stein, Chapter 9). These proteins appear to be membrane bound, having been localized to either the nuclear or the outer membrane, and they have been identified in terminally differentiated cells *in vivo* as well as *in vitro*. The presence of at least one of these proteins blocks proliferation in young, actively growing cells, suggesting that the limited life span phenotype is dominant over the immortal phenotype. The concept of two distinct growth blocks (designated as the G_0 and late G_1 blocks by E. Wang) should be useful in understanding the differences between quiescent and senescent cells. Whereas only growth factors are necessary to induce cells arrested at the first block to proliferate, transformation of cells (e.g., by SV40 virus) is required to induce senescent cells to proliferate. A difficulty still to be sorted out is the molecular basis for the relative ease of transforming mouse cells compared to human cells, which are transformed with difficulty.

Also still to be resolved is the role of specific extracellular factors related to inhibition of cell proliferation. The proteins described by E. Wang (Chapter 11) and J. Wang et al.

(Chapter 10) appear to be unrelated, as the former are large and have merely been shown to be secreted from senescent cells, whereas the latter is small and has growth inhibitory properties in cell cultures. Sequence analysis will eventually clarify whether there are any structural homologies among the various putative inhibitory proteins and between these proteins and already known proteins; for example, these could function as antagonists to growth factors.

An ongoing controversy has been the applicability of *in vitro* models to senescence *in vivo*. This has been addressed by Bruce (Chapter 14), Muggleton-Harris (Chapter 15), and Wice et al. (Chapter 16). The work described by these authors is encouraging in this regard. Antibodies to proteins found predominately in senescent cultured cells specifically stain nuclear proteins in nonproliferating cells *in vivo*. The remaining life span of cells in culture correlates inversely with the *in vivo* donor age. The messenger RNA for a putative inhibitor protein in senescent cells in culture can be isolated in significant amounts from terminally differentiated cells *in vivo*. Finally, specialized quiescent cells in culture express high levels of the appropriate proteins *in vitro*, but these cells will reenter the cell cycle when placed in high levels of serum.

Finally, the DNA sequence itself may play important roles. Howard (Chapter 12) has proposed that repetitive sequences in DNA, when unmasked, may be binding sites for regulatory proteins needed elsewhere for specific functions such as DNA replication. The structure of DNA and the ability of these sequences to bind proteins are easily modulated by methylation of the DNA (Shmookler-Reis et al., Chapter 13), thus affecting not only replication but also transcription. The interaction of all these factors and how they are specifically altered during senescence remain most challenging areas for investigation, currently limited primarily by our inability to ask appropriate questions.

This information is summarized in Figure 1, which pictures possible sites for blocking DNA replication in eukaryotic cells. The data discussed in this volume relate primarily to human cells, but presumably many analogies can be drawn with events in other eukaryotic cells. First, anything which interferes with the initial stimulation of cells by growth factors will block the signal transduction to the DNA, presumably required to stimulate the expression of G_1-specific genes. Examples include inactive receptors, low concentration of receptors, and growth factor antagonists. A secondary effect could be a failure to activate existing but inactive enzymes, e.g., DNA polymerase. Also important, but not shown on the figure, is the requirement for active transcription and translation systems to ensure that the necessary enzymes and other proteins can be synthesized. The synthesis of these proteins, while necessary, may not be sufficient. The proteins and the DNA must be assembled into a functional replication complex, represented in the cartoon by the small solid circle "traveling" around the cell cycle. The failure to form such a functional complex is represented by the small open circle "stuck" at the senescent block. This failure could be due to an immense variety of deficiencies, but it is limited here to a few possibilities such as (1) failure to activate cellular oncogenes, (2) failure to synthesize enzymes required for DNA synthesis (referred to as DNA enzymes in the cartoon), (3) failure to activate DNA polymerases, and (4) inability to form the complex because of inhibitory proteins. The G_0 and G_S genes refer to a putative collection of genes whose gene products regulate the ability to enter S phase after traversing the G_1 phase, and it is particularly important to identify such genes and characterize their gene products if we are to understand the control of proliferation in senescent cells.

This cartoon is an admittedly over-simplified and perhaps fanciful view of the regulation of cell proliferation, but hopefully the research results of the next few years will provide the molecular insights needed to modify this cartoon or replace it completely with a more realistic model.

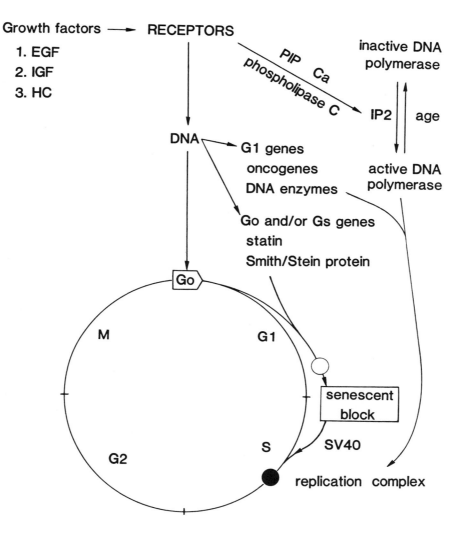

FIGURE 1. Possible gates for blocking cells from entry into S phase.

Index

INDEX